JOHNSON/EVINRUDE Outboards
1973-91 REPAIR MANUAL
60-235 Horsepower, 3-Cylinder V4 and V6

SELOC®

Managing Partners	Dean F. Morgantini, S.A.E.
	Barry L. Beck
Executive Editor	Kevin M. G. Maher, A.S.E.
Manager-Marine/Recreation	James R. Marotta, A.S.E.
Production Managers	Melinda Possinger
	Ronald Webb
Authors	Joan and Clarence Coles

Manufactured in USA
© 2003 Seloc Publishing
104 Willowbrook Lane
West Chester, PA 19382
ISBN 0-89330-010-1
10th Printing 4321098765

www.selocmarine.com
1-866-SELOC55

Other titles
Brought to you by

SELOC

Title	Part #
Chrysler Outboards, All Engines, 1962-84	018-7
Force Outboards, All Engines, 1984-96	024-1
Honda Outboards, All Engines 1988-98	1200
Johnson/Evinrude Outboards, 1-2 Cyl, 1956-70	007-1
Johnson/Evinrude Outboards, 1-2 Cyl, 1971-89	008-X
Johnson/Evinrude Outboards, 1-2 Cyl, 1990-95	026-8
Johnson/Evinrude Outboards, 3, 4 & 6 Cyl, 1973-91	010-1
Johnson/Evinrude Outboards, 3-4 Cyl, 1958-72	009-8
Johnson/Evinrude Outboards, 4, 6 & 8 Cyl, 1992-96	040-3
Kawasaki Personal Watercraft, 1973-91	032-2
Kawasaki Personal Watercraft, 1992-97	042-X
Marine Jet Drive, 1961-96	029-2
Mariner Outboards, 1-2 Cyl, 1977-89	015-2
Mariner Outboards, 3, 4 & 6 Cyl, 1977-89	016-0
Mercruiser Stern Drive, Type I, Alpha/MR, Bravo I & II, 1964-92	005-5
Mercruiser Stern Drive, Alpha I (Generation II), 1992-96	039-X
Mercruiser Stern Drive, Bravo I, II & III, 1992-96	046-2
Mercury Outboards, 1-2 Cyl, 1965-91	012-8
Mercury Outboards, 3-4 Cyl, 1965-92	013-6
Mercury Outboards, 6 Cyl, 1965-91	014-4
Mercury/Mariner Outboards, 1-2 Cyl, 1990-94	035-7
Mercury/Mariner Outboards, 3-4 Cyl, 1990-94	036-5
Mercury/Mariner Outboards, 6 Cyl, 1990-94	037-3
Mercury/Mariner Outboards, All Engines, 1995-99	1416
OMC Cobra Stern Drive, Cobra, King Cobra, Cobra SX, 1985-95	025-X
OMC Stern Drive, 1964-86	004-7
Polaris Personal Watercraft, 1992-97	045-4
Sea Doo/Bombardier Personal Watercraft, 1988-91	033-0
Sea Doo/Bombardier Personal Watercraft, 1992-97	043-8
Suzuki Outboards, All Engines, 1985-99	1600
Volvo/Penta Stern Drives 1968-91	011-X
Volvo/Penta Stern Drives, Volvo Engines, 1992-93	038-1
Volvo/Penta Stern Drives, GM and Ford Engines, 1992-95	041-1
Yamaha Outboards, 1-2 Cyl, 1984-91	021-7
Yamaha Outboards, 3 Cyl, 1984-91	022-5
Yamaha Outboards, 4 & 6 Cyl, 1984-91	023-3
Yamaha Outboards, All Engines, 1992-98	1706
Yamaha Personal Watercraft, 1987-91	034-9
Yamaha Personal Watercraft, 1992-97	044-6
Yanmar Inboard Diesels, 1975-98	7400

SAFETY NOTICE

Proper service and repair procedures are vital to the safe, reliable operation of all marine engines, as well as the personal safety of those performing repairs. This manual outlines procedures for servicing and repairing outboards using safe, effective methods. The procedures contain many NOTES, CAUTIONS and WARNINGS which should be followed, along with standard procedures, to eliminate the possibility of personal injury or improper service which could damage the vessel or compromise its safety.

It is important to note that repair procedures and techniques, tools and parts for servicing marine engines, as well as the skill and experience of the individual performing the work, vary widely. It is not possible to anticipate all of the conceivable ways or conditions under which these engines may be serviced, or to provide cautions as to all possible hazards that may result. Standard and accepted safety precautions and equipment should be used during cutting, grinding, chiseling, prying, or any other process that can cause material removal or projectiles.

Some procedures require the use of tools specially designed for a specific purpose. Before substituting another tool or procedure, you must be completely satisfied that neither your personal safety, nor the performance of the marine engine, will be compromised.

Although information in this manual is based on industry sources and is complete as possible at the time of publication, the possibility exists that some vehicle manufacturers made later changes which could not be included here. While striving for total accuracy, Seloc® cannot assume responsibility for any errors, changes or omissions that may occur in the compilation of this data.

PART NUMBERS

Part numbers listed in this reference are not recommendations by Seloc® for any product brand name. They are references that can be used with interchange manuals and aftermarket supplier catalogs to locate each brand supplier's discrete part number.

SPECIAL TOOLS

Special tools are recommended by the marine manufacturer to perform a specific task. Use has been kept to a minimum, but, where absolutely necessary, they are referred to in the text by the part number of the tool manufacturer. These tools can be purchased, under the appropriate part number, from your local dealer or regional distributor, or an equivalent tool can be purchased locally from a tool supplier or parts outlet. Before substituting any tool for the one recommended, read the SAFETY NOTICE at the top of this page.

ACKNOWLEDGMENTS

Seloc® expresses sincere appreciation to Johnson/Evinrude for their assistance in the production of this manual.

ALL RIGHTS RESERVED

No part of this publication may be reproduced, transmitted or stored in any form or by any means, electronic or mechanical, including photocopy, recording, or by information storage or retrieval system, without prior written permission from the publisher.

FOREWORD

This is a comprehensive tune-up and repair manual for Johnson/Evinrude outboards manufactured between 1973 and 1991. Competition, high-performance, and commercial units (including aftermarket equipment), are not covered. The book has been designed and written for the professional mechanic, the do-it-yourselfer, and the student developing his mechanical skills.

Professional Mechanics will find it to be an additional tool for use in their daily work on outboard units because of the many special techniques described.

Boating enthusiasts interested in performing their own work and in keeping their unit operating in the most efficient manner will find the step-by-step illustrated procedures used throughout the manual extremely valuable. In fact, many have said this book almost equals an experienced mechanic looking over their shoulder giving them advice.

Students and Instructors have found the chapters divided into practical areas of interest and work. Technical trade schools, from Florida to Michigan and west to California, as well as the U.S. Navy and Coast Guard, have adopted these manuals as a standard classroom text.

Troubleshooting sections have been included in many chapters to assist the individual performing the work in quickly and accurately isolating problems to a specific area without unnecessary expense and time-consuming work. As an added aid and one of the unique features of this book, many worn parts are illustrated to identify and clarify when an item should be replaced.

Illustrations and procedural steps are so closely related and identified with matching numbers that, in most cases, captions are not used. Exploded drawings show internal parts and their interrelationship with the major component.

Impeller K.t. 0389158

TABLE OF CONTENTS

1 SAFETY

INTRODUCTION	1-1
CLEANING, WAXING, AND POLISHING	1-1
CONTROLLING CORROSION	1-2
PROPELLERS	1-2
FUEL SYSTEM	1-7
LOADING	1-9
HORSEPOWER	1-10
FLOTATION	1-10
EMERGENCY EQUIPMENT	1-12
COMPASS	1-15
STEERING	1-17
ANCHORS	1-17
MISCELLANEOUS EQUIPMENT	1-18
BOATING ACCIDENT REPORTS	1-19
NAVIGATION	1-19

2 TUNING

INTRODUCTION	2-1
TUNE-UP SEQUENCE	2-2
COMPRESSION CHECK	2-3
SPARK PLUG INSPECTION	2-4
IGNITION SYSTEM	2-5
SYNCHRONIZING	2-5
BATTERY SERVICE	2-5
CARBURETOR ADJUSTMENTS	2-7
FUEL PUMPS	2-8
STARTER AND SOLENOID	2-8
INTERNAL WIRING HARNESS	2-9
WATER PUMP CHECK	2-10
PROPELLER	2-11
LOWER UNIT	2-11
BOAT TESTING	2-12

3 POWERHEAD

INTRODUCTION	3-1
Theory of Operation	3-1
CHAPTER ORGANIZATION	3-4
POWERHEAD DISASSEMBLING	3-5
HEAD SERVICE	3-6
REED SERVICE	3-7
Description	3-7
Cleaning and Service	3-9
Reed Assembling	3-10
BYPASS COVERS	3-10
EXHAUST COVER	3-10
Cleaning	3-11
TOP SEAL	3-12
Removal V4 and V6 Engine	3-12
Removal 3-Cylinder Engine	3-12
Assembling V4 and V6 Engine	3-12
BOTTOM SEAL	3-12
CENTERING PINS	3-13
MAIN BEARING BOLTS AND CRANKCASE SIDE BOLTS	3-14
CRANKCASE COVER	3-15
CONNECTING RODS AND PISTONS	3-15
Removal	3-16
Disassembly	3-17
Rod Inspection & Service	3-19
Piston & Ring Inspection and Service	3-21
Cleaning & Inspecting	3-22
Assembling	3-23
CRANKSHAFT	3-25
Removal	3-25
Cleaning & Inspection	3-27
Assembling	3-27
CYLINDER BLOCK SERVICE	3-28
Honing Procedures	3-29
POWERHEAD ASSEMBLING	3-29
PISTON & ROD ASSEMBLY Installation	3-33
CRANKCASE INSTALLATION NEEDLE MAIN & ROD BRGS	3-36
CRANKCASE COVER INSTALLATION	3-39
MAIN BRG. BOLT AND CRANKCASE SIDE BOLT INSTALLATION	3-41
EXHAUST COVER AND BYPASS COVER INSTALLATION	3-42
REED BOX INSTALLATION	3-43

3 POWERHEAD (CONT)

HEAD INSTALLATION	3-43
BREAK-IN PROCEDURES	3-45

4 FUEL

INTRODUCTION	4-1
GENERAL CARBURETION INFO.	4-1
FUEL SYSTEM	4-3
TROUBLESHOOTING	4-3
FUEL PUMP TESTS	4-5
Fuel Line Test	4-6
Rough Engine Idle	4-7
Excessive Fuel Consumption	4-7
Engine Surge	4-8
J/E CARBURETORS	4-8
Carburetor Installations	4-9

TYPE I CARBURETOR
65 HP -- 1972 AND 1973)
70 HP -- 1974 THRU 1978)   4-9
75 HP -- 1974 THRU 1978)

Removal	4-9
Disassembling	4-10
Cleaning & Inspecting	4-11
Assembling	4-13
Installation	4-15

TYPE II CARBURETOR)
All V4 & V6 powerheads thru)
1978 **EXCEPT** Model 85,)   4-16
115 &140 hp thru 1979)

TYPE III CARBURETOR)
All V4 and V6 powerheads)
1979 and on)
EXCEPT 85hp 1980 only;)   4-16
115hp, 120 hp & 140 hp '80 & on;)
200hp and 225hp 1986 & on)

Removal	4-17
Disassembling	4-18
Cleaning & Inspecting	4-19
Assembling	4-20
Installation	4-21

TYPE IV CARBURETOR)
V4 120hp and 140hp 1985 and on)   4-22
V6 200hp and 225hp 1986 and on)

Removal/Disassembling.	4-22
Cleaning & Inspecting	4-24
Assembling/Installation	4-24
PRIMER CHOKE SYSTEM	4-25
VRO (VARIABLE RATIO OIL)	4-26
FUEL TANK SERVICE	4-31

5 IGNITION

INTRODUCTION	5-1
Chapter Coverage	5-1
SPARK PLUG EVALUATION	5-2
POLARITY CHECK	5-4
WIRING HARNESS	5-4
GENERAL TROUBLESHOOTING	
ALL ENGINES ALL IGNITIONS	5-5

TYPE I CD FLYWHEEL MAGNETO
65 HP -- 1972 AND 1973)
70 HP -- 1974 THRU 1978)
75 HP -- 1974 THRU 1978)
85 HP -- 1973 THRU 1977)
115 HP -- 1973 THRU 1977)
140 HP -- 1977)   5-7
150 HP -- 1978)
175 HP -- 1977 AND 1978)
200 HP -- 1976 THRU 1978)
235 HP -- 1978)

Description	5-7
Theory of Operation	5-7

TROUBLESHOOTING
TYPE I CD FLYWHEEL MAGNETO
65 HP -- 1972 AND 1973)
70 HP -- 1974 THRU 1978)   5-10
75 HP -- 1974 THRU 1978)

TROUBLESHOOTING
TYPE I CD FLYWHEEL MAGNETO
85 HP -- 1973 THRU 1977)
115 HP -- 1973 THRU 1977)   5-14
135 HP -- 1973 THRU 1976)
140 HP -- 1977)

TROUBLESHOOTING
TYPE I CD FLYWHEEL MAGNETO
150 HP -- 1978)
175 HP -- 1977 AND 1978)   5-18
200 HP -- 1976 THRU 1978)
235 HP -- 1978)

TYPE II CAPACITOR DISCHARGE
70 HP AND 75 HP 1979 AND ON
AND ALL OTHERS COVERED IN THIS
MANUAL -- 1978 AND ON   5-25

Description	5-25
Theory of Operation	5-26
Checking the System	5-26

TROUBLESHOOTING
TYPE II CAPACITOR DISCHARGE
70 HP -- 1979 AND ON)   5-27
75 HP -- 1979 AND ON)

5 IGNITION (CONT)

TROUBLESHOOTING
TYPE II CAPACITOR DISCHARGE
85 HP -- 1978 THRU 1980)
90 HP -- 1981 AND ON)
100 HP -- 1979 AND 1980)
110 HP -- 1985 THRU 1989) 5-31
115 HP -- 1978 THRU 1984)
115 HP -- 1990 AND ON)
140 HP -- 1978 AND ON)

TROUBLESHOOTING
TYPE II CAPACITOR DISCHARGE
150 HP -- 1979 AND ON)
175 HP -- 1979 AND ON)
200 HP -- 1979 AND ON) 5-35
225 HP -- 1986 AND ON)
235 HP -- 1979 THRU 1985)

SERVICING TYPE I AND TYPE II
CD FLYWHEEL MAGNETO   5-42
  Stator & Charge Coil
    Replacement   5-42
  Timer Base & Sensor
    Replacement   5-44
  Power Pack Type I
    Replacement   5-45

TIMING POINTER ADJUSTMENT
ALL UNITS COVERED IN
THIS MANUAL   5-46

SYNCHRONIZING
ALL 3-CYLINDER UNITS
COVERED IN THIS MANUAL   5-46

SYNCHRONIZING
ALL V4 UNITS COVERED
IN THIS MANUAL   5-47

SYNCHRONIZING
ALL V6 UNITS
PRIOR TO 1986   5-48

SYNCHRONIZING
ALL V6 UNITS
1986 AND ON   5-49

6 ELECTRICAL

INTRODUCTION   6-1
BATTERIES   6-1
  Jumper Cables   6-5
  Dual Battery Installation 6-5
GAUGES AND HORNS   6-7
FUEL SYSTEM   6-8
  Fuel Gauge Troubleshooting 6-9
TACHOMETER   6-10
HORNS   6-10
ELECTRICAL SYSTEM
GENERAL INFORMATION   6-11
ALTERNATOR CIRCUIT   6-12
  Operation   6-12
ALTERNATOR WITH CD IGNITION
ALL ENGINES IN THIS MANUAL 6-14
  Troubleshooting   6-14
CHOKE CIRCUIT SERVICE   6-16
STARTER MOTOR CIRCUIT   6-17
  Troubleshooting   6-20
STARTER MOTOR DRIVE GEAR
ALL ENGINES IN THIS MANUAL 6-21
  Removal   6-21
  Disassembling   6-22
  Cleaning & Inspecting 6-22
  Assembling   6-23
PRESTOLITE SERVICE   6-23
  Removal   6-23
  Disassembling   6-24
  Cleaning & Inspecting 6-26
  Assembling   6-27
AMERICAN BOSCH SERVICE 6-28
  Removal   6-28
  Disassembling   6-29
  Cleaning & Inspecting 6-31
  Assembling   6-32
  Starter Motor Testing 6-35
  Starter Motor Installation 6-36

7 REMOTE CONTROLS

INTRODUCTION   7-1
SHIFT BOXES   7-1
  Description   7-1
HYDRO-ELECTRIC SHIFT BOX
65 HP -- 1972 ONLY   7-2
  Troubleshooting   7-2
  Disassembling   7-4
  Cleaning & Inspecting 7-7
  Assembling   7-8

SINGLE-LEVER REMOTE CONTROL
SHIFT BOX -- ALL ENGINES
COVERED IN THIS MANUAL
EXCEPT 65 HP -- 1972   7-10
  Description   7-10
  Troubleshooting   7-10
  Disassembling   7-11
  Assembling   7-13
SHIFT BOX REPAIR   7-15
  Disassembling   7-15
  Cleaning & Inspecting 7-17
  Assembling   7-17

8 LOWER UNIT

DESCRIPTION	8-1
CHAPTER COVERAGE	8-1
PROPELLER SERVICE	8-2
Removal	8-3
Installation	8-4
LOWER UNIT LUBRICATION	8-4
Draining	8-4
Filling	8-5

**ELECTRIC SHIFT
TWO SOLENOIDS
65 HP -- 1972**

	8-5
Description	8-5
Troubleshooting	8-6
Removal	8-8

WATER PUMP REMOVAL	8-9
Lower Unit Disassembling	8-10
"Frozen" Propeller Shaft	8-12
Continued Disassembling	8-12
Cleaning & Inspecting	8-16
Assembling	8-18
WATER PUMP INSTALLATION	8-28
Lower Unit Installation	8-31
Functional Check	8-32

**MECHANICAL SHIFT
HYDRAULIC ASSIST
SHIFT DISCONNECT UNDER
LOWER CARBURETOR
65 HP TO 135 HP -- 1973 THRU 1975
85 HP TO 200 HP -- 1976 AND 1977**

Description	8-33
Troubleshooting	8-33
Removal	8-36
WATER PUMP REMOVAL	8-36
Lower Unit Disassembling	8-37
"Frozen" Propeller Shaft	8-39
Continued Disassembling	8-39
Cleaning & Inspecting	8-48
Assembling	8-48
WATER PUMP INSTALLATION	8-59
Lower Unit Installation	8-62
Functional Check	8-63

**MECHANICAL SHIFT
SHIFT DISCONNECT UNDER
LOWER CARBURETOR
70 HP -- 1976 AND 1977
75 HP -- 1975 THRU 1977**

ALL UNITS 1978 AND ON

Description	8-63
Troubleshooting	8-63
Removal	8-65
WATER PUMP REMOVAL	8-67
Lower Unit Disassembling	8-67
"Frozen" Propeller Shaft	8-70
Continued Disassembling	8-70
Cleaning & Inspecting	8-74
Assembling	8-77
WATER PUMP INSTALLATION	8-85
Lower Unit Installation	8-88
Functional Check	8-88
"FROZEN" PROPELLER	8-89

9 POWER TILT/TRIM

SYSTEM DESCRIPTION	9-1
COMPONENT FEATURES	9-2
OPERATION	9-3
TROUBLESHOOTING MECHANICAL COMPONENTS	9-6
Problems & Corrections	9-8
Troubleshooting with Equipment	9-10
TROUBLESHOOTING ELECTRICAL SYSTEM ALL 1978, 1979, 1980 AND CIH 1981 ONLY	9-12
TROUBLESHOOTING ELECTRICAL SYSTEM MODELS 1981 AND ON EXCEPT CIH	9-14
SYSTEM SERVICE	9-17
Removal	9-17
Disassembling	9-18
Cleaning & Inspecting	9-21
Assembling & Installation	9-21
Trim Motor Test & Repair	9-25
Mount Tilt/Trim 1978-1980	9-28
Mount Tilt/Trim 1981 & On	9-28
UNIQUE TRIM/TILT SOME V4 UNITS PRIOR TO 1978	9-29

10 MAINTENANCE

INTRODUCTION	10-1
ENGINE MODEL NUMBERS	10-2
FIBERGLASS HULLS	10-3
ALUMINUM HULLS	10-3
BELOW WATERLINE SERVICE	10-4
SUBMERGED ENGINE SERVICE	10-5
WINTER STORAGE	10-7
LOWER UNIT SERVICE	10-9

10 MAINTENANCE (CONT)

Draining	10-10
Filling	10-11
BATTERY STORAGE	10-12
PRESEASON PREPARATION	10-13

11 OUTBOARD JET DRIVE

INTRODUCTION	11-1
Description & Operation	11-1
General Information	11-1
Model Identification Serial Numbers	11-1
REMOVAL & DISASSEMBLING	11-2
CLEANING & INSPECTING	11-6
Bearing Assembly	11-6
Driveshaft & Associated Parts	11-6
Reverse Gate	11-6
Jet Impeller	11-7
Water Pump	11-7
ASSEMBLING	11-9
Water Pump	11-10
Jet Drive Installation	11-12
ADJUSTMENTS	11-14
Cable Alignment and Free "Play"	11-14
Neutral Stop	11-15
Trim Adjustment	11-16
LUBRICATION	11-17
OUTBOARD JET CHART	11-18

APPENDIX

METRIC CONVERSION CHART	A-1
POWERHEAD SPECIFICATIONS	A-2
TORQUE SPECIFICATIONS	A-3
ENGINE SPECIFICATIONS AND TUNE-UP ADJUSTMENTS	A-4
GEAR OIL CAPACITIES	A-13
HOSE ROUTING PRIMER CHOKE	
V4 ENGINES	A-14
V6 ENGINES	A-15

WIRING IDENTIFICATION DRAWINGS	
65 hp -- 3-cylinder with alternator -- 1972	A-16
70 hp -- 1974-1975 without safety switch 75 hp -- 1975	A-17
70 hp and 75 hp -- 1976 & 77	A-18
70 hp and 75 hp -- 1978	A-19
60 hp 1985 & on	A-20
70 hp 1979 & On	A-20
75 hp 1979-82	A-20
85 hp, 115 hp, and 135 hp -- 1972-1976	A-21
85 hp & 115 hp -- 1977	A-22
140 hp -- 1977-84	A-22
85 hp - 1978-80	A-23
90 hp 1981 & On	A-23
100 hp 1979-80	A-23
110 hp -- 1985-89	A-23
115 hp -- 1978-84	A-23
115 hp -- 1990 & On	A-23
120 hp -- 1985 & On	A-23
120 hp -- 1986 & On	A-26
140 hp -- 1985 Only	A-23
140 hp -- 1986 & On	A-26
175 hp & 200 hp -- 1977-78	A-24
235 hp -- 1978	A-24
150 hp -- 1978-85	A-25
150 hp -- 1986 & On	A-27
175 hp -- 1979-85	A-25
175 hp -- 1986 & On	A-27
185 hp -- 1984-85	A-25
200 hp -- 1979-85	A-25
200 hp -- 1986 & On	A-27
225 hp -- 1986 & On	A-27
235 hp -- 1979-85	A-25

1
SAFETY

1-1 INTRODUCTION

Your boat probably represents a sizeable investment for you. In order to protect this investment and to receive the maximum amount of enjoyment from your boat it must be cared for properly while being used and when it is out of the water. Always store your boat with the bow higher than the stern and be sure to remove the transom drain plug and the inner hull drain plugs. If you use any type of cover to protect your boat, plastic, canvas, whatever, be sure to allow for some movement of air through the hull. Proper ventilation will assure evaporation of any condensation that may form due to changes in temperature and humidity.

1-2 CLEANING, WAXING, AND POLISHING

An outboard boat should be washed with clear water after each use to remove surface dirt and any salt deposits from use in salt water. Regular rinsing will extend the time between waxing and polishing. It will also give you "pride of ownership", by having a sharp looking piece of equipment. Elbow grease, a mild detergent, and a brush will be required to remove stubborn dirt, oil, and other unsightly deposits.

Stay away from harsh abrasives or strong chemical cleaners. A white buffing compound can be used to restore the original gloss to a scratched, dull, or faded area. The finish of your boat should be thoroughly cleaned, buffed, and polished at least once each season. Take care when buffing or polishing with a marine cleaner not to overheat the surface you are working, because you will burn it.

A small outboard engine mounted on an aluminum boat should be removed from the boat and stored separately. Under all circumstances, any outboard engine must **ALWAYS** be stored with the powerhead higher than the lower unit and exhaust system. This position will prevent water trapped in the lower unit from draining back through the exhaust ports into the powerhead.

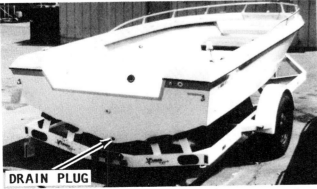

Whenever the boat is stored, for long or short periods, the bow should be slightly higher than the stern and the drain plug in the transom removed to ensure proper drainage of rain water.

Lower unit badly corroded because the zinc was not replaced. Once the zinc is destroyed, more costly parts will be damaged. Attention to the zinc condition is extremely important during boat operation in salt water.

A new zinc prior to installation. This inexpensive item will save corrosion on more valuable parts.

Most outboard engines have a flat area on the back side of the powerhead. When the engine is placed with the flat area on the powerhead and the lower unit resting on the floor, the engine will be in the proper altitude with the powerhead higher than the lower unit.

1-3 CONTROLLING CORROSION

Since man first started out on the water, corrosion on his craft has been his enemy. The first form was merely rot in the wood and then it was rust, followed by other forms of destructive corrosion in the more modern materials. One defense against corrosion is to use similar metals throughout the boat. Even though this is difficult to do in designing a new boat, particularily the undersides, similar metals should be used whenever and wherever possible.

A second defense against corrosion is to insulate dissimilar metals. This can be done by using an exterior coating of Sea Skin or by insulating them with plastic or rubber gaskets.

Using Zinc

The proper amount of zinc attached to a boat is extremely important. The use of too much zinc can cause wood burning by placing the metals close together and they become "hot". On the other hand, using too small a zinc plate will cause more rapid deterioration of the the metal you are trying to protect. If in doubt, consider the fact that it is far better to replace the zincs than to replace planking or other expensive metal parts from having an excess of zinc.

When installing zinc plates, there are two routes available. One is to install many different zincs on all metal parts and thus run the risk of wood burning. Another route, is to use one large zinc on the transom of the boat and then connect this zinc to every underwater metal part through internal bonding. Of the two choices, the one zinc on the transom is the better way to go.

Small outboard engines have a zinc plate attached to the cavitation plate. Therefore, the zinc remains with the engine at all times.

1-4 PROPELLERS

As you know, the propeller is actually what moves the boat through the water. This is how it is done. The propeller operates in water in much the manner as a wood screw does in wood. The propeller "bites" into the water as it rotates. Water passes between the blades and out to the rear in the shape of a cone. The propeller "biting" through the water in much the same manner as a wood auger is what propels the boat.

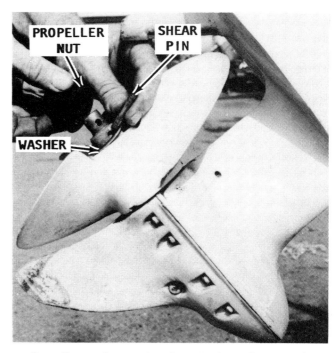

Propeller and associated parts in order, washer, shear-pin, and nut, ready for installation.

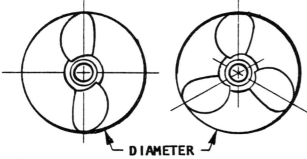

Diameter and pitch are the two basic dimensions of a propeller. The diameter is measured across the circumference of a circle scribed by the propeller blades, as shown.

Arrangement of propeller and associated parts, in order, for a small horsepower engine.

Diameter and Pitch

Only two dimensions of the propeller are of real interest to the boat owner: the diameter and the pitch. These two dimensions are stamped on the propeller hub and always appear in the same order: the diameter first and then the pitch. For instance, the number 15-19 stamped on the hub, would mean the propeller had a diameter of 15 inches with a pitch of 19.

The diameter is the measured distance from the tip of one blade to the tip of the other as shown in the accompanying illustration.

The pitch of a propeller is the angle at which the blades are attached to the hub. This figure is expressed in inches of water travel for each revolution of the propeller. In our example of a 15-19 propeller, the propeller should travel 19 inches through the water each time it revolves. If the propeller action was perfect and there was no slippage, then the pitch multiplied by the propeller rpms would be the boat speed.

Most outboard manufacturers equip their units with a standard propeller with a diameter and pitch they consider to be best suited to the engine and the boat. Such a propeller allows the engine to run as near to the rated rpm and horsepower (at full throttle) as possible for the boat design.

The blade area of the propeller determines its load-carrying capacity. A two-blade propeller is used for high-speed running under very light loads.

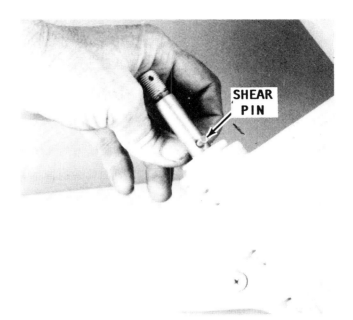

Shear-pin installed behind the propeller instead of in front of the propeller.

A four-blade propeller is installed in boats intended to operate at low speeds under very heavy loads such as tugs, barges, or large houseboats. The three-blade propeller is the happy medium covering the wide range between the high performance units and the load carrying workhorses.

Propeller Selection

There is no standard propeller that will do the proper job in very many cases. The list of sizes and weights of boats is almost endless. This fact coupled with the many boat-engine combinations makes the propeller selection for a specific purpose a difficult job. In fact, in many cases the propeller is changed after a few test runs. Proper selection is aided through the use of charts set up for various engines and boats. These charts should be studied and understood when buying a propeller. However, bear in mind, the charts are based on average boats

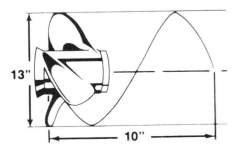
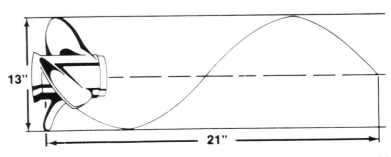

Diagram to explain the pitch dimension of a propeller. The pitch is the theoretical distance a propeller would travel through the water if there was no slippage.

1-4 SAFETY

with average loads, therefore, it may be necessary to make a change in size or pitch, in order to obtain the desired results for the hull design or load condition.

A wide range of pitch is available for each of the larger horsepower engines. The choice available for the smaller engines, up to about 25 hp, is restricted to one or two sizes. Remember, a low pitch takes a smaller bite of the water than the high pitch propeller. This means the low pitch propeller will travel less distance through the water per revolution. The low pitch will require less horsepower and will allow the engine to run faster and more efficiently.

It stands to reason, and it's true, that the high pitch propeller will require more horsepower, but will give faster boat speed if the engine is allowed to turn to its rated rpm.

If a higher-pitched propeller is installed on a boat, in an effort to get more speed, extra horsepower will be required. If the extra power is not available, the rpms will be reduced to a less efficient level and the actual boat speed will be less than if the lower-pitched propeller had been left installed.

All engine manufacturers design their units to operate with full throttle at, or slightly above, the rated rpm. If you run your engine at the rated rpm, you will increase spark plug life, receive better fuel economy, and obtain the best performance from your boat and engine. Therefore, take time to make the proper propeller selection for the rated rpm of your engine at full throttle with what you consider to be an average load. Your boat will then be correctly balanced between engine and propeller throughout the entire speed range.

A reliable tachometer must be used to measure engine speed at full throttle to ensure the engine will achieve full horsepower and operate efficiently and safely. To test for the correct propeller, make your run in a body of smooth water with the lower unit in forward gear at full throttle. Observe the tachometer at full throttle. **NEVER** run the engine at a high rpm when a flush attachment is installed. If the reading is above the manufacturer's recommended operating range, you must try propellers of greater pitch, until you find the one that allows the engine to operate continually within the recommended full throttle range.

If the engine is unable to deliver top performance and you feel it is properly tuned, then the propeller may not be to blame. Operating conditions have a marked effect on performance. For instance, an engine will lose rpm when run in very cold water. It will also lose rpm when run in salt water as compared with fresh water. A hot, low-barometer day will also cause your engine to lose power.

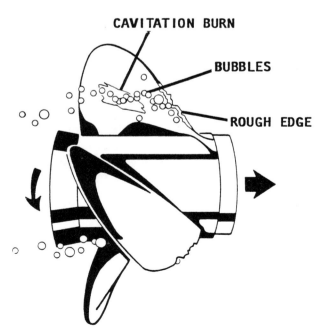

Cavitation (air bubbles) formed at the propeller. Manufacturers are constantly fighting this problem, as explained in the text.

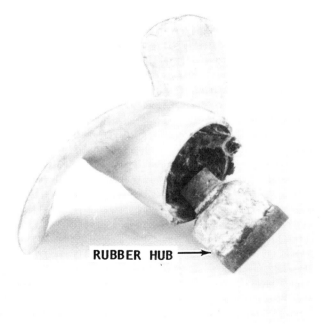

Example of a damaged propeller. This unit should have been replaced long before this amount of damage was sustained.

Ventilation

Ventilation is the forming of voids in the water just ahead of the propeller blades. Marine propulsion designers are constantly fighting the battle against the formation of these voids due to excessive blade tip speed and engine wear. The voids may be filled with air or water vapor, or they may actually be a partial vacuum. Ventilation may be caused by installing a piece of equipment too close to the lower unit, such as the knot indicator pickup, depth sounder, or bait tank pickup.

Vibration

Your propeller should be checked regularly to be sure all blades are in good condition. If any of the blades become bent or nicked, this condition will set up vibrations in the drive unit and the motor. If the vibration becomes very serious it will cause a loss of power, efficiency, and boat performance. If the vibration is allowed to continue over a period of time it can have a damaging effect on many of the operating parts.

Vibration in boats can never be completely eliminated, but it can be reduced by keeping all parts in good working condition and through proper maintenance and lubrication. Vibration can also be reduced in some cases by increasing the number of blades. For this reason, many racers use two-blade props and luxury cruisers have four- and five-blade props installed.

Shock Absorbers

The shock absorber in the propeller plays a very important role in protecting the shafting, gears, and engine against the shock of a blow, should the propeller strike an underwater object. The shock absorber allows the propeller to stop rotating at the instant of impact while the power train continues turning.

How much impact the propeller is able to withstand before causing the clutch hub to slip is calculated to be more than the force needed to propel the boat, but less than the amount that could damage any part of the power train. Under normal propulsion loads of moving the boat through the water, the hub will not slip. However, it will slip if the propeller strikes an object with a force that would be great enough to stop any part of the power train.

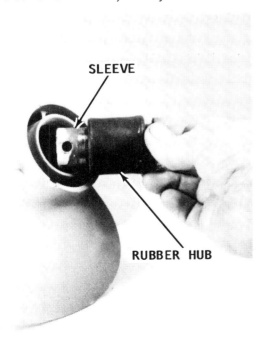

Rubber hub removed from a propeller. This hub was removed because the hub was slipping in the propeller.

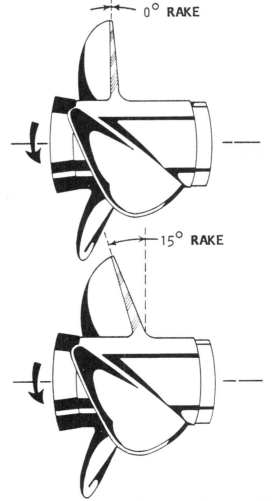

Illustration depicting the rake of a propeller, as explained in the text.

If the power train was to absorb an impact great enough to stop rotation, even for an instant, something would have to give and be damaged. If a propeller is subjected to repeated striking of underwater objects, it would eventually slip on its clutch hub under normal loads. If the propeller would start to slip, a new hub and shock absorber would have to be installed.

Propeller Rake

If a propeller blade is examined on a cut extending directly through the center of the hub, and if the blade is set vertical to the propeller hub, as shown in the accompanying illustration, the propeller is said to have a zero degree ($0°$) rake. As the blade slants back, the rake increases. Standard propellers have a rake angle from $0°$ to $15°$.

A higher rake angle generally improves propeller performance in a cavitating or ventilating situation. On lighter, faster boats, higher rake often will increase performance by holding the bow of the boat higher.

Progressive Pitch

Progressive pitch is a blade design innovation that improves performance when forward and rotational speed is high and/or the propeller breaks the surface of the water.

Progressive pitch starts low at the leading edge and progressively increases to the trailing edge, as shown in the accompanying illustration. The average pitch over the entire blade is the number assigned to that propeller. In the illustration of the progressive pitch, the average pitch assigned to the propeller would be 21.

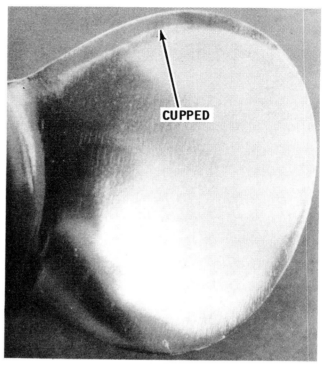

Propeller with a "cupped" leading edge. "Cupping" gives the propeller a better "hold" in the water.

Cupping

If the propeller is cast with a edge curl inward on the trailing edge, the blade is said to have a cup. In most cases, cupped blades improve performance. The cup helps the blades to **"HOLD"** and not break loose, when operating in a cavitating or ventilating situation. This action permits the engine to be trimmed out further, or to be mounted higher on the transom. This is especially true on high-performance boats. Either of these two adjustments will usually add to higher speed.

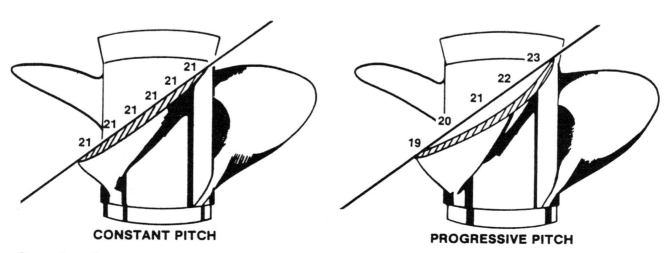

Comparison of a constant and progressive pitch propeller. Notice how the pitch of the progressive pitch propeller, right, changes to give the blade more thrust and therefore, the boat more speed.

The cup has the effect of adding to the propeller pitch. Cupping usually will reduce full-throttle engine speed about 150 to 300 rpm below the same pitch propeller without a cup to the blade. A propeller repair shop is able to increase or decrease the cup on the blades. This change, as explained, will alter engine rpm to meet specific operating demands. Cups are rapidly becoming standard on propellers.

In order for a cup to be the most effective, the cup should be completely concave (hollowed) and finished with a sharp corner. If the cup has any convex rounding, the effectiveness of the cup will be reduced.

Rotation

Propellers are manufactured as right-hand rotation (RH), and as left-hand rotation (LH). The standard propeller for outboards is RH rotation.

A right-hand propeller can easily be identified by observing it as shown in the accompanying illustration. Observe how the blade slants from the lower left toward the upper right. The left-hand propeller slants in the opposite direction, from upper left to lower right, as shown.

When the propeller is observed rotating from astern the boat, it will be rotating clockwise when the engine is in forward gear. The left-hand propeller will rotate counterclockwise.

Propeller Modification

If poor acceleration is experienced on hard-to-plane boats, OMC suggests a slight modification be performed. The modification involves drilling three 6 mm (7/32") holes through the outer shell in a precise pattern. The holes allow exhaust gasses to bleed onto the propeller blades causing controlled ventilation during the acceleration period. This action will allow the motor to turn at a higher rpm under acceleration, thus providing more power to plane the boat.

Layout the exact position of each hole 16 ±2mm (5/8" ±1/16") back from the inner lip and the same amount in a **CLOCKWISE** direction from the base of each blade, as shown in the accompanying illustration.

If an inner rib is located under the position of any one of the holes, another propeller must be used. If the holes are properly positioned, and the correct size is drilled, there will be no affect on top speed, maximum rpm, or ventilation in turns. Improper location or size holes will have no affect on performance, particularly in turns.

1-5 FUEL SYSTEM

With Built-in Fuel Tank

All parts of the fuel system should be selected and installed to provide maximum service and protection against leakage. Reinforced flexible sections should be installed in fuel lines where there is a lot of motion, such as at the engine connection. The flaring of copper tubing should be annealed after it is formed as a protection against hardening. **CAUTION:** Compression fittings should **NOT** be used because they are so easily overtightened, which places them under a strain and subjects them to fatigue.

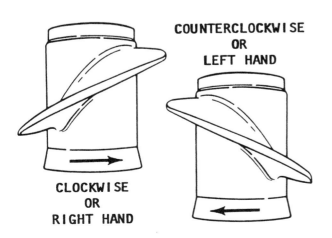

Right- and left-hand propellers showing how the angle of the blades is reversed. Right-hand propellers are by far the most popular.

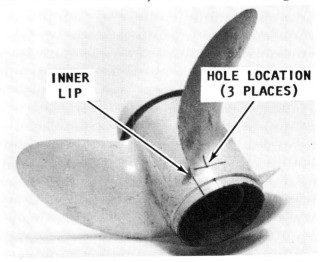

Layout for drilling three holes through the propeller shell for increased performance to provide more power to plane the boat, as explained in the text.

1-8 SAFETY

Such conditions will cause the fitting to leak after it is connected a second time.

The capacity of the fuel filter must be large enough to handle the demands of the engine as specified by the engine manufacturer.

A manually-operated valve should be installed if anti-siphon protection is not provided. This valve should be installed in the fuel line as close to the gas tank as possible. Such a valve will maintain anti-siphon protection between the tank and the engine.

The supporting surfaces and hold-downs must fasten the tank firmly and they should be insulated from the tank surfaces. This insulation material should be non-abrasive and nonabsorbent material. Fuel tanks installed in the forward portion of the boat should be especially well secured and protected because shock loads in this area can be as high as 20 to 25 g's ("g" equals force of gravity).

Static Electricity

In very simple terms, static electricity is called frictional electricity. It is generated by two dissimilar materials moving over each other. One form is gasoline flowing through a pipe or into the air. Another form is when you brush your hair or walk across a synthetic carpet and then touch a metal object. All of these actions cause an electrical charge. In most cases, static electricity is generated during very dry weather conditions, but when you are filling the fuel tank on a boat it can happen at any time.

Fuel Tank Grounding

One area of protection against the build-up of static electricity is to have the fuel

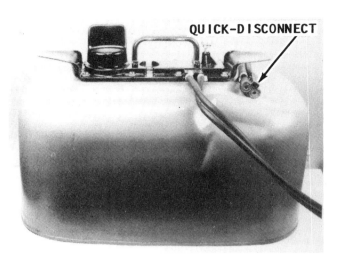

Old style pressure-type tank showing the fuel line to the engine and quick-disconnect fitting.

tank properly grounded (also known as bonding). A direct metal-to-metal contact from the fuel hose nozzle to the water in which the boat is floating. If the fill pipe is made of metal, and the fuel nozzle makes a good contact with the deck plate, then a good ground is made.

As an economy measure, some boats use rubber or plastic filler pipes because of compound bends in the pipe. Such a fill line does not give any kind of ground and if your boat has this type of installation and you do

A three-position valve permits fuel to be drawn from either tank or to be shut off completely. Such an arrangement prevents accidental siphoning of fuel from the tank.

Adding fuel to a six-gallon OMC fuel tank. Some fuel must be in the tank before oil is added to prevent the oil from accumulating on the tank bottom.

not want to replace the filler pipe with a metal one, then it is possible to connect the deck fitting to the tank with a copper wire. The wire should be 8 gauge or larger.

The fuel line from the tank to the engine should provide a continuous metal-to-metal contact for proper grounding. If any part of this line is plastic or other non-metallic material, then a copper wire must be connected to bridge the non-metal material. The power train provides a ground through the engine and drive shaft, to the propeller in the water.

Fiberglass fuel tanks pose problems of their own. One method of grounding is to run a copper wire around the tank from the fill pipe to the fuel line. However, such a wire does not ground the fuel in the tank. Manufacturers should imbed a wire in the fiberglass and it should be connected to the intake and the outlet fittings. This wire would avoid corrosion which could occur if a wire passed through the fuel. **CAUTION: It is not advisable to use a fiberglass fuel tank if a grounding wire was not installed.**

Anything you can feel as a "shock" is enough to set off an explosion. Did you know that under certain atmospheric conditions you can cause a static explosion yourself, particularly if you are wearing synthetic clothing. It is almost a certainty you could cause a static spark if you are **NOT** wearing insulated rubber-soled shoes.

As soon as the deck fitting is opened, fumes are released to the air. Therefore, to be safe you should ground yourself before opening the fill pipe deck fitting. One way to ground yourself is to dip your hand in the water overside to discharge the electricity in your body before opening the filler cap. Another method is to touch the engine block or any metal fitting on the dock which goes down into the water.

1-6 LOADING

In order to receive maximum enjoyment, with safety and performance, from your boat, take care not to exceed the load capacity given by the manufacturer. A plate attached to the hull indicates the U.S. Coast Guard capacity information in pounds for persons and gear. If the plate states the maximum person capacity to be 750 pounds and you assume each person to weigh an average of 150 lbs., then the boat could carry five persons safely. If you add another 250 lbs. for motor and gear, and the maximum weight capacity for persons and gear is 1,000 lbs. or more, then the five persons and gear would be within the limit.

Try to load the boat evenly port and starboard. If you place more weight on one side than on the other, the boat will list to the heavy side and make steering difficult. You will also get better performance by placing heavy supplies aft of the center to keep the bow light for more efficient planing.

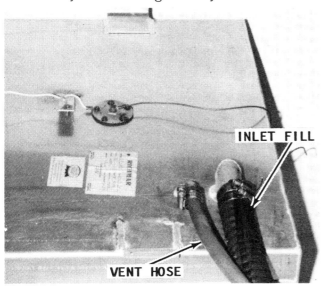

A fuel tank properly grounded to prevent static electricity. Static electricity could be extremely dangerous when taking on fuel.

U.S. Coast Guard plate affixed to all new boats. When the blanks are filled in, the plate will indicate the Coast Guard's recommendations for persons, gear, and horsepower to ensure safe operation of the boat. These recommendations should not be exceeded, as explained in the text.

Clarification

Much confusion arises from the terms, certification, requirements, approval, regulations, etc. Perhaps the following may clarify a couple of these points.

1- The Coast Guard does not approve boats in the same manner as they "Approve" life jackets. The Coast Guard applies a formula to inform the public of what is safe for a particular craft.

2- If a boat has to meet a particular regulation, it must have a Coast Guard certification plate. The public has been led to believe this indicates approval of the Coast Guard. Not so.

3- The certification plate means a willingness of the manufacturer to meet the Coast Guard regulations for that particular craft. The manufacturer may recall a boat if it fails to meet the Coast Guard requirements.

4- The Coast Guard certification plate, see accompanying illustration, may or may not be metal. The plate is a regulation for the manufacturer. It is only a warning plate and the public does not have to adhere to the restrictions set forth on it. Again, the plate sets forth information as to the Coast Guard's opinion for safety on that particular boat.

5- Coast Guard Approved equipment is equipment which has been approved by the Commandant of the U.S. Coast Guard and has been determined to be in compliance with Coast Guard specifications and regulations relating to the materials, construction, and performance of such equipment.

1-7 HORSEPOWER

The maximum horsepower engine for each individual boat should not be increased by any great amount without checking requirements from the Coast Guard in your area. The Coast Guard determines horsepower requirements based on the length, beam, and depth of the hull. **TAKE CARE NOT** to exceed the maximum horsepower listed on the plate or the warranty and possibly the insurance on the boat may become void.

1-8 FLOTATION

If your boat is less than 20 ft. overall, a Coast Guard or BIA (Boating Industry of America) now changed to NMMA (National Marine Manufacturers Association) requirement is that the boat must have buoyant material built into the hull (usually foam) to keep it from sinking if it should become swamped. Coast Guard requirements are mandatory but the NMMA is voluntary.

"Kept from sinking" is defined as the ability of the flotation material to keep the boat from sinking when filled with water

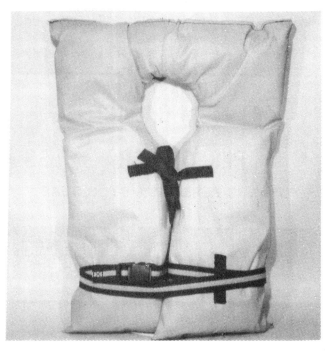

*Type I PFD Coast Guard Approved life jacket. This type flotation device provides the greatest amount of buoyancy. **NEVER** use them for cushions or other purposes.*

A Type IV PFD cushion device intended to be thrown to a person in the water. If air can be squeezed out of the cushion it is no longer fit for service as a PFD.

and with passengers clinging to the hull. One restriction is that the total weight of the motor, passengers, and equipment aboard does not exceed the maximum load capacity listed on the plate.

Life Preservers —Personal Flotation Devices (PFDs)

The Coast Guard requires at least one Coast Guard approved life-saving device be carried on board all motorboats for each person on board. Devices approved are identified by a tag indicating Coast Guard approval. Such devices may be life preservers, buoyant vests, ring buoys, or buoyant cushions. Cushions used for seating are serviceable if air cannot be squeezed out of it. Once air is released when the cushion is squeezed, it is no longer fit as a flotation device. New foam cushions dipped in a rubberized material are almost indestructible.

Life preservers have been classified by the Coast Guard into five type categories. All PFDs presently acceptable on recreational boats fall into one of these five designations. All PFDs **MUST** be U.S. Coast Guard approved, in good and serviceable condition, and of an appropriate size for the persons who intend to wear them. Wearable PFDs **MUST** be readily accessible and throwable devices **MUST** be immediately available for use.

Type I PFD has the greatest required buoyancy and is designed to turn most **UNCONSCIOUS** persons in the water from a face down position to a vertical or slightly backward position. The adult size device provides a minimum buoyancy of 22 pounds and the child size provides a minimum buoyancy of 11 pounds. The Type I PFD provides the greatest protection to its wearer and is most effective for all waters and conditions.

Type II PFD is designed to turn its wearer in a vertical or slightly backward position in the water. The turning action is not as pronounced as with a Type I. The device will not turn as many different type persons under the same conditions as the Type I. An adult size device provides a minimum buoyancy of 15½ pounds, the medium child size provides a minimum of 11 pounds, and the infant and small child sizes provide a minimum buoyancy of 7 pounds.

Type III PFD is designed to permit the wearer to place himself (herself) in a vertical or slightly backward position. The Type III device has the same buoyancy as the Type II PFD but it has little or no turning ability. Many of the Type III PFD are designed to be particularly useful when water skiing, sailing, hunting, fishing, or engaging in other water sports. Several of this type will also provide increased hypothermia protection.

Type IV PFD is designed to be thrown to a person in the water and grasped and held by the user until rescued. It is **NOT** designed to be worn. The most common Type IV PFD is a ring buoy or a buoyant cushion.

Type V PFD is any PFD approved for restricted use.

Coast Guard regulations state, in general terms, that on all boats less than 16 ft. overall, one Type I, II, III, or IV device shall be carried on board for each person in the boat. On boats over 26 ft., one Type I, II, or III device shall be carried on board for each person in the boat **plus** one Type IV device.

It is an accepted fact that most boating people own life preservers, but too few actually wear them. There is little or no excuse for not wearing one because the modern comfortable designs available today do not subtract from an individual's boating pleasure. Make a life jacket available to

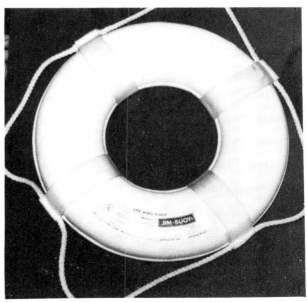

Type IV PFD ring buoy designed to be thrown. On ocean cruisers, this type device usually has a weighted pole with flag, attached to the buoy.

your crew and advise each member to wear it. If you are a crew member ask your skipper to issue you one, especially when boating in rough weather, cold water, or when running at high speed. Naturally, a life jacket should be a must for non-swimmers any time they are out on the water in a boat.

1-9 EMERGENCY EQUIPMENT

Visual Distress Signals
The Regulation

Since January 1, 1981, Coast Guard Regulations require all recreation boats when used on coastal waters, which includes the Great Lakes, the territorial seas and those waters directly connected to the Great Lakes and the territorial seas, up to a point where the waters are less than two miles wide, and boats owned in the United States when operating on the high seas to be equipped with visual distress signals.

The only exceptions are during daytime (sunrise to sunset) for:

Recreational boats less than 16 ft. (5 meters) in length.

Boats participating in organized events such as races, regattas or marine parades.

Open sailboats not equipped with propulsion machinery and less than 26 ft. (8 meters) in length.

Manually propelled boats.

The above listed boats need to carry night signals when used on these waters at night.

Pyrotechnic visual distress signaling devices **MUST** be Coast Guard Approved, in serviceable condition and stowed to be readily accessible. If they are marked with a date showing the serviceable life, this date must not have passed. Launchers, produced before Jan. 1, 1981, intended for use with approved signals are not required to be Coast Guard Approved.

USCG Approved pyrotechnic visual distress signals and associated devices include:

Pyrotechnic red flares, hand held or aerial.

Pyrotechnic orange smoke, hand held or floating.

Launchers for aerial red meteors or parachute flares.

Internationally accepted distress signals.

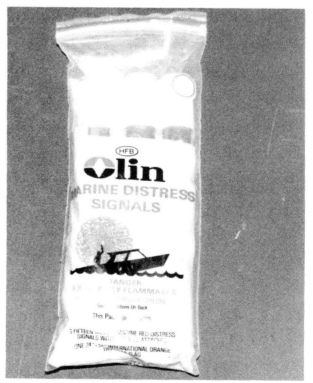

Moisture-protected flares should be carried on board for use as a distress signal.

EMERGENCY EQUIPMENT   1-13

Non-pyrotechnic visual distress signaling devices must carry the manufacturer's certification that they meet Coast Guard requirements. They must be in serviceable condition and stowed so as to be readily accessible.

This group includes:

Orange distress flag at least 3 x 3 feet with a black square and ball on an orange background.

Electric distress light -- not a flashlight but an approved electric distress light which **MUST** automatically flash the international **SOS** distress signal (. . . - - . . .) four to six times each minute.

Types and Quantities

The following variety and combination of devices may be carried in order to meet the requirements.

1- Three hand-held red flares (day and night).

2- One electric distress light (night only).

3- One hand-held red flare and two parachute flares (day and night).

4- One hand-held orange smoke signal, two floating orange smoke signals (day) and one electric distress light (day and night).

If young children are frequently aboard your boat, careful selection and proper stowage of visual distress signals becomes especially important. If you elect to carry pyrotechnic devices, you should select those in tough packaging and not easy to ignite should the devices fall into the hands of children.

Coast Guard Approved pyrotechnic devices carry an expiration date. This date can **NOT** exceed 42 months from the date of manufacture and at such time the device can no longer be counted toward the minimum requirements.

SPECIAL WORDS

In some states the launchers for meteors and parachute flares may be considered a firearm. Therefore, check with your state authorities before acquiring such a launcher.

First Aid Kits

The first-aid kit is similar to an insurance policy or life jacket. You hope you don't have to use it but if needed, you want it there. It is only natural to overlook this essential item because, let's face it, who likes to think of unpleasantness when planning to have only a good time. However, the prudent skipper is prepared ahead of time, and is thus able to handle the emergency without a lot of fuss.

Good commercial first-aid kits are available such as the Johnson and Johnson "Marine First-Aid Kit". With a very modest expenditure, a well-stocked and adequate kit can be prepared at home.

Any kit should include instruments, supplies, and a set of instructions for their use. Instruments should be protected in a watertight case and should include: scissors, tweezers, tourniquet, thermometer, safety

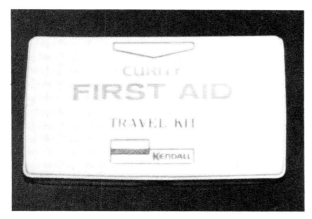

An adequately stocked first-aid kit should be on board for the safety of crew and guests.

A sounding device should be mounted close to the helmsman for use in sounding an emergency alarm.

pins, eye-washing cup, and a hot water bottle. The supplies in the kit should include: assorted bandages in addition to the various sizes of "band-aids", adhesive tape, absorbent cotton, applicators, petroleum jelly, antiseptic (liquid and ointment), local ointment, aspirin, eye ointment, antihistamine, ammonia inhalent, sea-sickness pills, antacid pills, and a laxative. You may want to consult your family physician about including antibiotics. Be sure your kit contains a first-aid manual because even though you have taken the Red Cross course, you may be the patient and have to rely on an untrained crew for care.

Fire Extinguishers

All fire extinguishers must bear Underwriters Laboratory (UL) "Marine Type" approved labels. With the UL certification, the extinguisher does not have to have a Coast Guard approval number. The Coast Guard classifies fire extinguishers according to their size and type.

Type B-I or B-II Designed for extinguishing flammable liquids. Required on all motorboats.

The Coast Guard considers a boat having one or more of the following conditions as a "boat of closed construction" subject to fire extinguisher regulations.

1- Inboard engine or engines.
2- Closed compartments under thwarts and seats wherein portable fuel tanks may be stored.
3- Double bottoms not sealed to the hull or which are not completely filled with flotation materials.
4- Closed living spaces.
5- Closed stowage compartments in which combustible or flammable material is stored.
6- Permanently installed fuel tanks.

Detailed classification of fire extinguishers is by agent and size:

B-I contains 1-1/4 gallons foam, or 4 pounds carbon dioxide, or 2 pounds dry chemical agent, or 2-1/2 pounds Halon.

B-II contains 2-1/2 gallons foam, or 15 pounds carbon dioxide, or 10 pounds dry chemical agent, or 10 pounds Halon.

The class of motorboat dictates how many fire extinguishers are required on board. One B-II unit can be substituted for two B-I extinguishers. When the engine compartment of a motorboat is equipped with a fixed (built-in) extinguishing system, one less portable B-I unit is required.

Dry chemical fire extinguishers without

A suitable fire extinguisher should be mounted close to the helmsman for emergency use.

At least one gallon of emergency fuel should be kept on board in an approved container.

gauges or indicating devices must be weighed and tagged every 6 months. If the gross weight of a carbon dioxide (CO_2) fire extinguisher is reduced by more than 10% of the net weight, the extinguisher is not acceptable and must be recharged.

READ labels on fire extinguishers. If the extinguisher is U.L. listed, it is approved for marine use.

DOUBLE the number of fire extinguishers recommended by the Coast Guard, because their requirements are a bare **MINIMUM** for safe operation. Your boat, family, and crew, must certainly be worth much more than "bare minimum".

1-10 COMPASS

Selection

The safety of the boat and her crew may depend on her compass. In many areas weather conditions can change so rapidly that within minutes a skipper may find himself "socked-in" by a fog bank, a rain squall, or just poor visibility. Under these conditions, he may have no other means of keeping to his desired course except with the compass. When crossing an open body of water, his compass may be the only means of making an accurate landfall.

During thick weather when you can neither see nor hear the expected aids to navigation, attempting to run out the time on a given course can disrupt the pleasure of the cruise. The skipper gains little comfort in a chain of soundings that does not match those given on the chart for the expected area. Any stranding, even for a short time, can be an unnerving experience.

A pilot will not knowingly accept a cheap parachute. A good boater should not accept a bargain in lifejackets, fire extinguishers, or compass. Take the time and spend the few extra dollars to purchase a compass to fit your expected needs. Regardless of what the salesman may tell you, postpone buying until you have had the chance to check more than one make and model.

Lift each compass, tilt and turn it, simulating expected motions of the boat. The compass card should have a smooth and stable reaction.

The card of a good quality compass will come to rest without oscillations about the lubber's line. Reasonable movement in your hand, comparable to the rolling and pitching

The compass is a delicate instrument and deserves respect. It should be mounted securely and in position where it can be easily observed by the helmsman.

of the boat, should not materially affect the reading.

Installation

Proper installation of the compass does not happen by accident. Make a critical check of the proposed location to be sure compass placement will permit the helmsman to use it with comfort and accuracy. First, the compass should be placed directly in front of the helmsman and in such a position that it can be viewed without body stress as he sits or stands in a posture of relaxed alertness. The compass should be in the helmsman's zone of comfort. If the compass is too far away, he may have to bend forward to watch it; too close and he must rear backward for relief.

Do not hesitate to spend a few extra dollars for a good reliable compass. If in doubt, seek advice from fellow boaters.

Second, give some thought to comfort in heavy weather and poor visibilty conditions during the day and night. In some cases, the compass position may be partially determined by the location of the wheel, shift lever, and throttle handle.

Third, inspect the compass site to be sure the instrument will be at least two feet from any engine indicators, bilge vapor detectors, magnetic instruments, or any steel or iron objects. If the compass cannot be placed at least two feet (six feet would be better) from one of these influences, then either the compass or the other object must be moved, if first order accuracy is to be expected.

Once the compass location appears to be satisfactory, give the compass a test before installation. Hidden influences may be concealed under the cabin top, forward of the cabin aft bulkhead, within the cockpit ceiling, or in a wood-covered stanchion.

Move the compass around in the area of the proposed location. Keep an eye on the card. A magnetic influence is the only thing that will make the card turn. You can quickly find any such influence with the compass. If the influence can not be moved away or replaced by one of non-magnetic material, test to determine whether it is merely magnetic, a small piece of iron or steel, or some magnetized steel. Bring the north pole of the compass near the object, then shift and bring the south pole near it. Both the north and south poles will be attracted if the compass is demagnetized. If the object attracts one pole and repels the other, then the compass is magnetized. If your compass needs to be demagnetized, take it to a shop equipped to do the job **PROPERLY.**

After you have moved the compass around in the proposed mounting area, hold it down or tape it in position. Test everything you feel might affect the compass and cause a deviation from a true reading. Rotate the wheel from hard over to hard over. Switch on and off all the lights, radios, radio direction finder, radio telephone, depth finder and the shipboard intercom, if one is installed. Sound the electric whistle, turn on the windshield wipers, start the engine (with water circulating through the engine), work the throttle, and move the gear shift lever. If the boat has an auxiliary generator, start it.

If the card moves during any one of these tests, the compass should be relocated. Naturally, if something like the windshield wipers cause a slight deviation, it may be necessary for you to make a different deviation table to use only when certain pieces of equipment is operating. Bear in mind, following a course that is only off a degree or two for several hours can make considerable difference at the end, putting you on a reef, rock, or shoal.

"Innocent" objects close to the compass, such as diet coke in an aluminum can, may cause serious problems and lead to disaster, as these three photos and the accompanying text illustrate.

Check to be sure the intended compass site is solid. Vibration will increase pivot wear.

Now, you are ready to mount the compass. To prevent an error on all courses, the line through the lubber line and the compass card pivot must be exactly parallel to the keel of the boat. You can establish the fore-and-aft line of the boat with a stout cord or string. Use care to transfer this line to the compass site. If necessary, shim the base of the compass until the stile-type lubber line (the one affixed to the case and not gimbaled) is vertical when the boat is on an even keel. Drill the holes and mount the compass.

Magnetic Items After Installation

Many times an owner will install an expensive stereo system in the cabin of his boat. It is not uncommon for the speakers to be mounted on the aft bulkhead up against the overhead (ceiling). In almost every case, this position places one of the speakers in very close proximity to the compass, mounted above the ceiling.

As we all know, a magnet is used in the operation of the speaker. Therefore, it is very likely that the speaker, mounted almost under the compass in the cabin will have a very pronounced affect on the compass accuracy.

Consider the following test and the accompanying photographs as prove of the statements made.

First, the compass was read as 190 degrees while the boat was secure in her slip.

Next a full can of diet coke in an **aluminum** can was placed on one side and the compass read as 204 degrees, a good 14 degrees off.

Next, the full can was moved to the opposite side of the compass and again a reading was observed. This time as 189 degrees, 11 degrees off from the original reading.

Finally the contents of the can were consumed, the can placed on both sides of the compass with **NO** affect on the compass reading.

Two very important conclusions can be drawn from these tests.

1- Something must have been in the contents of the can to affect the compass so drastically.

2- Keep even "innocent" things clear of the compass to avoid any possible error in the boat's heading.

REMEMBER, a boat moving through the water at 10 knots on a compass error of just 5 degrees will be almost 1.5 miles off course in only **ONE** hour. At night, or in thick weather, this could very possibly put the boat on a reef, rock, or shoal, with disastrous results.

1-11 STEERING

USCG or BIA certification of a steering system means that all materials, equipment, and installation of the steering parts meet or exceed specific standards for strength, type, and maneuverability. Avoid sharp bends when routing the cable. Check to be sure the pulleys turn freely and all fittings are secure.

1-12 ANCHORS

One of the most important pieces of equipment in the boat next to the power plant is the ground tackle carried. The engine makes the boat go and the anchor and its line are what hold it in place when the boat is not secured to a dock or on the beach.

*The weight of the anchor **MUST** be adequate to secure the boat without dragging.*

The anchor must be of suitable size, type, and weight to give the skipper peace of mind when his boat is at anchor. Under certain conditions, a second, smaller, lighter anchor may help to keep the boat in a favorable position during a non-emergency daytime situation.

In order for the anchor to hold properly, a piece of chain must be attached to the anchor and then the nylon anchor line attached to the chain. The amount of chain should equal or exceed the length of the boat. Such a piece of chain will ensure that the anchor stock will lay in an approximate horizontal position and permit the flutes to dig into the bottom and hold.

1-13 MISCELLANEOUS EQUIPMENT

In addition to the equipment you are legally required to carry in the boat and those previously mentioned, some extra items will add to your boating pleasure and safety. Practical suggestions would include: a bailing device (bucket, pump, etc.), boat hook, fenders, spare propeller, spare engine parts, tools, an auxiliary means of propulsion (paddle or oars), spare can of gasoline, flashlight, and extra warm clothing. The area of your boating activity, weather conditions, length of stay aboard your boat, and the specific purpose will all contribute to the kind and amount of stores you put aboard. When it comes to personal gear, heed the advice of veteran boaters who say, "Decide on how little you think you can get by with, then cut it in half".

Bilge Pumps

Automatic bilge pumps should be equipped with an overriding manual switch. They should also have an indicator in the operator's position to advise the helmsman when the pump is operating. Select a pump that will stabilize its temperature within the manufacturer's specified limits when it is operated continuously. The pump motor should be a sealed or arcless type, suitable for a marine atmosphere. Place the bilge pump inlets so excess bilge water can be removed at all normal boat trims. The intakes should be properly screened to prevent the pump from sucking up debris from the bilge. Intake tubing should be of a high quality and stiff enough to resist kinking and not collapse under maximum pump suction condition if the intake becomes blocked.

To test operation of the bilge pump, operate the pump switch. If the motor does not run, disconnect the leads to the motor. Connect a voltmeter to the leads and see if voltage is indicated. If voltage is not indicated, then the problem must be in a blown fuse, defective switch, or some other area of the electrical system.

If the meter indicates voltage is present at the leads, then remove, disassemble, and inspect the bilge pump. Clean it, reassemble, connect the leads, and operate the switch again. If the motor still fails to run, the pump must be replaced.

To test the bilge pump switch, first disconnect the leads from the pump and connect them to a test light or ohmmeter. Next, hold the switch firmly against the mounting location in order to make a good ground. Now, tilt the opposite end of the switch upward until it is activated as indicated by the test light coming on or the ohmmeter showing continuity. Finally, lower the switch slowly toward the mounting

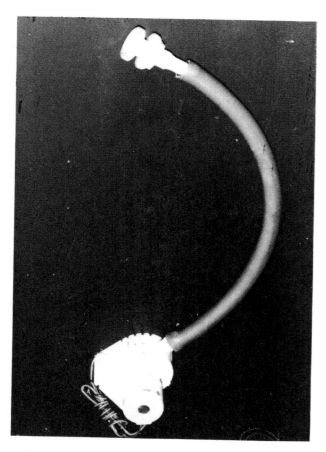

The bilge pump line must be cleaned frequently to ensure the entire bilge pump system will function properly in an emergency.

position until it is deactivated. Measure the distance between the point the switch was activated and the point it was deactivated. For proper service, the switch should deactivate between 1/2-inch and 1/4-inch from the planned mounting position. **CAUTION: The switch must never be mounted lower than the bilge pump pickup.**

1-14 BOATING ACCIDENT REPORTS

New federal and state regulations require an accident report to be filed with the nearest State boating authority within 48 hours if a person is lost, disappears, or is injured to the degree of needing medical treatment beyond first aid.

Accidents involving only property or equipment damage **MUST** be reported within 10 days, if the damage is in excess of $500.00. Some states require reporting of accidents with property damage less than $500.00 or a total boat loss. A **$1,000.00 PENALTY** may be assessed for failure to submit the report.

WORD OF ADVICE

Take time to make a copy of the report to keep for your records or for the insurance company. Once the report is filed, the Coast Guard will not give out a copy, even to the person who filed the report.

The report must give details of the accident and include:

1- The date, time, and exact location of the occurrence.

2- The name of each person who died, was lost, or injured.

3- The number and name of the vessel.

4- The names and addresses of the owner and operator.

If the operator cannot file the report for any reason, each person on board **MUST** notify the authorities, or determine that the report has been filed.

1-15 NAVIGATION

Buoys

In the United States, a buoyage system is used as an assist to all boaters of all size craft to navigate our coastal waters and our navigable rivers in safety. When properly read and understood, these buoys and markers will permit the boater to cruise with comparative confidence and enable her/him to avoid reefs, rocks, shoals, and other hazards.

In the spring of 1983, the Coast Guard began making modifications to U.S. aids to navigation in support of an agreement sponsored by the International Association of Lighthouse Authorities (IALA) and signed by representatives from most of the maritime nations of the world. The primary purpose of the modifications is to improve safety by making buoyage systems around the world more alike and less confusing.

The modifications mentioned moved ahead rapidly in the mid and late 1980's and were scheduled to be completed in the early 1990's.

Lights

All boats are required to show lights at night. The kind and number vary with the type and size of the boat. Inland and Great Lakes Rules require that powerboats under 26 ft. carry a combination red (port) and green (starboard) 20-pt. light, visible at least a mile, and a white 32-pt. (all-around) stern light, visible for 2 miles. Powerboats 26 to 65 ft. must carry a white 32-pt. red light astern, a white 20-pt. light forward, and separate 10-pt. red and green side lights.

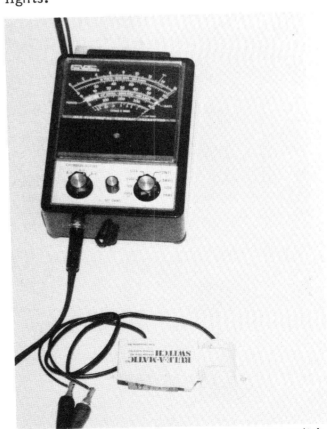

Hookup for testing an automatic bilge pump switch.

SAFETY

Waterway Rules

On the water, certain basic safe-operating practices must be followed. You should learn and practice them, for to **know**, is to be able to handle your boat with confidence and safety. Knowledge of what to do, and not do, will add a great deal to the enjoyment you will receive from your boating investment.

Rules of the Road

The best advice possible and a Coast Guard requirement for boats over 39' 4" (12 meters) since 1981, is to obtain an official copy of the "Rules of the Road", which includes Inland Waterways, Western Rivers, and the Great Lakes for study and ready reference.

The following two paragraphs give a **VERY** brief condensed and abbreviated -- almost a synopsis of the rules and should not be considered in any way as covering the entire subject.

Powered boats must yield the right-of-way to all boats without motors, except when being overtaken. When meeting another boat head-on, keep to starboard, unless you are too far to port to make this practical. When overtaking another boat, the right-of-way belongs to the boat being overtaken. If your boat is being passed, you must maintain course and speed.

When two boats approach at an angle and there is danger of collision, the boat to port must give way to the boat to starboard. Always keep to starboard in a narrow channel or canal. Boats underway must stay clear of vessels fishing with nets, lines, or trawls. (Fishing boats are not allowed to fish in channels or to obstruct navigation.)

Port Side Odd Numbers	Starboard Side Even Numbers
	Light Rhythms FIXED ━━━━━ FLASHING ▪▪▪▪▪ OCCULTING ━ ━ ━ QUICK FLASHING ▪▪▪▪▪▪ EQ INT ━ ━ ━
G "9" Fl G 4sec	R "8" Fl R 4sec
Lighted Buoy (Green Light Only)	Lighted Buoy (Red Light Only)
or	or
C "7" Can Buoy (Unlighted)	N "6" Nun Buoy (Unlighted)
SG G "1" Daymark	TR R "2" Daymark

MODIFICATIONS: Port hand aids will be green with green lights. All starboard hand aids will have red lights.

Preferred Channel to Starboard	Preferred Channel to Port
	Light Rhythm ▪ ▪▪ ▪▪▪ ▪ ▪▪ Composite Group Flashing (2 + 1)
GR "M" CGpFl G	RG "D" CGpFl R
Lighted Buoy (Green Light Only)	Lighted Buoy (Red Light Only)
or	or
GR C "F" Can Buoy (Unlighted)	RG N "L" Nun Buoy (Unlighted)
JG GR "A" Daymark	JR RG "B" Daymark

MODIFICATIONS: Green will replace black. Light rhythm will be changed to Composite Gp Fl (2 + 1).

2
TUNING

2-1 INTRODUCTION

The efficiency, reliability, fuel economy and enjoyment available from engine performance are all directly dependent on having it tuned properly. The importance of performing service work in the sequence detailed in this chapter cannot be over emphasized. Before making any adjustments, check the Specifications in the Appendix. **NEVER** rely on memory when making critical adjustments.

Before beginning to tune any engine, check to be sure the engine has satisfactory compression. An engine with worn or broken piston rings, burned pistons, or badly scored cylinder walls, cannot be made to perform properly no matter how much time and expense is spent on the tune-up. Poor compression must be corrected or the tune-up will not give the desired results.

The opposite of poor compression would be to consider good compression as evidence of a satisfactory cylinder. However, this is not necessarily the case, when working on an outboard engine. As the professional mechanic has discovered, many times the compression check will indicate a satisfactory cylinder, but after the heads are pulled and an inspection made, the cylinder will require service.

A practical maintenance program that is followed throughout the year, is one of the best methods of ensuring the engine will give satisfactory performance at any time.

The extent of the engine tune-up is usually dependent on the time lapse since the last service. A complete tune-up of the entire engine would entail almost all of the work outlined in this manual. A logical sequence of steps will be presented in general terms. If additional information or

Damaged piston, probably caused by inaccurate fuel mixture, or improper timing.

Operating the engine in a test tank with the cowling removed in preparation to making adjustments.

detailed service work is required, the chapter containing the instructions will be referenced.

Each year higher compression ratios are built into modern outboard engines and the electrical systems become more complex, especially with electronic (capacitor discharge) units. Therefore, the need for reliable, authoritative, and detailed instructions becomes more critical. The information in this chapter and the referenced chapters fulfill that requirement.

2-2 TUNE-UP SEQUENCE

If twenty different mechanics were asked the question, "What constitutes a major and minor tune-up?", it is entirely possible twenty different answers would be given. As the terms are used in this manual and other Seloc outboard books, the following work is normally performed for a minor and major tune-up.

Minor Tune-up
Lubricate engine.
Drain and replace gear oil.
Adjust carburetor.
Clean exterior surface of engine.
Tank test engine for fine adjustments.
Check synchronization and timing.

Major Tune-up
Remove heads.
Clean carbon from pistons and cylinders.
Clean and overhaul carburetor.
Clean and overhaul fuel pump.
Test ignition system.
Lubricate engine.
Drain and replace gear oil.
Clean exterior surface of engine.
Tank test engine for fine adjustments.
Check synchronization and timing

During a major tune-up, a definite sequence of service work should be followed to return the engine to the maximum performance desired. This type of work should not be confused with attempting to locate problem areas of "why" the engine is not performing satisfactorily. This work is classified as "troubleshooting". In many cases, these two areas will overlap, because many times a minor or major tune-up will correct the malfunction and return the system to normal operation.

The following list is a suggested sequence of tasks to perform during the tune-up service work. The tasks are merely listed here. Generally procedures are given in subsequent sections of this chapter. For more detailed instructions, see the referenced chapter.

1- Perform a compression check of each cylinder. See Chapter 3.

2- Inspect the spark plugs to determine their condition. Test for adequate spark at the plug. See Chapter 5.

The time, effort, and expense of a tune-up will not restore an engine to satisfactory performance, if the pistons are damaged.

A boat and lower unit covered with marine growth. Such a condition is a serious hinderance to satisfactory performance.

3- Start the engine in a body of water and check the water flow through the engine. See Chapter 8.
4- Check the gear oil in the lower unit. See Chapter 8.
5- Check the carburetor adjustments and the need for an overhaul. See Chapter 4.
6- Check the fuel pump for adequate performance and delivery. See Chapter 4.
7- Make a general inspection of the ignition system. See Chapter 5.
8- Test the starter motor and the solenoid. See Chapter 6.
9- Check the internal wiring.
10- Check the synchronization and timing. See Chapter 5.

2-3 COMPRESSION CHECK

A compression check is extremely important, because an engine with low or uneven compression between cylinders **CANNOT** be tuned to operate satisfactorily. Therefore, it is essential that any compression problem be corrected before proceeding with the tune-up procedure. See Chapter 3.

If the powerhead shows any indication of overheating, such as discolored or scorched paint, especially in the area of the No. 1 cylinder (top on 3-cylinder units and top starboard bank on V4's and V6's), inspect the cylinders visually thru the transfer ports for possible scoring. A more thorough inspection can be made if the heads are removed. It is possible for a cylinder with satisfactory compression to be scored slightly. Also, check the water pump. The overheating condition may be caused by a faulty water pump.

An overheating condition may also be caused by running the engine out of the water. For unknown reasons, many operators have the misconception that running an engine for a short period of time without the lower unit submerged in water, can be done without harm. **FALSE!** Such a practice will result in an overheated condition in a matter of **SECONDS**.

Water must circulate through the lower unit to the engine any time the engine is run to prevent damage to the water pump and an overheating condition. Just five seconds without water will damage the water pump and cause the engine to overheat.

Checking Compression

Remove the spark plug wires. **ALWAYS** grasp the molded cap and pull it loose with a

Removing the spark plugs for inspection. Worn plugs are one of the major contributing factors to poor engine performance.

A compression check should be taken in each cylinder before spending time and money on tune-up work. Without adequate compression, efforts in other areas to regain engine performance will be wasted.

2-4 TUNING

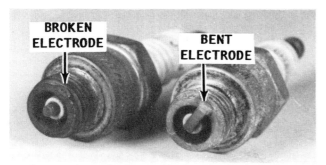

*Damaged spark plugs. Notice the broken electrode on the left plug. The broken part **MUST** be found and removed before returning the engine to service.*

twisting motion to prevent damage to the connection. Remove the spark plugs and keep them in **ORDER** by cylinder for evaluation later. Ground the spark plug leads to the engine to render the ignition system inoperative while performing the compression check.

Insert a compression gauge into the No. 1 (top starboard bank V4's and V6's or the top 3-cylinder) spark plug opening. Crank the engine with the starter thru at least 4 complete strokes with the throttle at the wide-open position, or until the highest possible reading is observed on the gauge. Record the reading. Repeat the test and record the compression for each cylinder. A variation between cylinders is far more important than the actual readings. A variation of more than 5 psi between cylinders indicates the lower compression cylinder may be defective. The problem may be worn, broken, or sticking piston rings, scored pistons or worn cylinders. These problems may only be determined after the heads have been removed. Removing the heads on an outboard engine is not that big a deal and may save many hours of frustration and the cost of purchasing unnecessary parts to correct a faulty condition.

2-4 SPARK PLUG INSPECTION

Inspect each spark plug for badly worn electrodes, glazed, broken, blistered, or lead fouled insulators. Replace all of the plugs, if one shows signs of excessive wear.

Make an evaluation of the cylinder performance by comparing the spark condition with those shown in Chapter 5. Check each spark plug to be sure they are all of the same manufacturer and have the same heat range rating.

Inspect the threads in the spark plug opening of the heads and clean the threads before installing the plug. If the threads are damaged, the heads should be removed and a Heli-coil insert installed. If an attempt is made to drill out the opening with the heads in place, some of the filings may fall into the cylinder and cause damage to the cylinder wall during operation. Because the heads are made of aluminum, the filings cannot be removed with a magnet.

When purchasing new spark plugs, **ALWAYS** ask the marine dealer if there has been a spark plug change for the engine being serviced.

Crank the engine through several revolutions to blow out any material which might have become dislodged during cleaning.

Install the spark plugs and tighten them to the torque value given in the Appendix. **ALWAYS** use a new gasket and wipe the seats in the block clean. The gasket must

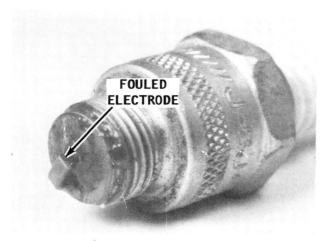

A foul spark plug. The condition of this plug indicates problems in the cylinder that should be corrected.

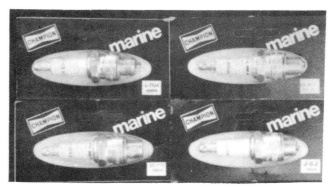

*Today, numerous type spark plugs are available for service. **ALWAYS** check with your local marine dealer to be sure you are purchasing the proper plugs for the engine being serviced.*

be fully compressed on clean seats to complete the heat transfer process and to provide a gas tight seal in the cylinder. If the torque value is too high, the heat will dissipate too rapidly. Conversely, if the torque value is too low, heat will not dissipate fast enough.

2-5 IGNITION SYSTEM

All engines covered in this manual are equipped with a magneto capacitor discharge ignition system, commonly referred to as a "CD" system. Chapter 5, Ignition, is divided into six separate sections covering the various changes to the CD system. Therefore, if engine performance is less than expected, and the ignition is diagnosed as the problem area, refer to Chapter 5 for detailed troubleshooting and service procedures. To synchronize the ignition system with the fuel system, see Chapter 5.

2-6 SYNCHRONIZING

Synchronizing engines covered in this manual consists of adjusting the carburetor linkage to allow throttle movement precisely with ignition advance. Detailed synchronizing procedures are outlined in Chapter 5.

2-7 BATTERY SERVICE

A battery may or may not be required for engine operation once the engine is running. To clarify: With a magneto ignition system, a battery is only used to crank the engine for starting purposes. Once the engine is running properly, theoretically the

The battery MUST be located near the engine in a well-ventilated area. It must be secured in such a manner that absolutely no movement is possible in any direction under the most violent action of the boat.

battery could be removed without affecting engine operation. This statement is in theory to illustrate a point. Engines equipped with an alternator or generator still require the battery as a final storage for the current being generated while the engine is operating. If the battery is completely dead, and enough muscle power is available, the engine may be hand started with a pull cord and operate efficiently.

Engines covered in this manual require a 70-ampere or higher rated battery.

If a battery is used for starting, inspect and service the battery, cables and connections. Check for signs of corrosion. Inspect the battery case for cracks or bulges, dirt, acid, and electrolyte leakage. Check the electrolyte level in each cell.

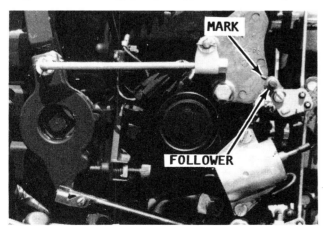

The fuel and ignition systems on any engine MUST be properly synchronized before maximum performance can be obtained from the unit.

A Check of the electrolyte in the battery should be a regular task on the maintenance schedule on any boat.

Fill each cell to the proper level with distilled water or water passed thru a demineralizer.

Clean the top of the battery. The top of a 12-volt battery should be kept especially clean of acid film and dirt, because of the high voltage between the battery terminals. For best results, first wash the battery with a diluted ammonia or baking soda solution to neutralize any acid present. Flush the solution off the battery with clean water. Keep the vent plugs tight to prevent the neutralizing solution or water from entering the cells.

Check to be sure the battery is fastened securely in position. The hold-down device should be tight enough to prevent any movement of the battery in the holder, but not so tight as to place a strain on the battery case.

If the battery posts or cable terminals are corroded, the cables should be cleaned separately with a baking soda solution and a wire brush. Apply a thin coating of Multi-purpose Lubricant to the posts and cable clamps before making the connections. The lubricant will help to prevent corrosion.

If the battery has remained under-charged, check for high resistance in the charging circuit. If the battery appears to be using too much water, the battery may be defective, or it may be too small for the job.

Jumper Cables

If booster batteries are used for starting an engine the jumper cables must be connected correctly and in the proper sequence to prevent damage to either battery, or the alternator diodes.

ALWAYS connect a cable from the positive terminals of the dead battery to the positive terminal of the good battery **FIRST**. **NEXT**, connect one end of the other cable to the negative terminals of the good battery and the other end of the **ENGINE** for a good ground. By making the ground connection on the engine, if there is an arc when you make the connection it will not be near the battery. An arc near the battery could cause an explosion, destroying the battery and causing serious personal injury.

If it is necessary to use a fast-charger on a dead battery, **ALWAYS** disconnect one of the boat cables from the battery first, to prevent burning out the diodes in the alternator.

NEVER use a fast charger as a booster to start the engine because the diodes in the generator will be **DAMAGED**.

Alternator Charging

When the battery is partially discharged, the ammeter should change from discharge to charge between 800 to 1000 rpm for all models. If the battery is fully-charged, the rpm will be a little higher.

With the engine running, increase the throttle to approximately 5200 rpm. The ammeter reading should be approximately equal to the amperage rating of the alternator installed. With a fully-charged battery,

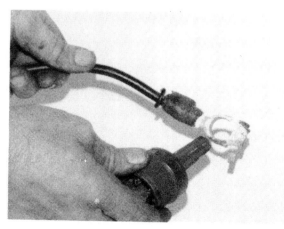

An inexpensive brush should be purchased and used to clean the battery terminals. Clean terminals will ensure a proper connection.

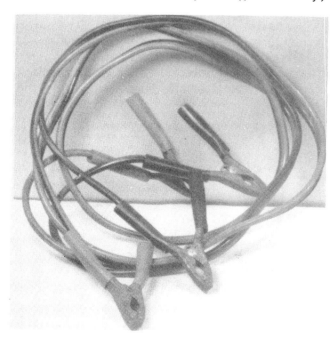

*Common set of jumper cables for using a second battery to crank and start the engine. **EXTREME** care should be exercised when using a second battery, as explained in the text.*

CARBURETOR ADJUSTMENTS 2-7

the ammeter reading will be a bit lower due to the self-regulating characteristics of the generating systems. Before disconnecting the ammeter, reconnect the red harness lead to the positive battery terminal and install the wing nut.

2-8 CARBURETOR ADJUSTMENTS

The carburetors used on the engines covered in this manual do not have either a high-speed nor a low-speed adjustment. Fixed non-adjustable orifices are installed. Detailed service procedures for the carburetor and these orifices are given in Chapter 4.

Fuel and Fuel Tanks

Take time to check the fuel tank and all of the fuel lines, fittings, couplings, valves, flexible tank fill and vent. If the fuel was not drained at the end of the previous season, make a careful inspection for gum formation. When gasoline is allowed to stand for long periods of time, particularly in the presence of copper, gummy deposits form. This gum can clog the filters, lines, and passageway in the carburetor.

If the condition of the fuel is in doubt, drain, clean, and fill the tank with fresh fuel.

An OMC six-gallon fuel tank with the fuel line connected through a quick-disconnect fitting. Such a fitting is handy when the tank is removed from the boat for filling.

Fuel pressure at the carburetor should be checked whenever a lack of fuel volume at the carburetor is suspected.

Repairs and Adjustments

For detailed procedures to disassemble, clean, assemble, and adjust the carburetor, see the appropriate section in Chapter 4 for the carburetor type on the engine being serviced.

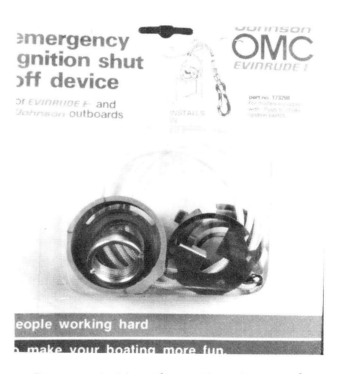

Emergency ignition safety device. One end of the cord is secured to the ignition key and the other end attached to the helmsman's clothing. Should the man at the wheel be accidentally thrown overboard, the key will be ejected from the ignition and the engine immediately shut down.

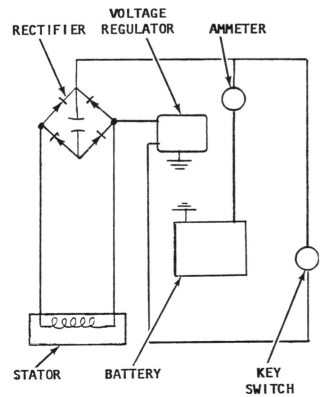

Wiring diagram for an alternator charging circuit.

2-9 FUEL PUMPS

Many times, a defective fuel pump diaphragm is mistakenly diagnosed as a problem in the ignition system. The most common problem is a tiny pin-hole in the diaphragm. Such a small hole will permit gas to enter the crankcase and wet foul the spark plug of the cylinder to which the fuel pump is attached on 3-cylinder units. This fouling will occur at idle-speed. On V4 units, the fuel pump is attached to the No. 4 (bottom port bank) cylinder. Two fuel pumps are installed on the V6 powerhead. Vacuum for these pumps is taken from No. 4 and No. 6 cylinders. During high-speed operation, gas quantity is limited, the plug is not foul and will therefore fire in a satisfactory manner.

If the fuel pump fails to perform properly, an insufficient fuel supply will be delivered to the carburetor. This lack of fuel will cause the engine to run lean, lose rpm or cause piston scoring.

When a fuel pressure gauge is added to the system, it should be installed at the end of the fuel line leading to the upper carburetor. To ensure maximum performance, the fuel pressure must be 2 psi or more at full throttle.

Tune-up Task

All fuel pumps are equipped with a fuel filter. The filter may be cleaned by first removing the cap, then the filter element, cleaning the parts and drying them with compressed air, and finally installing them in their original position.

A late model fuel pump installed on a 3-cylinder engine. The only service possible on this particular pump is to change the filter.

A fuel pump pressure test should be made any time the engine fails to perform satisfactorily at high speed.

NEVER use liquid Neoprene on fuel line fittings. Always use Permatex when making fuel line connections. Permatex is available at almost all marine and hardware stores.

All fuel pumps installed on engines covered in this manual are the "throwaway" type and must be replaced as a unit. For fuel pump service, see Chapter 4.

2-10 STARTER AND SOLENOID

Starter Motor Test

Check to be sure the battery has a 70-ampere rating and is fully charged. Would you believe, many starter motors are needlessly disassembled, when the battery is actually the culprit.

Lubricate the pinion gear and screw shaft with No. 10 oil.

Connect one lead of a voltmeter to the positive terminal of the starter motor. Connect the other meter lead to a good ground on the engine. Check the battery voltage under load by turning the ignition switch to the **START** position and observing the voltmeter reading.

OMC fuel conditioner added to the fuel will keep it fresh for up to one full year.

STARTER MOTOR 2-9

If the reading is 9-1/2 volts or greater, and the starter motor fails to operate, repair or replace the starter motor. See Chapter 6.

Solenoid Test

An ohmmeter is the only instrument required to effectively test a solenoid. Test the ohmmeter by connecting the red and black leads together. Adjust the pointer to the right side of the scale.

On all Johnson/Evinrude engines the case of the solenoid does **NOT** provide a suitable ground to the engine. Hundreds of solenoids have been discarded because of the erroneous belief the case is providing a ground and the unit should function when 12-volts is applied. Not so! One terminal of the solenoid is connected to a 12-volt source. The other terminal is connected via a white wire to a cutout switch on top of the engine. This cutout switch provides a safety to break the ground to the solenoid in the event the engine starts at a high rpm. Therefore, the solenoid ground is made and broken by the cutout switch.

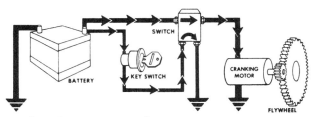
Functional diagram of a typical cranking system.

NEVER connect the battery leads to the large terminals of the solenoid, or the test meter will be damaged. Connect each lead of the test meter to each of the large terminals on the solenoid.

Using battery jumper leads, connect the positive lead from the positive terminal of the battery to the small **"S"** terminal of the solenoid. Connect the negative lead to the negative battery terminal and the **"I"** terminal of the solenoid. If the meter pointer hand moves into the **OK** block, the solenoid is serviceable. If the pointer fails to reach the **OK** block, the solenoid must be replaced.

2-11 INTERNAL WIRING HARNESS

An internal wiring harness is only used on the larger horsepower engines covered in this manual. If the engine is equipped with a wiring harness, the following checks and test will apply.

Starter motor installed on all 3-cylinder engines covered in this manual.

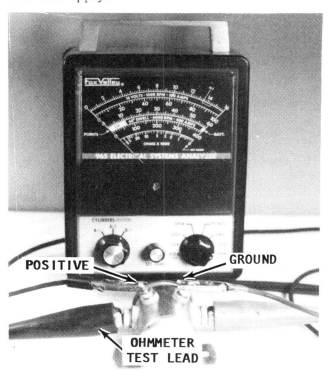
Proper hook-up of an ohmmeter in preparation for testing a starter solenoid.

2-10 TUNING

Check the internal wiring harness if problems have been encountered with any of the electrical components. Check for frayed or chafed insulation and/or loose connections between wires and terminal connections.

Short Test (See Wiring Diagram for engine being serviced in the Appendix.)

Disconnect the internal wiring harness from the electrical components. Use a magneto analyzer, set on Scale No. 3 and check for continuity between any of the wires in the harness. Use Scale No. 3 and check for continuity between any wire and a good ground. If continuity exists, the harness **MUST** be repaired or replaced.

Resistance Test (See the Wiring Diagram in the Appendix.)

Use a magneto analyzer, set on Scale No. 2. Clip the small red and black leads together. Turn the meter adjustment knob for Scale No. 2 until the meter pointer aligns with the set position on the left side of the **"OK"** block on Scale No. 2. Separate the small red and black leads. Use the Wiring Diagram in the Appendix, and check each wire for resistance between the harness connection and the terminal ends. If resistance exists (meter reading outside the **"OK"** block) the harness **MUST** be repaired or replaced.

2-12 WATER PUMP CHECK

FIRST A GOOD WORD: The water pump **MUST** be in very good condition for the engine to deliver satisfactory service. The pump performs an extremely important function by supplying enough water to properly cool the engine. Therefore, in most cases, it is advisable to replace the complete water pump assembly at least once a year, or anytime the lower unit is disassembled for service.

Water pump housing on a V4 engine. Notice how the pump is installed on top of the lower unit.

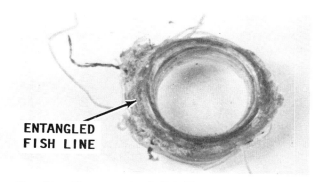

Considerable amount of fish line entangled around the propeller shaft. Some of the fish line actually melted, giving it the appearance of a washer.

Sometimes during adjustment procedures, it is necessary to run the engine with a flush device attached to the lower unit. **NEVER** operate the engine over 1000 rpm with a flush device attached, because the engine may **"RUNAWAY"** due to the no-load condition on the propeller. A "runaway" engine could be severely damaged. As the name implies, the flush device is primarily used to flush the engine after use in salt water or contaminated fresh water. Regular use of the flush device will prevent salt or silt deposits from accumulating in the water passageway. During and immediately after flushing, keep the motor in an upright position until all of the water has drained from the drive shaft housing. This will prevent water from entering the power heads by way of the drive shaft housing and the exhaust ports, during the flush. It will also prevent residual water from being trapped in the driveshaft housing and other passageways.

To test the water pump, the lower unit **MUST** be placed in a test tank or the boat moved into a body of water. The pump must now work to supply a volume to the engine.

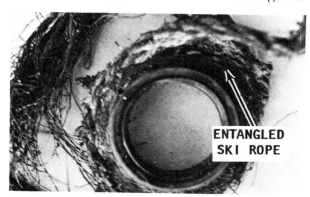

Ski rope entangled around the propeller shaft. Some of the rope has actually melted and fused together. In this case, the rope cut through the lower unit oil seal, allowing the lubricant to escape.

Lack of adequate water supply from the water pump thru the engine will cause any number of power head failures, such as stuck rings, scored cylinder walls, burned pistons, etc.

2-13 PROPELLER

Check the propeller blades for nicks, cracks, or bent condition. If the propeller is damaged, the local marine dealer can make repairs or send it out to a shop specializing in such work.

Remove the cotter key, propeller nut, and propeller from the shaft. Check the propeller shaft seal to be sure it is not leaking. Check the area just forward of the seal to be sure a fish line is not wrapped around the shaft.

Operation At Recommended RPM

Check with the local OMC dealer, or a propeller shop for the recommended size and pitch for a particular size engine, boat, and intended operation. The correct propeller should be installed on the engine to enable operation at recommended rpm.

Two rpm ranges are usually given. The lower rpm is recommended for large, heavy slow boats, or for commercial applications. The higher rpm is recommended for light, fast boats. The wide rpm range will result in greater satisfaction because of maximum performance and greater fuel economy. If the engine speed is above the recommended rpm, try a higher pitch propeller or the same pitch cupped. See Chapter 1 for explanation of propeller terms, pitch, diameter, cupped, etc.

For a dual engine installation, the next higher pitch propeller may prove the most satisfactory condition for water skiing.

2-14 LOWER UNIT

NEVER remove the vent or filler plugs when the lower unit is hot. Expanded lubricant would be released through the plug hole. Check the lubricant level after the unit has been allowed to cool. Add only OMC approved gear lubricant. **NEVER** use regular automotive-type grease in the lower unit, because it expands and foams too much. Outboard lower units do not have provisions to accommodate such expansion.

If the lubricant appears milky brown, indicating the presence of water, a check should be made to determine how the water entered. If large amounts of lubricant must be added to bring the lubricant up to the full mark, a thorough inspection should be made to find the cause of the lubricant loss.

Draining Lower Unit

The fill/drain plug on Johnson/Evinrude lower units may be located towards the bottom of the unit on the port side, starboard side, or on the leading edge of the lower unit.

Remove the drain plug and then remove the vent plug located just above the anti-cavitation plate.

Checking the water circulation with the engine operating in an adequate test tank. Notice the water discharge, indicating the water pump and passages are in satisfactory condition.

Example of a damaged propeller. This unit should have been replaced long before this amount of damage was sustained.

2-12 TUNING

Filling Lower Unit

Position the drive unit approximately vertical and without a list to either port or starboard. Insert the lubricant tube into the **FILL/DRAIN** hole at the bottom plug hole, and inject lubricant until the excess begins to come out the **VENT** hole. Install the **VENT** plug first then replace the **FILL** plug with **NEW** gaskets. Check to be sure the gaskets are properly positioned to prevent water from entering the housing. Many times some of the gear lubricant is lost during installation of the plugs. Therefore, if the vent plug is removed again, and more lubricant added very **SLOWLY** using a small-spout oil can to allow air to pass out the opening, there is no doubt but what the unit will be filled to capacity.

For detailed lower unit service procedures, see Chapter 8. For lower unit lubrication capacities, see the Appendix.

Repairs and Adjustments

For detailed procedures to disassemble, clean, assemble, and adjust the carburetor, see the appropriate section in Chapter 4 for the carburetor type on the engine being serviced.

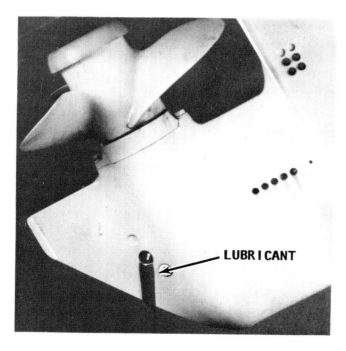

Draining lubricant from the lower unit. The gear oil in the lower unit should be checked on a daily basis during the season of operation. The oil should be drained and replenished with new oil every 100 hours.

2-15 BOAT TESTING

Hook and Rocker

Before testing the boat, check the boat bottom carefully for marine growth or evidence of a "hook" or a "rocker" in the bottom. Either one of these conditions will greatly reduce performance.

Performance

Mount the motor on the boat. Install the remote control cables and check for proper adjustment.

Make an effort to test the boat with what might be considered an average gross load. The boat should ride on an even keel, without a list to port or starboard. Adjust the motor tilt angle, if necessary, to permit the boat to ride slightly higher than the stern. If heavy supplies are stowed aft of the center, the bow will be light and the boat will "plane" more efficiently. For this test the boat must be operated in a body of water.

Check the engine rpm at full throttle. The rpm should be within the Specifications in the Appendix. All OMC engine model serial number identification plates indicate the horsepower rating and rpm range for the engine. If the rpm is not within specified range, a propeller change may be in order. A higher pitch propeller will decrease rpm, and a lower pitch propeller will increase rpm.

For maximum low speed engine performance, the idle mixture and the idle rpm should be readjusted under actual operating conditions.

Lower unit drain plug with built-in magnet. Such a magnet will pick up small metallic particles before they can cause damage to expensive parts.

3
POWERHEAD

3-1 INTRODUCTION

The carburetion and ignition principles of two-cycle engine operation **MUST** be understood in order to perform a proper tune-up on an outboard motor. Therefore, it would be well worth the time to study the principles of two-cycle engines as outlined in this section.

A Polaroid, or equivalent instant-type camera, is an extremely useful item providing the means of accurately recording the arrangement of parts and wire connections **BEFORE** the disassembly work begins. Such a record is most valuable during the assembly work.

Tags are handy to identify wires after they are disconnected to ensure they will be connected to the same terminal from which they were removed. These tags may also be used for parts where marks or other means of identification are not possible.

THEORY OF OPERATION

The two-cycle engine differs in several ways from a conventional four-cycle (automobile) engine.

1- The method by which the fuel-air mixture is delivered to the combustion chamber.
2- The complete lubrication system.
3- In most cases, the ignition system.
4- The frequency of the power stroke.

These differences will be discussed briefly and compared with four-cycle engine operation.

Intake/Exhaust

Two-cycle engines utilize an arrangement of port openings to admit fuel to the combustion chamber and to purge the exhaust gases after burning has been completed. The ports are located in a precise pattern in order for them to be open and closed off at an exact moment by the piston as it moves up and down in the cylinder. The exhaust port is located slightly higher than the fuel intake port. This arrangement opens the exhaust port first, as the piston starts downward, and therefore, the exhaust phase begins a fraction of a second before the intake phase.

Actually, the intake and exhaust ports are spaced so closely together that both open almost simultaneously. For this reason, the pistons of most two-cycle engines have a deflector-type top. This design of the piston top serves two purposes very effectively.

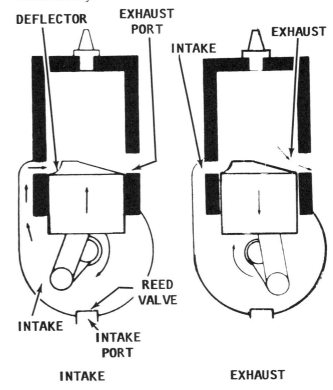

Drawing to depict the intake and exhaust cycles of a two-cycle engine.

First, it creates turbulence when the incoming charge of fuel enters the combustion chamber. This turbulence results in more complete burning of the fuel than if the piston top were flat. The second effect of the deflector-type piston crown is to force the exhaust gases from the cylinder more rapidly.

This system of intake and exhaust is in marked contrast to individual valve arrangement employed on four-cycle engines.

Lubrication

A two-cycle engine is lubricated by mixing oil with the fuel. Therefore, various parts are lubricated as the fuel mixture passes through the crankcase and the cylinder. Four-cycle engines have a crankcase containing oil. This oil is pumped through a circulating system and returned to the crankcase to begin the routing again.

Power Stroke

The combustion cycle of a two-cycle engine has four distinct phases.
1- Intake
2- Compression
3- Power
4- Exhaust

Three phases of the cycle are accomplished with each stroke of the piston, and the fourth phase, the power stroke occurs with each revolution of the crankshaft. Compare this system with a four-cycle engine. A stroke of the piston is required to accomplish each phase of the cycle and the power stroke occurs on every other revolution of the crankshaft. Stated another way, two revolutions of the four-cycle engine crankshaft are required to complete one full cycle, the four phases.

Physical Laws

The two-cycle engine is able to function because of two very simple physical laws.

One: Gases will flow from an area of high pressure to an area of lower pressure. A tire blowout is an example of this principle. The high-pressure air escapes rapidly if the tube is punctured.

Two: If a gas is compressed into a smaller area, the pressure increases, and if a gas expands into a larger area, the pressure is decreased.

If these two laws are kept in mind, the operation of the two-cycle engine will be easier understood.

Actual Operation

Beginning with the piston approaching top dead center on the compression stroke: The intake and exhaust ports are closed by the piston; the reed valve is open; the spark plug fires; the compressed fuel-air mixture

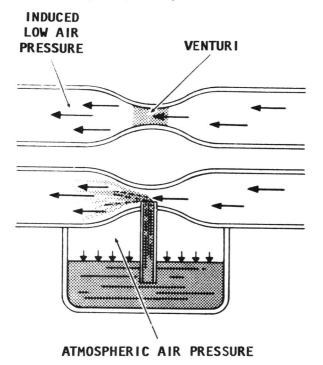

Air flow principle for a modern carburetor.

Adding OMC approved oil into the fuel tank.

is ignited; and the power stroke begins. The reed valve was open because as the piston moved upward, the crankcase volume increased, which reduced the crankcase pressure to less than the outside atmosphere.

As the piston moves downward on the power stroke, the combustion chamber is filled with burning gases. As the exhaust port is uncovered, the gases, which are under great pressure, escape rapidly through the exhaust ports. The piston continues its downward movement. Pressure within the crankcase increases, closing the reed valves against their seats. The crankcase then becomes a sealed chamber. The air-fuel mixture is compressed ready for delivery to the combustion chamber. As the piston continues to move downward, the intake port is uncovered. Fresh fuel rushes through the intake port into the combustion chamber striking the top of the piston where it is deflected along the cylinder wall. The reed valve remains closed until the piston moves upward again.

When the piston begins to move upward on the compression stroke, the reed valve opens because the crankcase volume has been increased, reducing crankcase pressure to less than the outside atmosphere. The intake and exhaust ports are closed and the fresh fuel charge is compressed inside the combustion chamber.

Pressure in the crankcase decreases as the piston moves upward and a fresh charge of air flows through the carburetor picking up fuel. As the piston approaches top dead center, the spark plug ignites the air-fuel mixture, the power stroke begins and one complete cycle has been completed.

Cross Fuel Flow Principle

OMC pistons are a deflector dome type. The design is necessary to deflect the fuel charge up and around the combustion chamber. The fresh fuel mixture enters the combustion chamber through the intake ports and flows across the top of the piston. The piston design contributes to clearing the combustion chamber, because the incoming fuel pushes the burned gases out the exhaust ports.

Loop Scavenging

All three-cylinder engine powerheads have what is commonly known as a loop scavenging system. The piston dome is relatively flat on top with just a small amount of crown. Pressurized fuel in the crankcase is forced up through the skirt of the piston and out through irregular shaped openings cut in the skirt. After the fuel is forced out the piston skirt openings it is transferred upward through long deep

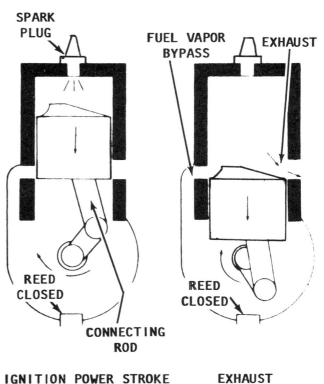

Complete piston cycle of a two-cycle engine, depicting intake, power, and exhaust.

grooves molded into the cylinder wall. The fuel then enters the combustion portion of the cylinder and is compressed, as the piston moves upward.

This particular powerhead does not have intake cover plates, because the intake passage is molded into the cylinder wall as described in the previous paragraph. Therefore, if these engines are being serviced, disregard the sections covering intake cover plates.

Timing

The exact time of spark plug firing depends on engine speed. At low speed the spark is retarded -- fires later than when the piston is at or beyond top dead center. Therefore, the timing is advanced as the throttle is advanced.

At high speed, the spark is advanced -- fires earlier than when the piston is at top dead center.

Procedures for making the timing and synchronization adjustment will be found in Chapter 5.

Summary

More than one phase of the cycle occurs simultaneously during operation of a two-cycle engine. On the downward stroke, power occurs above the piston while the ports are closed. When the ports open, exhaust begins and intake follows. Below the piston, fresh air-fuel mixture is compressed in the crankcase.

On the upward stroke, exhaust and intake continue as long as the ports are open. Compression begins when the ports are closed and continues until the spark plug ignites the air-fuel mixture. Below the piston, a fresh air-fuel mixture is drawn into the crankcase ready to be compressed during the next cycle.

3-2 CHAPTER ORGANIZATION

This chapter is divided into 24 main service sections. Each section covers a particular area of service and outlines complete instructions for the work to be performed. Because of the many countless number of outboard units in the field, it would be impractical and almost impossible to give detailed procedures for removal and installation of each bolt, carburetor, starter, and other "buildup" type units.

Therefore, the sections, for the particular powerhead work to be performed, begin with the preliminary access tasks completed. As an example, disassembly of the powerhead begins with the necessary hood, cowling, and accessories, removed.

The information is presented in a logical sequence for complete powerhead overhaul.

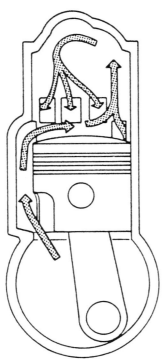

Drawing to depict fuel flow of the "loop charge" while the piston is on the down stroke.

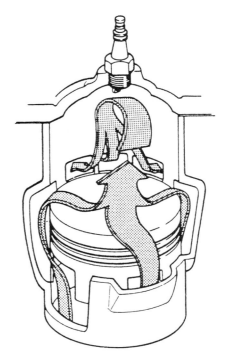

Drawing to depict exhaust leaving the cylinder and fuel entering through the three ports in the piston.

DISASSEMBLING 3-5

The instructions can be followed generally for almost any size horsepower engine. In rare cases, where the procedures differ depending on the model being serviced, separate steps are included. One example is the different types of crankshaft installations for the V4, V6, and 3-cylinder engines.

The illustrations accompanying the text are from different size units and the captions clearly identify which model is covered.

Exploded drawings, showing principle parts, for the various size powerheads are included at the end of the chapter. These drawings will prove to be most helpful to establish the relationship of the various parts to each other.

Special tools may be called out in certain instances. These tools may be purchased from the local Johnson/Evinrude dealer or directly from Customer Service Department, Outboard Marine Corporation (OMC), Waukegan, Illinois, 60085.

The chapter ends with Break-in Procedures, Section 3-24, to be performed after the powerhead has been assembled, all accessories installed, and the powerhead mounted on the exhaust housing.

Torque Values

All torque values must be met when they are specified. Many of the outboard castings and other parts are made of aluminum. The torque values are given to prevent stretching the bolts, but more importantly to protect the threads in the aluminum. It is extremely important to tighten the connecting rods to the proper torque value to ensure proper service. The head bolts are probably the next most important torque value.

Powerhead Components

Service procedures for the carburetors, fuel pumps, starter, and other powerhead components are given in their respective chapters of this manual. See the Table of Contents.

Reed Installation

All reeds on Johnson/Evinrude engines covered in this manual are installed just behind the carburetor behind the intake manifold.

Cleanliness

Make a determined effort to keep parts and the work area as clean as possible. Parts **MUST** be cleaned and thoroughly inspected before they are assembled, installed, or adjusted. Use proper lubricants, or their equivalent, whenever they are recommended.

Keep rods and rod caps together as a set to ensure they will be installed as a pair and in the proper sequence.

Needle bearings **MUST** remain as a complete set. **NEVER** mix needles from one set with another. If only one needle is damaged, the complete set **MUST** be replaced.

3-3 POWERHEAD DISASSEMBLING

Preliminary Work

Before the powerhead can be disassembled, the battery must be disconnected; fuel lines disconnected; and the carburetor, alternator, starter, flywheel, and ignition components, all removed. Take time to identify the hose between the intake manifold and the bypass covers, also the fuel primer hose, to ensure these hoses are connected and routed properly during installation. You may elect to follow the practice of many professional mechanics by taking a series of photographs of the engine with the flywheel removed: one from the top, and a couple from the sides showing the wiring and arrangement of parts.

If in doubt as to how these items are to be removed, refer to the appropriate chapter.

Remove the three bolts on each side of the powerhead, the two in the front, and the two nuts from the studs at the rear of the powerhead. If servicing a V6 unit, four studs are located at the rear. The powerhead is then removed from the exhaust housing.

BAD NEWS

If the unit is several years old, or if it has been operated in salt water, or has not had proper maintenance, or shelter, or any number of other factors, then separating the powerhead from the exhaust housing may not be a simple task. An air hammer may be required on the studs to shake the corrosion loose; heat may have to be applied to the casting to expand it slightly; or other devices employed in order to remove the powerhead. One very serious condition

Cylinder block water passages corroded, preventing proper circulation of coolant water.

would be the driveshaft "frozen" with the crankshaft. In this case, a circular plug-type hole must be drilled and a torch used to cut the driveshaft. Let's assume the powerhead will come free on the first attempt.

The following procedures pickup the work after these preliminary tasks have been completed.

3-4 HEAD SERVICE

Usually the head/s is removed and an examination of the cylinders made to determine the extent of overhaul required. However, if the head/s has not been removed, back out all of the head bolts and lift the head/s free of the powerhead.

A thermostat is installed in the head of all 3-cylinder and V6 engines. On V4 engines, the thermostat is installed in a Bakelite housing between the two heads just below the exhaust housing plate. The thermostat has three parts, a top, middle, and bottom. A hose connects the thermostat with each head. In addition to the thermostat, the engine has a thermostat

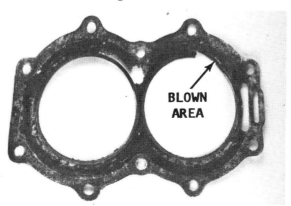

Blown head gasket, possibly caused by an overheating condition. The water can then find its way into the cylinder and cause further damage. The head gasket should be replaced as soon as possible, the engine run, and engine cleaner injected through the carburetor.

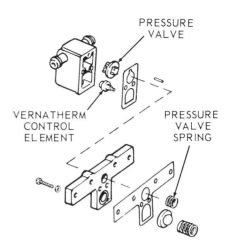

Thermostat installed on late model powerheads, mounted between the heads on V4 engines only.

bypass valve. These items are easily removed, inspected and cleaned.

Normally, if a thermostat is not functioning properly, it is almost always stuck in the open position. An engine operating at too low a temperature is almost as much a problem as an engine running too hot.

Therefore, during a major overhaul, good shop practice dictates to replace the thermostat and eliminate this area as a possible problem at a later date.

Lay a piece of fine sandpaper or emery paper on a flat surface (such as a piece of glass) with the abrasive side facing up. With the machined face of the head on the sandpaper, move the head in a circular motion to dress the surface. This procedure will also indicate any "high" or "low" spots.

Check the spark plug opening/s to be sure the threads are not damaged. Most marine dealers can insert a heli-coil into a spark plug opening if the threads have been damaged.

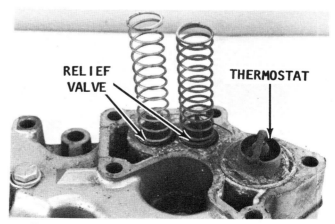

Thermostat installation on a 3-cylinder engine. Both springs are the same weight and control the relief valves.

REED SERVICE 3-7

A 3-cylinder head (left) and a V4 head (right). The water temperature sending unit wire is identified.

On many engines, a sending unit is installed in the head to warn the operator if the engine begins to run too hot. The light on the dash can be checked by turning the ignition switch to the **ON** position, and then ground the wire to the sending unit. The light should come on. If it does not, replace the bulb and repeat the test. On later model engines the light was replaced with a horn mounted in the shift box. This horn sounds if the engine overheats.

Head installation procedures are given in Section 3-23, Head Installation.

3-5 REED SERVICE

DESCRIPTION

Fuel Delivery

The fuel delivery may be one of several types. All three-cylinder engines are equipped with three carburetors, one for each cylinder. V4 engines have two carburetors, each with double barrels, each barrel serving a single cylinder. V6 engines have three double barrel carburetors, each barrel serving a single cylinder.

Six reed boxes mounted in the reed plate of a V6 powerhead.

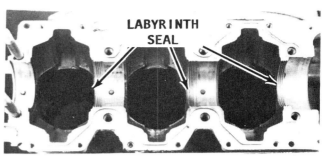

A 3-cylinder engine block with the labyrinth seals to hold pressure and vacuum within each cylinder.

Cylinder Sealing

The three-cylinder engines covered in this manual are equipped with an upper seal and bearing, a lower seal and bearing, and two main bearings between. A labyrinth seal is used at the two center main bearings and just above the bottom seal, to provide an effective seal between the cylinders.

On the V4 and V6 engines covered in this manual, a top and bottom seal is used with sealing rings around the crankshaft in the center between each cylinder, in each bank, to seal for pressure and vacuum. The V4 engines use 6 rings and the V6 engines use 8 rings.

Reed Arrangement

On a three-cylinder engine, three sets of reeds are use, one for each cylinder. On a four-cylinder powerhead, four sets of reeds are installed, and the V6 powerhead has six sets of reeds. These reeds are installed on a reed plate with a reed box. Fuel is delivered to the reeds as described in the paragraph above, Fuel Delivery. The reed arrangement operates in much the same manner as the reed in a saxophone or other wind instrument. At rest, the reed is closed and seals the opening to which it is attached. In the case of an outboard engine, this opening is between the crankcase and the carburetor. The reeds are mounted in the intake manifold, just behind the carburetor.

V4 engine block with sealing rings to hold pressure and vacuum within each cylinder.

Actual Operation

The piston creates vacuum and pressure in the crankcase as it moves up and down in the cylinder. As the piston moves upward, a vacuum is created in the crankcase. This vacuum "lifts" the reed off its seat, allowing fuel to pass. On the compression stroke, when the piston moves downward, pressure is created and the reed is forced closed.

Reed Designs

A wide range of reed boxes may be found on an outboard unit, due to the varying designs of the engines. All installations employ the same principle and there is no difference in their operation.

Broken Reed

A broken reed is usually caused by metal fatigue over a long period of time. The failure may also be due to the reed flexing too far because the reed stop has not been adjusted properly or the stop has become distorted. If the reed is broken, the loose piece **MUST** be located and removed, Illustration #1, before the engine is returned to service. The piece of reed may have found its way into the crankcase, into the passage leading to the cylinder, or in the cylinder. If the broken piece cannot be located, the powerhead must be completely disassembled until it is located and removed.

An excellent check for a broken reed on an operating engine is to hold an ordinary business card approximately 2" (9.08 cm) in

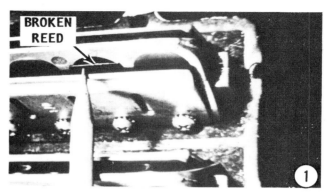

front of the carburetor. Under normal operating conditions, a very small amount of fine mist will be noticeable, but if fuel begins to appear rapidly on the card from the carburetor, one of the reeds is broken and causing the backflow through the carburetor and onto the card, Illustration #2.

A broken reed will cause the engine to operate roughly and with a "pop" back through the carburetor.

Reed Stops

If the reed stops have become distorted, the most effective corrective action is to replace the complete reed box as an assembly.

Reed to Reed Box Check

The specified clearance of the reed from the base plate, when the reed is at rest, is 0.010" (0.254 mm) at the tip of the reed, Illustration #3.

An alternate method of checking the reed clearance is to hold the reed up to the sunlight and look through the back side.

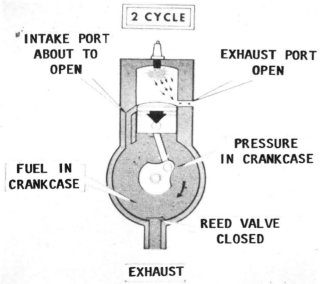

Diagram to illustrate operation of a two-cycle outboard engine.

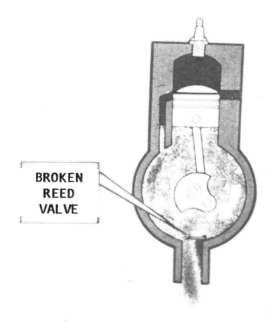

REED SERVICE 3-9

Some air space should be visible, but not a great amount. If in doubt, check the reed at the tip with a feeler gauge. The maximum clearance should not exceed 0.010" (0.254 mm). If the clearance is excessive, the reed box must be replaced as a complete assembly.

The reeds must **NEVER** be turned over in an attempt to correct a problem. Such action would cause the reed to flex in the opposite direction and the reed would break in a very short time.

V-Type Reed Boxes

As the name implies, these reed boxes are shaped in a "V" with a set of reeds and stops on both arms of the "V". If a problem develops with this type reed box, the complete assembly **MUST** be replaced -- reeds, box, and stops. The assembly may be purchased as a complete unit and the cost will usually not exceed the time, effort, and problems encountered in an attempt to replace only one part.

CLEANING AND SERVICE

Always handle the reeds with the utmost care. Rough treatment will result in the reeds becoming distorted and will affect their performance.

Wash the reeds in solvent, and blow them dry with compressed air from the **BACK SIDE ONLY**. Do not blow air through the reed from the front side. Such action would

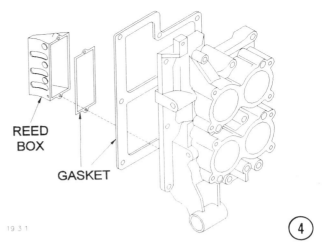

cause the reed to open and fly up against the reed stop. Wipe the front of the reed dry with a lint free cloth.

Clean the reed box thoroughly by removing any old gasket material.

Secure the reed blocks to the reed plate with the screws tightened to a torque value of 25-35 ft lbs (3-4Nm).

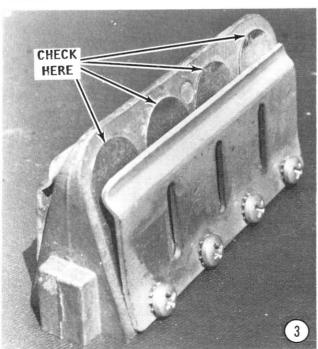

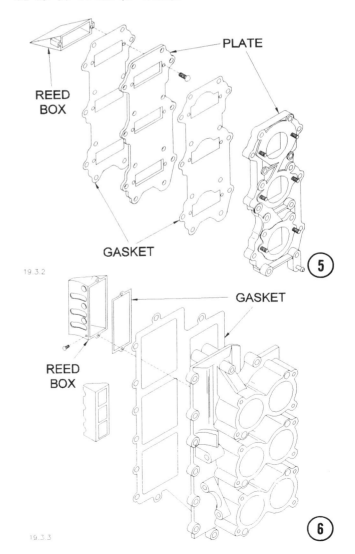

Check for chipped or broken reeds. Observe that the reeds are not preloaded or standing open. Satisfactory reeds will not adhere to the reed block surface, but still there is not more than 0.010" (0.254 mm) clearance between the reed and the block surface.

Check the reed location over the reed block, or plate openings, to be sure the reed is centered.

The reed assemblies are then ready for installation.

REED ASSEMBLING

Position a **NEW** gasket in place on the reed plate, Illustration #4 for V4 engines, Illustration #5 for the 3-cylinder engines, and Illustration #6 for the V6 engines. (Illustrations are on previous page.) Install the reed box to the reed plate by threading the screws through the reed plate from the back side and into the reed box. For installation of the reed box to the powerhead, see Section 3-22.

3-6 BYPASS COVERS

On three-cylinder engines the powerhead does not contain bypass covers. The bypass covers the area the fuel travels from the crankcase up the side of the powerhead and into the cylinder.

Seldom does a bypass cover cause any problem.

During a normal overhaul, the bypass covers should be removed, cleaned, and new gaskets installed, Illustration #7. Identify the covers to ensure installation in the same location from which they are removed. This identification is important because the fuel pump or starter solenoid may be attached to the bypass cover.

INSTALLATION

Procedures to install the bypass covers to the powerhead will be found in Section 3-21, Exhaust Cover and Bypass Cover Installation.

3-7 EXHAUST COVER

The exhaust covers are one of the most neglected items on any outboard engine. Seldom are they checked and serviced. Many times an engine may be overhauled and returned to service without the exhaust covers ever having been removed.

The exhaust manifold is located on the port side of three-cylinder engines. On V4 and V6 engines, the exhaust manifold is located at the rear of the powerhead between the two cylinder heads. See next five illustrations for arrangement of the gaskets and plates.

One reason the exhaust covers are not removed is because the attaching bolts usually become corroded in place. This means

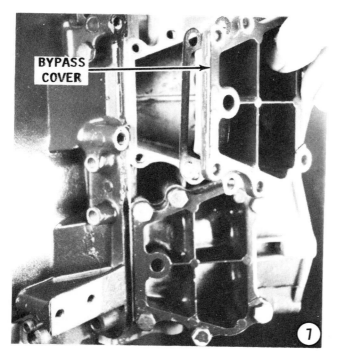

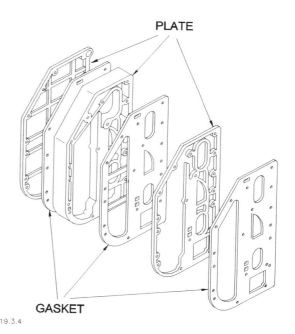

Exhaust plate and gasket arrangement for 3-cylinder engines, 1973 thru 1982.

EXHAUST COVER 3-11

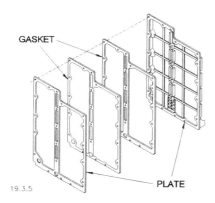

Exhaust plate and gasket arrangement for 85 hp and 115 hp engines, 1973 thru 1977.

they are very difficult to remove, but the work should be done. Heat applied to the bolt head and around the exhaust cover will help in removal. However, some bolts may still be broken. If the bolt is broken it must be drilled out and the hole tapped with new threads.

The exhaust covers are installed over the exhaust ports to allow the exhaust to leave the powerhead and be transferred to the exhaust housing. If the cover was the only item over the exhaust ports, they would become so hot from the exhaust gases they might cause a fire or a person would be severely burned if they came in contact with the cover.

Therefore, an inner plate is installed to help dissipate the exhaust heat. Two gaskets are installed -- one on either side of the inner plate. Water is channeled to circulate between the exhaust cover and the inner plate. This circulating water cools the exhaust cover and prevents it from becoming a hazard.

A thorough cleaning of the inner plate behind the exhaust covers should be performed during a major engine overhaul. If the integrity of the exhaust cover assembly is in doubt, replace the inner plate.

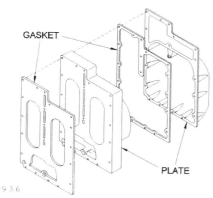

Exhaust plate and gasket arrangement for all V4 powerheads except the 85 hp and 115 hp units -- 1973 thru 1977.

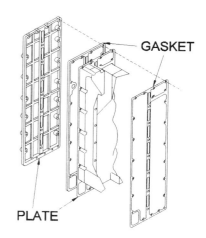

Exhaust plate and gasket arrangement for the 175 hp and 200 hp engines 1977.

On powerheads equipped with the heat/electric choke, a baffle is installed on the inside surface of the inner plate. This baffle is heated from the engine exhaust gases. Air passing through the baffle heats the choke and allows the choke to open as engine temperature rises.

CLEANING

Clean any gasket material from the cover and inner plate surfaces. Check to be sure the water passages in the cover and plate are clean to permit adequate passage of cooling water.

Inspect the inlet and outlet holes in the powerhead to be sure they are clean and free of corrosion. The openings in the powerhead may be cleaned with a small size screwdriver.

Clean the area around the exhaust ports and in the webs running up to the exhaust ports. Carbon has a habit of forming in this area.

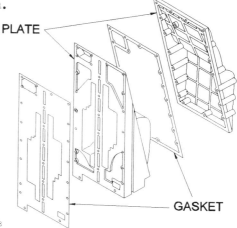

Exhaust plate and gasket arrangement for all V6 powerheads except the 175 hp -- 1977 and the 200 hp -- 1976 and 1977.

3-12 POWERHEAD

INSTALLATION

Procedures to install the exhaust covers will be found in Section 3-21, Exhaust Cover and Bypass Cover Installation.

3-8 TOP SEAL

The top seal simply prevents oil from escaping around the bearing. This seal also acts as a barrier against dirt or other foreign matter from reaching the bearing.

REMOVAL V4 AND V6 ENGINE

Remove the bolts securing the upper cap in place. **CAREFULLY** work the bearing cap up and free of the block. Drive the seal from the bearing cap using a punch and hammer. Remove the O-rings from the outside edge of the bearing cap.

REMOVAL THREE-CYLINDER ENGINE

In order to replace the top seal on the three-cylinder engine, the crankcase cover must be removed, the crankshaft raised, and the top bearing removed, before the seal can be removed. See section 3-18 to accomplish this task.

ASSEMBLING V4 AND V6 ENGINE

Install a **NEW** seal into the bearing cap, with the hard side facing **UP**, Illustration #8. Position a **NEW** O-ring in place around the

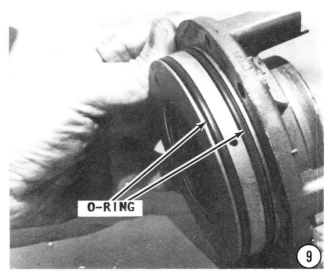

bearing cap, Illustration #9. Some models have one O-ring, and others have two. Do not attempt to install two rings if there is only one groove. Coat the O-ring/s with oil. Set the assembly aside for later installation, see Section 3-18.

3-9 BOTTOM SEAL

The bottom seal has equal importance as the top seal. This seal is installed to maintain vacuum and pressure in the lower half of the crankcase for the lower cylinder.

The lower seal on all engines covered in this manual is pressed into the lower bearing cap.

Seal Installed in Bearing Cap
V4 and V6 Engines

The seal on these engines is installed inside the lower bearing cap. Its function is to prevent exhaust fumes from entering the crankcase, and maintain pressure and vacuum inside the crankcase.

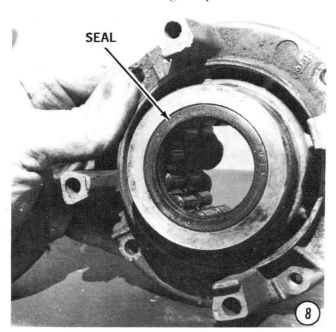

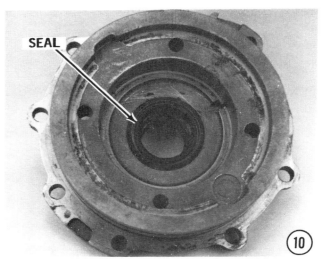

Removal

If the seal is to be replaced without overhauling the powerhead, proceed as follows:

Remove the outside perimeter bolts on the bearing cap. Remove the four bolts closer to the crankshaft. Use a pair of screwdrivers and work the bearing cap free of the lower bearing. Remove the O-rings from the outside edge of the bearing cap. Use a punch and drive the seal out of the bearing cap.

Install the new seal with the hard side **DOWN**, Illustration #10. Drive the seal into place using the correct size socket and a hammer. Install the outside O-ring/s around the cap, Illustration #11. Some models use one O-ring and others use two. If the powerhead is to be overhauled, set the assembly aside for later installation.

If the powerhead is not to be overhauled, ee installation procedures in Section 3-18.

al in a Cap
Three-cylinder Engines

The seal on these engines is installed in a The cap is bolted to the bottom of the ⸱rhead. The four bolts securing the cap   be removed before the crankcase cover   be removed. The seal can be punched out d new ones installed without difficulty. Remove the O-rings from the cap.

If the powerhead is not to be overhauled, install the seal with the hard side **DOWN**. Drive the seal and O-ring into place using the correct size socket and hammer, Illustration #12. Align the holes in the cap with the holes in the powerhead. Secure the cap in place with the attaching bolts.

If the powerhead is to be overhauled, see Section 3-18 to install the bottom seal and cap.

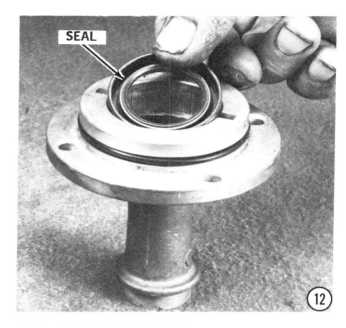

3-10 CENTERING PINS

All Johnson/Evinrude outboard engines have at least one, and in most cases two, centering pins installed through the crankcase cover. These pins index into matching holes in the powerhead block when the crankcase cover is installed. These pins center the crankcase cover on the powerhead block.

The centering pins are tapered. The pins must be carefully checked to determine how they are to be removed from the cover, Illustration #13. In most cases the pin is

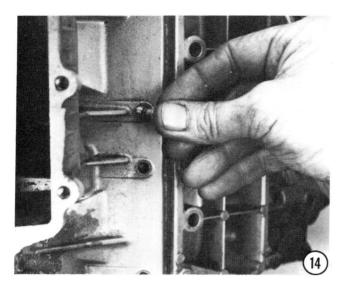

removed by using a center punch and tapping the pin towards the carburetor or intake manifold side of the crankcase.

When removing a centering pin, hold the punch securely onto the pin head, then strike the punch a good hard forceful blow. **DO NOT** keep beating on the end of the pin, because such action would round the pin head until it would not be possible to drive it out of the cover.

Centering pins are the first item to be installed in the cover when replacing the crankcase cover.

3-11 MAIN BEARING BOLTS AND CRANKCASE SIDE BOLTS

The main bearing bolts are installed

through the crankcase cover into the powerhead block. Most engines have two bolts installed for the top main bearing, two for the center main bearing, and two for the lower main bearing.

In many cases the upper and lower main bearing bolts are **DIFFERENT** lengths. Therefore, take time to tag and identify the bolts to ensure they will be installed in the same location from which they were removed.

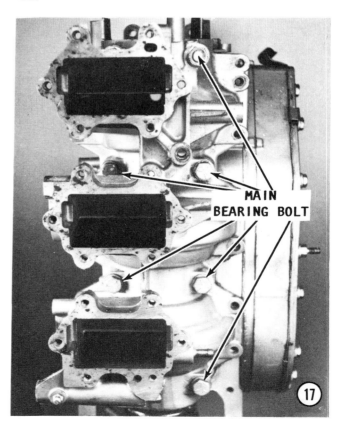

The crankcase side bolts are installed along the edge of the crankcase cover to secure the cover to the cylinder block. These bolts usually have a 7/16" head and all must be removed before the crankcase cover can be removed. Remove the crankcase side bolts, Illustration #14.

Remove all the main bearing bolts. Be sure to remove any bolts or Allen screws from the intake manifold in the reed box areas. Naturally, all main bearing bolts must be removed before the crankcase cover will come free, Illustration #15 for V4 engines, Illustration #16 for V6 engines, and Illustration #17 for 3-cylinder engines. (These illustration are on the previous page.)

INSTALLATION

Main bearing bolt and the crankcase side bolt installation is given in Section 3-20, Cylinder Block Assembling, under Main Bearing and Crankcase Side Bolt Installation.

3-12 CRANKCASE COVER ALL ENGINES

REMOVAL

Remove all bolts securing the upper bearing cap and all bolts securing the lower bearing cap in place, if this has not been done.

After all side bolts and main bearing bolts have been removed, use a soft-headed mallet and tap on the bottom side of the crankshaft. A soft, hollow sound should be heard indicating the cover has broken loose from the crankcase. If this sound is not heard, check to be sure all the side bolts and main bearing bolts have been removed. **NEVER** pry between the cover and the crankcase or the cover will surely be distorted. If the cover is distorted, it will fail to make a proper seal when it is installed. Such an installation would damage both crankcase halves and render them unfit for service -- an expensive replacement.

Once the crankshaft has been tapped, as described, and the proper sound heard, the cover will be jarred loose and may be removed.

CLEANING AND INSPECTING

Wash the cover with solvent, and then dry it thoroughly. Check the mating surface to the cylinder block for damage that may affect the seal.

Inspect the sealing ring grooves at the center main bearing area to be sure they are clean and not damaged in any manner, Illustration #18.

INSTALLATION

Installation procedures for the crankcase cover are given in Section 3-19, Crankcase Cover Installation.

3-13 CONNECTING RODS AND PISTONS

The connecting rods and their rod caps are a **MATCHED set**. They absolutely **MUST** be identified, kept, and installed as a set. Under no circumstances should the connecting rod and caps be interchanged. Therefore, on a multiple piston engine, **TAKE TIME AND CARE** to tag each rod and rod cap; to keep them together as a set while they are on the bench; and to install them into the same cylinder from which they were removed as a set.

The connecting rod and its cap are manufactured as a set -- as a single unit. After the complete rod and cap have been mated, two holes are drilled through the side of the cap and rod, and the cap is then fractured from the rod. Therefore, the cap must always be installed with its original rod. The cap half of the break can **ONLY** be matched with the other half of the break on the **ORIGINAL** rod, Illustration #19.

Inspect the rod and the rod cap before removing the cap from the crankshaft. Under normal conditions, a line or a dot is

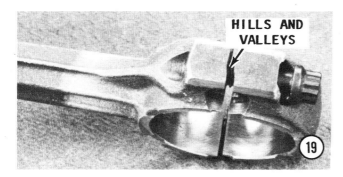

visible on the side of the rod and the cap, Illustration #20. This identification is an assist to assemble the parts together and in the proper location.

Observe into the block and notice how the rods have a "trough". Also notice the hole in the rod in the area where the wrist pin passes through the piston. This hole **MUST ALWAYS** face toward the **UPPER** end of the powerhead during installation. A cutaway area at the lower end of the rod serves to "throw" lubrication onto the main bearing. If the hole at the upper end faces correctly, this cutaway section will also face in the proper direction and the bearing will be adequately lubricated.

REMOVAL

To remove the rod bolts from the cap, it is recommended to loosen each bolt just a little at-a-time and alternately, Illustration #21. This procedure will prevent one bolt from being completely removed while the other is still tightened to its recommended torque value. Such action may very likely warp the cap.

Remove the bolts as described in the previous paragraph, and then **CAREFULLY** remove the rod cap.

Remove the needle bearings and cages from around the crankshaft, Illustration #22. Count the needle bearings and insert them into a separate container -- one container

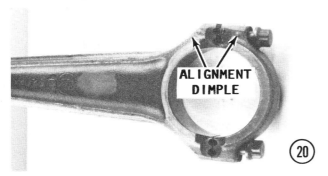

for each rod, with the container clearly identified to ensure they will be installed with the proper rod at the crankshaft journal from which they were removed.

Tap the piston out of the cylinder from the crankshaft side. Immediately attach the proper rod cap to the rod and hold it in place with the rod bolts, Illustration #23. The few minutes involved in securing the cap with the rod will ensure the matched cap remains with its mating rod during the cleaning and assembling work.

Fill the piston skirt with a rag, towel, shop cloths, or other suitable material, Illustration #24. The rag will prevent the rod from coming in contact with the piston skirt

RODS AND PISTONS 3-17

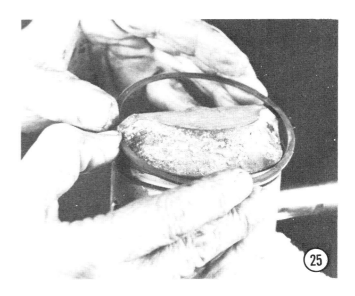

while it is laying on the bench. If the rod is allowed to strike the piston skirt, the skirt may become distorted.

Identify the rod to ensure it will be installed into the cylinder from which it was removed.

Remove and identify the other rod caps, needle bearings and cages, and rods with pistons, in the same manner.

Remove the rings from the pistons. This is accomplished by placing the rod in a vise with the piston skirt resting on the top surface of the vise. Expand each ring with your fingers or a pair of reverse pliers enough to slip the ring free of the piston, Illustration #25.

DISASSEMBLY

Before separating the piston from the rod, notice the location of the piston in relation to the rod. Observe the hole in the rod trough on one side of the rod near the wrist pin opening. This hole must face toward the **TOP** of the engine during installation.

Slanted Piston V4 and V6 Engines

Observe the slanted edge and the sharp edge of the dome-type piston. The slanted edge **MUST** face toward the exhaust side of the cylinder and the sharp edge toward the intake side during installation, Illustration #26.

Loop-Charged Pistons
Three-Cylinder Engines Only

If servicing a three-cylinder loop-charged powerhead, carefully observe the hole in the rod near the wrist pin and the relationship of the irregular cutouts in the piston skirt, Illustration #27. Only in this position will this relationship exist. The rod and piston **MUST** be assembled in this manner or the engine will run **VERY** poorly.

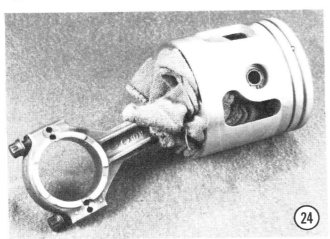

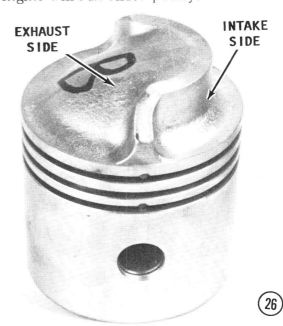

3-18 POWERHEAD

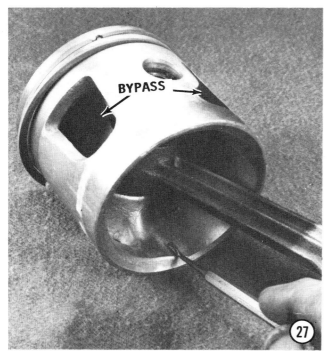

When the rod is installed to the piston, the relationship of the rod can only be one way. The rod holes must face upward and the piston must face as described in the previous paragraph.

All Engines

Observe into the piston skirt. On most model pistons, notice the **"L"** stamped on the boss through which the wrist pin passes. The letter mark identifies the "loose" side of the piston and indicates the side of the piston from which the wrist pin must be driven out without damaging the piston. Some pistons may have the full word **"LOOSE"** stamped on the inside of the piston skirt, Illustration **#28**.

If the piston does not have the **"L"** or the word **"LOOSE"** stamped, the wrist pin may be driven out in either direction.

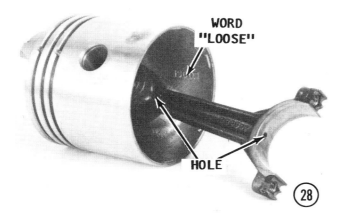

Remove the retaining clips from each end of the wrist pin. Some clips are spring wire type and may be worked free of the piston using a screwdriver, Illustration **#29**. Other model pistons have a Truarc snap ring. This type of ring can only be successfully removed using a pair of Truarc pliers, Illustration **#30**.

It may be necessary to heat the piston in a container of boiling water, Illustration **#31**, in order to press the wrist pin free.

Place the piston in an arbor press using the **PROPER** size cradle for the piston being serviced, and with the **LOOSE** side of the piston facing **UPWARD**.

The wrist pin must be driven out **FROM** the loose side. This may not seem reasonable, but there is a very simple explanation. By placing the piston in the arbor press cradle with the tight side down, and the arbor ram pushing from the loose side, the piston has good support and will not be distorted. If the piston is placed in the arbor press with the loose side down, the piston would be distorted and unfit for further service.

All rods have a wrist pin bearing. This bearing is the caged needle bearing type.

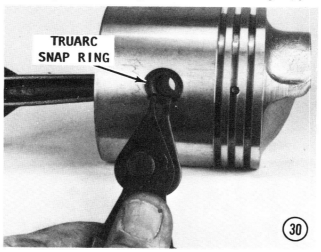

RODS AND PISTONS 3-19

TAKE CARE not to lose any of the bearings when the wrist pin is driven free of the piston.

Alternate Removal Method

If the piston does not have the **"L"** or the word **"LOOSE"** stamped, the wrist pin may be driven out in either direction.

Remove the retaining clips from each end of the wrist pin. Some clips are spring wire type and may be worked free of the piston using a screwdriver. Other model pistons have a Truarc snap ring. This type of ring can only be successfully removed using a pair of Truarc pliers.

It may be necessary to heat the piston in a container of boiling water in order to press the wrist pin free.

If an arbor press or cradle is not available, proceed as follows: heat the piston in a container of very hot water for about ten minutes (heating the piston will cause the metal to expand ever so slightly, but ease the task of driving the pin out); assume a sitting position in a chair, on a box, whatever. Next, lay a couple towels over your legs. Hold your legs tightly together to form a cradle for the piston above your knees. Set the piston between your legs with the **LOOSE** side of the piston facing upward, Illustration #32. Now, drive the wrist pin free using a drift pin with a shoulder. The drift pin will fit into the hole through the wrist pin and the shoulder will ride on the edge of the wrist pin. Use sharp hard blows with a hammer. Your legs will absorb the shock without damaging the piston. If this method is used on a regular basis during the busy season, your legs will develop black-and-blue areas, but no problem, the marks will disappear in a few days.

ROD INSPECTION AND SERVICE

All rods for engines covered in this manual have needle bearings. The needles should be replaced anytime a major overhaul is performed. It is not necessary to replace the cages, but a complete **NEW** set of needles should be purchased and installed.

Place each connecting rod on a surface plate and check the alignment. If light can be seen under any portion of the machined surfaces, or if the rod has a slight wobble on the plate, or if a 0.002" (.05 mm) feeler gauge can be inserted between the machined surface and the surface plate, the rod is bent and unfit for further service.

Inspect the connecting rod bearings for rust or signs of bearing failure. **NEVER** intermix new and used bearings. If even one bearing in a set needs to be replaced, all bearings at that location **MUST** be replaced.

Inspect the bearing surface of the rod and the rod cap for rust and pitting.

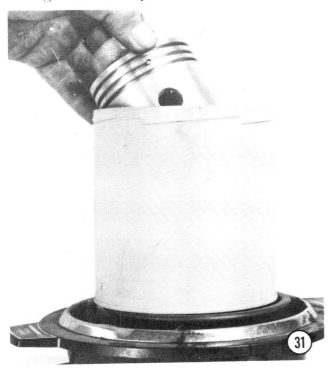

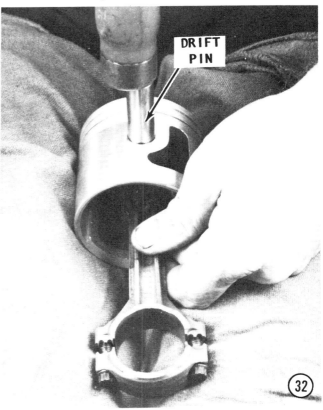

Inspect the bearing surface of the rod and the rod cap for water marks. Water marks are caused by the bearing surface being subjected to water contamination, which causes "etching". The etching resembles the size of the bearing as shown in the accompanying illustration.

Inspect the bearing surface of the rod and rod cap for signs of spalling. Spalling is the loss of bearing surface, and resembles flaking or chipping. The spalling condition will be most evident on the thrust portion of the connecting rod in line with the I-beam. Bearing surface damage is usually caused by improper lubrication.

Check the bearing surface of the rod and rod cap for signs of chatter marks. This condition is identified by a rough bearing surface resembling a tiny washboard. The condition is caused by a combination of low-speed low-load operation in cold water, and is aggravated by inadequate lubrication and improper fuel. Under these conditions, the crankshaft journal is hammered by the connecting rod. As ignition occurs in the cylinder, the piston pushes the connecting rod with tremendous force, and this force is transferred to the connecting rod journal.

Since there is little or no load on the crankshaft, it bounces away from the connecting rod. The crankshaft then remains immobile for a split second, until the piston travel causes the connecting rod to catch up to the waiting crankshaft journal, then hammers it.

In some instances, the connecting rod crankpin bore becomes highly polished.

While the engine is running, a "whirr" and/or "chirp" sound may be heard when the engine is accelerated rapidly from idle speed to about 1500 rpm, then quickly returned to idle. If chatter marks are discovered, the crankshaft and the connecting rods should be replaced.

Rod, rod cap, wrist pin, and wrist pin bearing, after removal.

Inspect the bearing surface of the rod and rod cap for signs of uneven wear and possible overheating. Uneven wear is usually caused by a bent connecting rod. Overheating is identified as a bluish bearing surface color and is caused by inadequate lubrication or operating the engine at excessively high rpm.

Inspect the needle bearings. A bluish color indicates the bearing became very hot and the complete set for the rod **MUST** be replaced, no question.

Service the connecting rod bearing surfaces according to the following procedures and precautions:

a- Align the etched marks on the knob side of the connecting rod with the etched marks on the connecting rod cap.

b- Tighten the connecting rod cap attaching bolts securely.

c- Use **ONLY** crocus cloth to clean bearing surface at the crankshaft end of the connecting rod. **NEVER** use any other type of abrasive cloth.

d- Insert the crocus cloth in a slotted 3/8" diameter shaft. Chuck the shaft in a drill press and operate the press at high speed and at the same time, keep the connecting rod at a 90° angle to the slotted shaft.

e- Clean the connecting rod **ONLY** enough to remove marks. **DO NOT** continue once the marks have disappeared.

f- Clean the piston pin end of the connecting rod using the method described in Steps **d** and **e**, but using 320 grit Carborundum cloth instead of crocus cloth.

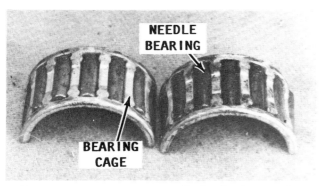

Needle bearings and cages unfit for further service.

Testing two rods at the wrist pin end for warpage.

RODS AND PISTONS 3-21

Testing two rods at the rod cap end for warpage.

g- Thoroughly wash the connecting rods to remove abrasive grit. After washing, check the bearing surfaces a second time.

h- If the connecting rod cannot be cleaned properly, it should be replaced.

i- Lubricate the bearing surfaces of the connecting rods with light-weight oil to prevent corrosion.

PISTON AND RING INSPECTION AND SERVICE

Inspect each piston for evidence of scoring, cracks, metal damage, cracked piston pin boss, or worn pin boss, Illustration #33. Be especially critical during inspection if the engine has been submerged, Illustration #34.

When installing flat dome pistons, the top ring is tapered and fits into a tapered groove in the piston. Care must be exercised when installing rings to the pistons to ensure the proper ring is used with the proper piston in the proper ring groove, Illustration #35.

Carefully check each wrist pin to be sure it is not the least bit bent. If a wrist pin is bent, the pin and piston MUST be replaced

as a set, because the pin will have damaged the boss when it was removed.

Check the wrist pin needle bearings. If any one of the needles is damaged, has a flat spot, or is unfit for further service for any reason, the complete set MUST be replaced.

Grasp each end of the ring with either a ring expander or your thumbnails; open the ring and remove it from the piston, Illustration #36. Many times, the ring may be difficult to remove because it is "frozen" in the piston ring groove. In such a case, use a screwdriver and pry the ring free. The ring may break, but if it is difficult to remove, it MUST be replaced.

OBSERVE the pin in each ring groove of the piston. The ends of the ring MUST

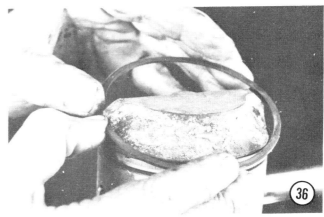

straddle this pin. The pin prevents the ring from rotating while the engine is operating. This fact is the direct opposite of a four-cycle engine where the ring must rotate. In a two-cycle engine, if the ring is permitted to rotate, at one point, the opening between the ring ends would align with either the intake or exhaust port in the cylinder. At that time, the ring would expand very slightly, catch on the edge of the port, and **BREAK**.

Therefore, when checking the condition of the piston, **ALWAYS** check the pin in each groove to be sure it is tight, Illustration #37. If one pin is the least bit loose, the piston **MUST** be replaced, without question. Never attempt to replace the pin, it is **NEVER** successful.

CLEANING AND INSPECTING

Check the piston ring grooves for wear, burns, distortion, or loose locating pins. During an overhaul, the rings should be replaced to ensure lasting repair and proper engine performance after the work has been completed.

Clean the piston dome, ring grooves, and the piston skirt. Clean the piston skirt with a crocus cloth.

Clean carbon deposits from the top of the piston using a soft wire brush, carbon removal solution, or by sand blasting. If a wire brush is used, **TAKE CARE** not to burr or round machined edges.

Wear a pair of good gloves for protection against sharp edges, and clean the piston ring grooves using the recessed end of the

*Cleaning the piston ring grooves. An automotive type ring groove cleaner should **NEVER** be used.*

proper broken ring as a tool. **NEVER** use a rectangular ring to clean the groove for a tapered ring, or use a tapered ring to clean the groove for a rectangular ring.

NEVER use an automotive-type ring groove cleaner to clean piston ring grooves, because this type of tool could loosen the piston ring locating pins. **TAKE CARE** not to burr or round the machined edges. Inspect the piston ring locating pins to be sure they are tight.

Oversize Pistons and Rings

Scored cylinder blocks can be saved for further service by reboring and installing oversize pistons and piston rings. **ONE MORE WORD:** Oversize pistons and rings are not available for all engines. At the time of this printing, the sizes listed in the Appendix were available. Check with the parts department at your local dealer for the model engine you are servicing, and to be sure the factory has not deleted a size from their stock. The pistons should always be ordered and received **BEFORE** the block is rebored, to ensure a proper piston fit for each cylinder, after the work is accomplished.

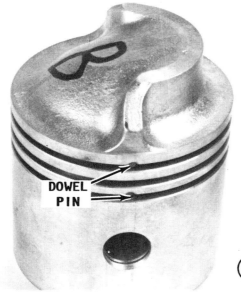

Using a micrometer to check the roundness of the piston.

ASSEMBLING

CRITICAL WORDS

Two conditions absolutely **MUST** exist when the piston and rod assembly are installed into the cylinder block.

For V4 and V6 engine installation, the slanted side of the piston must face toward the exhaust side of the cylinder, Illustration **#38**.

For three-cylinder engine installation, make sure the openings in the pistons are aligned with the cutaways in the cylinder wall, Illustration **#39**.

The hole in the rod near the wrist pin opening and at the lower end of the rod must face **UPWARD**.

Therefore, the rod and piston **MUST** be assembled correctly in order for the assembly to be properly installed into the cylinder. Soak the piston in a container of very hot water for about ten minutes, Illustration **#40**. Heating the piston will cause it to expand ever so slightly, but enough to allow the wrist pin to be pressed through without difficulty.

Before pressing the wrist pin into place, hold the piston and rod near the cylinder block and check to be sure both will be facing in the right direction when they are installed.

Pack the wrist pin needle bearing cage with needle bearing grease, or a good grade of petroleum jelly. Load the bearing cage with needles and insert it into the end of the rod, Illustration **#41**.

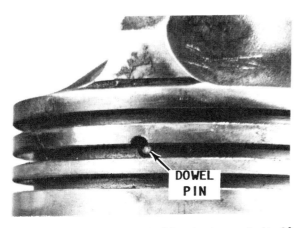

Heat and vibration caused this pin to work itself into the piston until the head became flush with the surface of the ring groove. This condition will allow the ring to rotate during engine operation. The ring end would soon catch on one of the cylinder ports and break.

Slide the rod into the piston boss. On V4 and V6 engine installation, check a second time to be sure the slanted side of the piston is facing toward the exhaust side of the cylinder and the hole in the rod is facing upward.

For a three-cylinder engine installation, check a second time to be sure the hole in the piston aligns with the cutaway in the cylinder wall.

Place the piston and rod in the arbor press with the **LOOSE** or stamped **"L"** side of the piston facing **UPWARD**, Illustration #42. Press the wrist pin through the piston and rod. Continue to press the wrist pin through until the groove in the wrist pin for the lock ring is visible on both ends of the pin, Illustration #43. Remove the assembly from the arbor press. Install the retaining ring onto each end of the wrist pin. Some models have a wire ring, and others have a Truarc ring. Use a pair of Truarc pliers to install the Truarc ring.

Fill the piston skirt with a rag, towel, shop cloths, or other suitable material, Illustration #44. The rag will prevent the rod from coming in contact with the piston skirt while it is laying on the bench. If the rod is allowed to strike the piston skirt, the skirt may become distorted.

Assemble the other pistons, rods, and wrist pins in the same manner. Fill the skirt with rags as protection until the assembly is installed.

Alternate Assembling Method

If an arbor press is not available, the piston may be assembled to the rod in much the same manner as described for disassembling.

First, soak the piston in a container of very hot water for about ten minutes, Illustration #40, previous page.

For a V4 engine installation, before pressing the wrist pin into place, hold the piston and rod near the cylinder block and check to be sure both will be facing in the right direction when they are installed.

For a three-cylinder installation, check to be sure the hole in the piston will align with the cutaway in the cylinder wall.

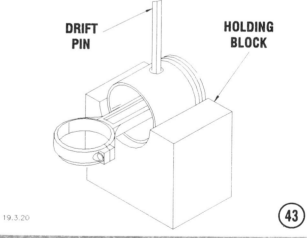

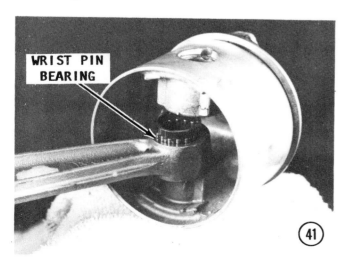

CRANKSHAFT 3-25

Pack the wrist pin needle bearing cage with needle bearing grease, or a good grade of petroleum jelly. Load the bearing cage with needles and insert it into the end of the rod, Illustration #41, previous page.

Slide the rod into the piston boss and check a second time to be sure the slanted side of the piston is facing toward the exhaust side of the cylinder and the hole in the rod is facing upward.

Now, assume a sitting position and lay a couple towels over your lap. Hold your legs tightly together to form a cradle for the piston above your knees. Set the piston between your legs with the **LOOSE** side of the piston facing upward. Now, drive the wrist pin through the piston using a drift pin with a shoulder. The drift pin will fit into the hole through the wrist pin and the shoulder will ride on the end of the wrist pin. Use sharp hard blows with a hammer. Your legs will absorb the shock without damaging the piston. If this method is used on a regular basis during the busy season, your legs will develop black-and-blue areas, but no problem, the marks will disappear in a few days.

Continue to drive the wrist pin through the piston, Illustration #45, until the groove in the wrist pin for the lockring is visible at both ends. Install the retaining spring wire or Truarc ring onto each end of the wrist pin.

Fill the piston skirt with a rag, towel, shop cloths, or other suitable material, Illustration #44, previous page. The rag will prevent the rod from coming in contact with the piston skirt while it is laying on the bench. If the rod is allowed to strike the piston skirt, the skirt may become distorted.

Assemble the other pistons, rods, and wrist pins in the same manner. Fill the skirt with rags as protection until the assembly is installed.

INSTALLATION

Piston and rod assembly installation procedures will be found in Section 3-17, Piston and Rod Assembly Installation.

3-14 CRANKSHAFT

REMOVAL

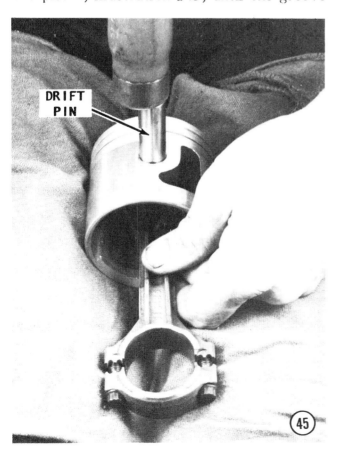

FIRST, THESE WORDS

The three-cylinder engines covered in this manual are equipped with an upper seal and bearing, a lower seal and bearing, and two main bearings between. A labyrinth seal is used at the two center main bearings and just above the bottom seal, to provide an effective seal between the cylinders.

On the V4 and V6 engines covered in this manual, a top and bottom seal is used with sealing rings around the crankshaft in the center between each cylinder in each bank to seal for pressure and vacuum. All models have a "six-ring" or "eight-ring" arrangement, which eliminates the necessity of having the labyrinth seals. It is not possible to exchange a two-ring crankshaft for a three-ring crankshaft or the other way around. The same type crankshaft must **ALWAYS** be installed as a replacement.

The rings on the V4 and V6 engines are brittle and break easily if twisted during removal of the crankshaft. Therefore, **EXERCISE CARE** when removing the crankshaft to prevent damaging the rings.

Check to be sure the bolts have been removed securing the upper and lower bearing caps.

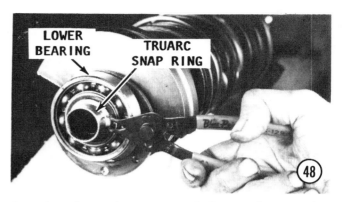

Lift the crankshaft assembly from the block, Illustration **#46**. On some models, it may be necessary to use a soft-headed mallet and tap on the bottom side of the crankshaft to jar it loose. As the crankshaft is lifted, **TAKE CARE** to work the center main bearing/s loose, and the sealing rings, if servicing a V4 engine. The center bearing/s are a split bearing held together with a snap wire ring.

The bottom half of the bearing may be stuck in the cylinder block. Therefore, the crankshaft and the center main bearing must be worked free of the block together.

Crankshaft Bearings

On the V4 engines, observe how the center main bearing has a hole in the outside circumference. On three-cylinder engines, the two center main bearings and the top main bearing also have a hole. On the V6 engines, the two center main bearings and the top and bottom bearings are all held in place by a pin indexing into a hole in the bearing. Notice the locating pin in the cylinder block, Illustration **#47**. The purpose of this arrangement is to prevent the bearing shell from rotating. During assembling, the holes in the bearings **MUST** index with the pins in the block. Notice that the hole is not in the center of the bearing. If the bearing is to be removed from the crankshaft, make a suitable mark or identification to **ENSURE** the bearing is installed in the same position from which it was removed. **ONLY** in this manner can the hole be properly indexed with the pin.

Labyrinth Seal

If servicing a three-cylinder engine, notice the grooves in the block on one side of the center main bearing. Observe the grooves in the crankcase cover. This arrangement of grooves forms what is commonly known as a "labyrinth" seal. The grooves fill with oil and/or fuel creating a seal between the cylinders. Also notice that the top bearing and the two center main bearings all have a locating pin.

Lower Bearing

The powerheads of engines covered in this manual have a lower roller bearing that is pressed onto the crankshaft. If this bearing requires replacement, it is removed by first removing the snap ring, Illustration **#48**, and then "pulling" the bearing from the crankshaft. A clamp-type puller is required to remove this bearing, Illustration **#49**. The bearing need not be removed for cleaning and inspection. Remove this bearing **ONLY** if the determination has been made that it is unfit for further service.

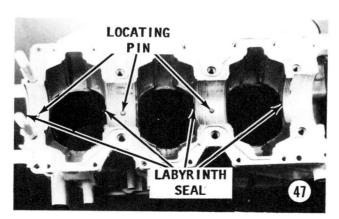

CRANKSHAFT

CLEANING AND INSPECTION

Inspect the splines for signs of abnormal wear. Check the crankshaft for straightness. Inspect the crankshaft oil seal surfaces to be sure they are not grooved, pitted or scratched. Replace the crankshaft if it is severely damaged or worn. Check all crankshaft bearing surfaces for rust, water marks, chatter marks, uneven wear or overheating. Clean the crankshaft surfaces with crocus cloth. Carefully check the grooves around the crankshaft for the sealing rings. Check the surfaces of block halves to be sure the sealing rings will affect an effective seal.

Clean the crankshaft and crankshaft bearing with solvent. Dry the parts, but **NOT** the bearing, with compressed air. Check the crankshaft surfaces a second time. Replace the crankshaft if the surfaces cannot be cleaned properly for satisfactory service. If the crankshaft is to be installed for service, lubricate the surfaces with light oil.

On the V4 engine, the center main bearing has a spring steel wire securing the two halves together. The three-cylinder engine has two center main bearings. Remove the wire, and then the outer sleeve, then the needle bearings, Illustration #50. **TAKE CARE** not to lose any of the needles. The outer shell is a fractured break type unit. Therefore, the two halves of the shell **MUST** absolutely be kept as a set.

Check the crankshaft bearing surfaces to be sure they are not pitted or show any signs of rust or corrosion. If the bearing surfaces are pitted or rusted, the crankshaft and bearings must be replaced.

During an engine overhaul to this degree, it is a good practice to remove the seal from the top main bearing. If the same type of seal is used in the bottom main bearing, remove that seal also.

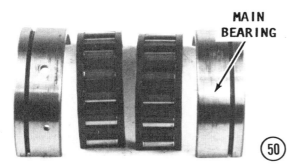
MAIN BEARING
50

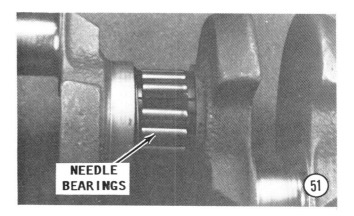
NEEDLE BEARINGS
51

ASSEMBLING

REMINDER WORDS

The V4 engines have a single center main bearing. The three-cylinder and V6 engines have two center main bearings.

Insert the proper number of needle bearings into the center main bearing cage, Illustration #51. Install the outer sleeve over the bearing cage. Check to be sure the two halves of the outer sleeve are matched. Again, these two halves are manufactured as a single unit and then broken. Therefore, the hills and valleys of the break absolutely **MUST** match during installation. Double check to be sure the marks made during bearing removal match to ensure the hole in the bearing will index with the pin in the block. Remember, the hole in the bearing is **NOT** in the center, therefore the bearing can only be installed properly **ONE WAY**, Illustration #52.

Snap the retaining ring into place around the bearing.

To install a new lower bearing, place the bearing onto the shaft and press it into place using an arbor press. If an arbor press is not available a socket large enough to fit over the crankshaft could be used to drive

BEARING SHELL
52

the bearing into place, Illustration #53.

Install the Truarc ring to secure the bearing in place, Illustration #54.

On V4 and V6 engines, the upper bearing is pressed into the upper bearing cap, Illustration #55. On the three-cylinder engines, the upper bearing is simply a "snug" fit onto the crankshaft, Illustration #56.

Rotate the installed bearings to be sure there is no evidence of binding or rough spots. The crankshaft is now ready for installation.

INSTALLATION

Installation procedures are given in Section 3-18, Crankshaft Installation.

3-15 CYLINDER BLOCK SERVICE

Inspect the cylinder block and cylinder bores for cracks or other damage. Remove carbon with a fine wire brush on a shaft attached to an electric drill or use a carbon remover solution.

Use an inside micrometer or telescopic gauge and micrometer to check the cylinders for wear. Check the bore for out-of-round and/or oversize bore. If the bore is tapered, out-of-round or worn more than 0.003" - 0.004" (0.076 mm - 0.102 mm) the cylinders should be rebored and oversize pistons and rings installed.

GOOD WORDS

Oversize piston weight is approximately the same as a standard size piston. Therefore, it is **NOT** necessary to rebore all cylinders in a block just because one cylinder requires reboring.

Hone the cylinder walls lightly to seat the new piston rings, as outlined in the

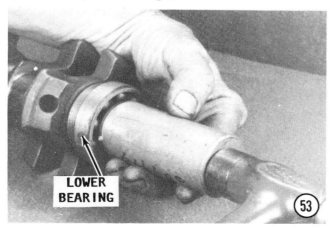

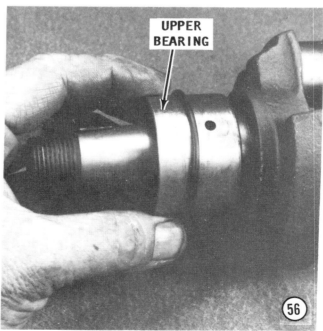

CYLINDER BLOCK 3-29

Honing Procedures Section in this chapter. If the cylinders have been scored, but are not out-of-round or the sleeve is rough, clean the surface of the cylinder with a cylinder hone as described in Honing Procedures, next section.

SPECIAL WORD

If only one cylinder is damaged, a cylinder sleeve may be installed on some models, but the cost is very high. Installation of the sleeve will bring the powerhead back to standard and permit reboring of that cylinder at a later date.

HONING PROCEDURES

To ensure satisfactory engine performance and long life following the overhaul work, the honing work should be performed with patience, skill, and in the following sequence:

a- Follow the hone manufacturer's recommendations for use of the hone and for cleaning and lubricating during the honing operation, Illustration **#57**.

b- Pump a continuous flow of honing oil into the work area. If pumping is not practical, use an oil can. Apply the oil generously and frequently on both the stones and work surface.

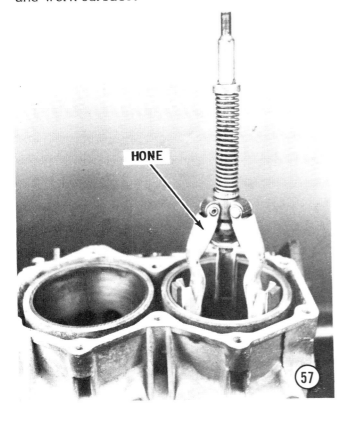

SPECIAL WORDS

If honing a three-cylinder powerhead block, use a long-size hone as a precaution against the hone becoming stuck in one of the oblong openings intake ports, in the cylinder wall, Illustration **#58**, and breaking. If the special long hone is not available, the block should be taken to a commercial shop equipped to do the job properly.

c- Begin the stroking at the top of the cylinder. Maintain a firm stone pressure against the cylinder wall to assure fast stock removal and accurate results.

d- Expand the stones as necessary to compensate for stock removal and stone wear. The best cross-hatch pattern is obtained using a stroke rate of 30 complete cycles per minute. Again, use the honing oil generously.

e- Hone the cylinder walls **ONLY** enough to de-glaze the walls.

f- After the honing operation has been completed, clean the cylinder bores with hot water and detergent. Scrub the walls with a stiff bristle brush and rinse thoroughly with hot water. The cylinders **MUST** be cleaned well as a prevention against any abrasive material remaining in the cylinder bore. Such material will cause rapid wear of new piston rings, the cylinder bore, and the bearings.

g- After cleaning, swab the bores several times with engine oil and a clean cloth, and then wipe them dry with a clean cloth. **NEVER** use kerosene or gasoline to clean the cylinders.

h- Clean the remainder of the cylinder block to remove any excess material spread during the honing operation.

3-16 POWERHEAD ASSEMBLING

SPECIAL WORD

The cylinder block assembling work

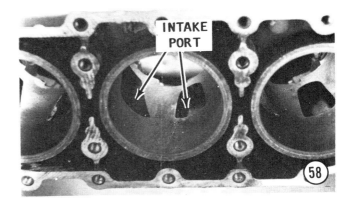

3-30 POWERHEAD

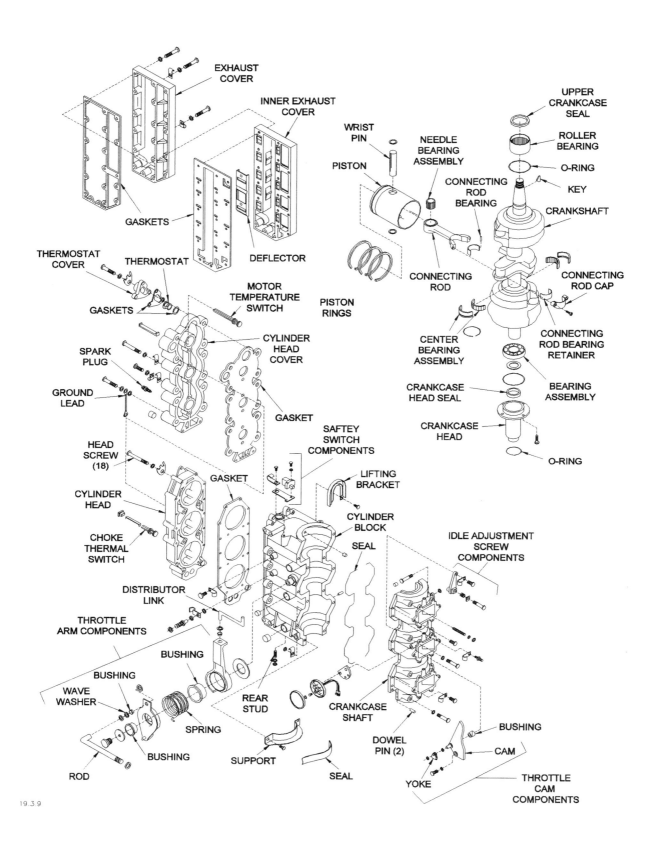

Exploded drawing of the complete powerhead for a three-cylinder engine.

CYLINDER BLOCK 3-31

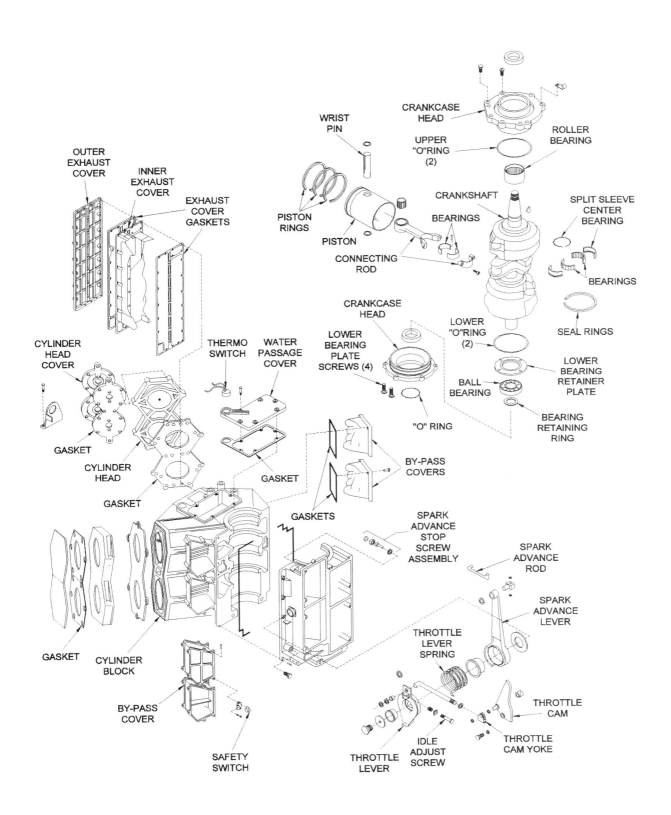

Exploded drawing of the complete powerhead for a late model V4 engine.

3-32 POWERHEAD

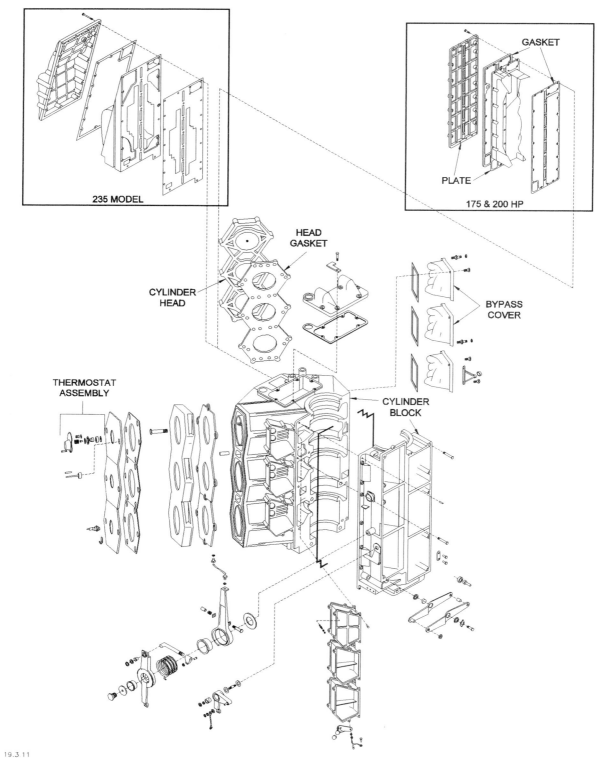

Exploded drawing of a complete powerhead for a late model V6 engine.

ASSEMBLING 3-33

should proceed quickly and without interruptions. If the work is partially completed and then left for any period of time, sealant may become hard, parts may be moved and their identity for a particular cylinder lost, or an important step may be bypassed, overlooked, or forgotten.

The following sections pickup the work of assembling the cylinder block **AFTER** the various parts have been serviced and assembled. Procedures for each area are found in this chapter under separate headings.

3-17 PISTON AND ROD ASSEMBLY INSTALLATION

WORDS OF ADVICE

If new rings are to be installed, each ring from the package **MUST** be checked in the cylinder. Errors happen. Men and machines can make mistakes. The wrong size ring can be included in a package with the proper part number.

Therefore, the end gap of **EACH** ring must be checked, one at-a-time. If the ring end gap is too **LARGE** -- when the powerhead reaches operating temperature, the ring will fail to bear

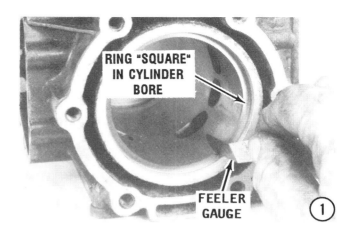

against the cylinder wall properly -- resulting in loss of compression. If the ring end gap is too **SMALL** -- when the powerhead reaches operating temperature, the ring will expand and have excessive friction against the cylinder wall resulting in rapid cylinder wall wear.

Turn the ring sideways and lower it a couple of inches into the proper cylinder bore. Now, make the ring "square" in the cylinder bore, using a piston to position the ring properly, Illustration #1. Next, use a feeler gauge and measure the distance (the gap), between the ends of the ring. The maximum and minimum allowable ring gap is .07-.17" (1.8-4.3mm) for units prior to 1988 and .19-.30" (4.8-7.6mm) for units 1989 and on.

Turn the piston upside down and slide it in and out of the cylinder, Illustration #2. The piston should slide without any evidence of binding.

Several different methods are possible to install the piston and rod assembly into the cylinder. The following procedures are outlined for the do-it-yourselfer, working at home without the advantage of special tools.

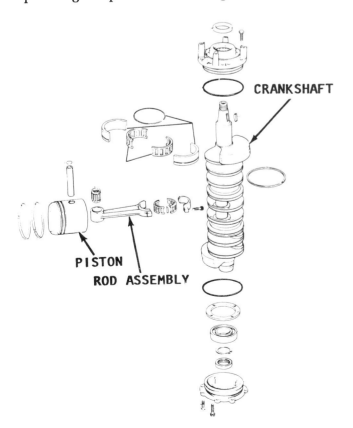

Typical V6 crankshaft and related parts.

First, purchase a special hose clamp with a strip of metal inside the clamp, Illustration #3. This piece of metal on the inside allows the outside portion of the clamp to slide on the inner strip without causing the ring to rotate.

Actually, to our knowledge a Mercruiser dealer is the only place such a clamp may be purchased. At the Mercruiser marine dealer, ask for an exhaust bellows hose clamp. The design of this hose clamp prevents the clamp and the piston ring from turning as the clamp is tightened. **DO NOT** attempt to use an ordinary hose clamp from an automotive parts house because such a clamp will cause the piston ring to rotate as the clamp is tightened. The ring **MUST NOT** rotate, because the ring ends must remain on either side of the dowel pin in the ring groove.

Next, coat the inside surface of the cylinder with a film of light-weight oil. Coat the exterior surface of the piston with the oil.

The powerhead of some 1973 and 1974 engines, covered in this manual, were equipped with three-ring pistons. Later three-cylinder, V4, and V6 models used two-ring pistons. When the change was made from the three-ring to the two-ring type, the area from the top groove and the piston dome was too small, causing failure of the piston. Therefore, when new pistons are purchased, check the first ring "land" to be sure it is a reasonable distance from the dome. Illustration #4 shows an old piston (left) and a new piston (right) with the wider distance from the top groove to the dome. When installing flat dome pistons, the top ring is tapered and fits into a tapered groove in the piston. Care must be exercised when installing rings to the pistons to ensure the proper

ring is used with the proper piston in the proper ring groove.

TAKE TIME

Take just a minute to notice how the piston rings are manufactured. Each end of the ring has a small cutout on the inside circumference, Illustration #5. Now, visualize the ring installed in the piston groove. The ring ends must straddle the pin installed in each piston groove. As the ring is tightened around the piston, the ends will begin to come together. When the piston is installed into the cylinder bore, the two ends of the ring will come together and the cutout edge will be up against the pin. For this reason, **CARE** must be exercised when installing the rings onto the piston and when the piston is installed into the cylinder.

Install only the bottom ring into the bottom piston groove. Do not expand the ring any further than necessary, as a precaution against breaking it.

Install the ring into the piston groove with the ends of the ring straddling the pin in the groove. Notice how the ring pins are staggered from one groove to the next, by 180°. The ring ends **MUST** straddle the pin to prevent the ring from rotating during engine operation. In a two-cycle engine, if the ring is permitted to rotate, at one point the opening between the ring ends would align with either the intake or exhaust port

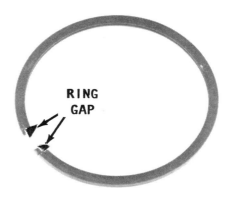

ROD AND PISTON INSTALLATION

in the cylinder, the ring would expand very slightly, catch on the edge of the port, and **BREAK**.

Install the hose clamp over the piston and bottom ring, Illustration #6. Tighten the hose clamp with one hand and at the same time rotate the clamp back-and-forth slightly with the other hand. This "rocking" motion of the clamp as it is tightened will convince you the ring ends are properly positioned on either side of the pin. Continue to tighten the clamp, and "rocking" the clamp until the clamp is against the piston skirt. At this point, the ring ends will be together and the cutout on each ring end will be against the pin.

CAREFULLY insert the rod and the piston skirt down into the cylinder. Push the piston into the cylinder until the bottom ring clamp, just installed, touches the top of the cylinder, Illustration #7. Watch to be sure the rod does not hang-up on one of the cylinder ports.

GOOD WORDS

The following four areas must be checked at this point in the assembling work:

a- The piston and rod are being installed into the same cylinder from which they were removed.

b- The hole in the rod is facing **UPWARD**.

c- The slanted side of the piston is **TOWARD** the exhaust side of the cylinder.

d- The ends of the bottom ring straddle the pin in the piston groove.

e- On the three-cylinder powerhead using the flat dome piston, check to be sure the cutouts in the piston align with "windows" in the cylinder wall.

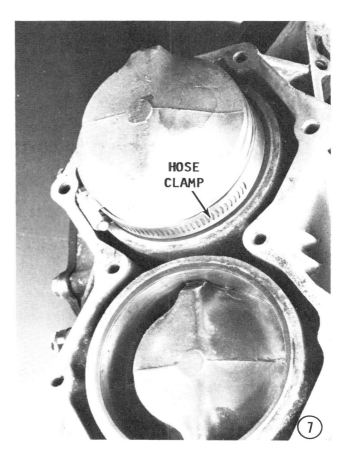

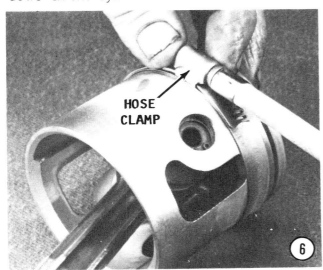

Tap the piston with the end of a wooden tool handle, or plastic mallet, Illustration #8, until the ring enters the cylinder. Remove the hose clamp.

Install the remaining rings in the same manner, one at-a-time, making sure the ends of each ring straddle the pin in the piston groove.

After the last ring has been installed and the clamp removed, tap the piston into the bore until the crown is about even with the cylinder block surface.

Install the other pistons in exactly the same manner.

Turn the cylinder block upside down with the top of the block to your **LEFT**. Remove the bolts and rod caps from each rod. Set each rod cap in a definite position to ensure each will be installed onto the rod from which it was removed.

V4 and V6 Engines

Pull on the rod to move the piston all the way into the cylinder. Install one of the rod bolts to each rod. Wrap one end of a rubber band around each rod bolt and the other end around something on the block to hold the rods out of the way during crankshaft installation.

Three-Cylinder Engines

Leave the pistons towards the top of the cylinders to allow room for crankshaft installation, Illustration **#9**.

3-18 CRANKSHAFT INSTALLATION NEEDLE MAIN AND ROD BEARINGS

The following procedures outline steps to install a crankshaft with a single main bearing; the lower bearings pressed onto the crankshaft; and the upper bearing pressed into the upper bearing cap, as for the V4 and V6 engine. The section also contains procedures for installation of the three-cylinder crankshaft with two center main bearings; the lower bearing pressed onto the crankshaft; and the upper bearing simply pushed onto the crankshaft. A sub-heading clearly identifies the procedures for the type of crankshaft being serviced. If the procedure applies to both type of crankshafts, this fact will also be indicated in the heading.

Three-Cylinder Crankshaft

Observe the pin installed in each main bearing recess. Notice the hole in each

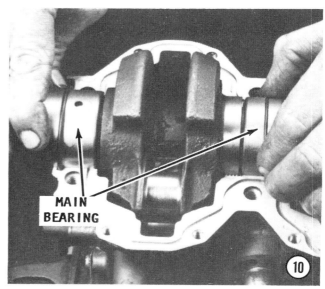

main bearing outer shell. During installation, the hole in each bearing shell **MUST** index over the pin in the cylinder block.

Hold the crankshaft over the cylinder block with the upper end to your **LEFT**. Now, lower the crankshaft into the block, Illustration **#10**, and at the same time, align the hole in each bearing to enable the pin in the block to index with the hole. Rotate each bearing slightly until all pins are properly indexed with the matching bearing hole. Once all pins are indexed, the crankshaft will be properly seated.

Bearing Cap Installation

Slide the lower bearing cap into place and just start the mounting bolts, Illustration **#11**. **DO NOT** tighten them at this time. Starting the bolts will hold the crankshaft in place in the block while the work continues.

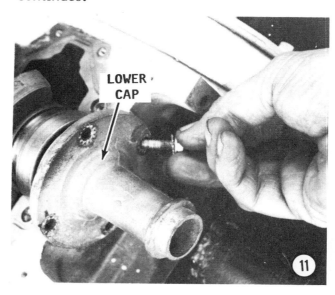

CRANKSHAFT INSTALLATION

GOOD WORDS

The reason for installing the bearing caps at this time is to prevent the crankshaft from lifting as it is turned during rod bearing and rod cap installation. If servicing a three-cylinder engine, drill two 7/16" holes in a small piece of wood. Now, slide the block of wood down the two studs at the upper bearing. Secure the block of wood in place with the nuts, Illustration #12. This block of wood will prevent the crankshaft from lifting as it is turned during rod cap installation.

V4 and V6 Crankshaft

Lower the crankshaft into place in the block with the center main bearing pin indexed into the hole in the block. At the same time as the crankshaft is being lowered, work the sealing rings around the crankshaft into the recesses in the block, Illustration #13. Rotate the rings until the openings are staggered across the crankshaft. **USE CARE** because the rings are brittle and will break if twisted or distorted. After the crankshaft is in place, it is not possible to rotate the rings. Check to be sure the pins in the block are indexed into the holes in the center main bearings. It should **NOT** be possible to rotate the bearings after the crankshaft is in place.

Slide the lower and upper bearing caps into place and just start the mounting bolts.

GOOD WORDS

On V4 and V6 engines, the bearing cap at both ends of the crankshaft can only align

with the bolt hole pattern in the powerhead **ONE WAY**. Continue to rotate the caps until all eight bolt holes are properly aligned. **DO NOT** tighten the bolts at this time. Starting the bolts will hold the crankshaft in place in the block while the work continues.

The lower bearing retainer plate has two ears which index into two recesses in the lower bearing cap. Therefore, rotate the retainer until the ears on the retainer plate index with the two recesses in the bearing cap, then install two 1/4x28 guide pins through the lower bearing cap and into the retainer plate, Illustration #14. These guide pins will hold the bearing cap and retainer plate in place while the work continues and until the retaining bolts are installed.

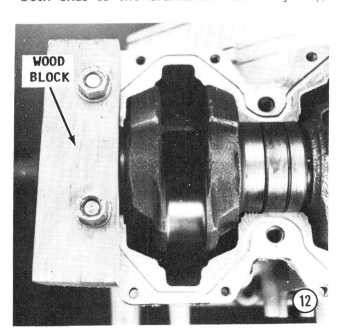

3-38 POWERHEAD

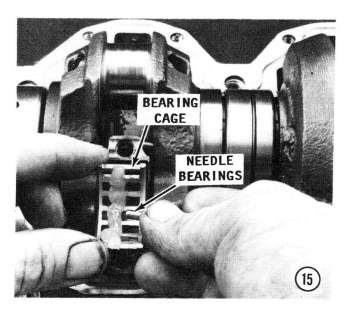

(15)

All Engines

Apply needle bearing grease to each bearing cage. Coat the rod half of the bearing area with needle bearing grease. Needle bearing grease **MUST** be used because other types of grease will not thin out and dissipate. The grease must dissipate to allow the gasoline and oil mixture to enter and lubricate the bearing. If needle bearing grease is not available, use a good grade of petroleum jelly (Vasoline).

Insert the proper number of needle bearings into each cage, Illustration #15. Set the bearing cage into the bottom half of the rod, Illustration #16, next page. With your fingers on each side of the rod, pull up on the rod and bring the rod up to the bottom side of the crankshaft. Place one needle bearing on each side of the crankshaft, Illustration #17. Using needle bearing

(17)

grease load the other cage and install the needle bearings into the cage. Lower the cage onto the crankshaft journal.

Install the proper rod cap to the rod with the identifying mark or dimple properly aligned to ensure the cap is being installed in the same position from which it was removed, Illustration #18. Tighten the rod bolts fingertight, and then just a bit more, Illustration #19.

Use a "scratchall", pick, or similar tool and move it back-and-forth on the outside surface of the rod and cap, Illustration #20, next page. Make the movement across the mating line of the rod and cap. The tool should not catch on the rod or on the cap. The rod cap must seat squarely with the rod. If not, tap the cap until the "scratchall" will move back-and-forth on the rod and cap across the mating line without any feeling

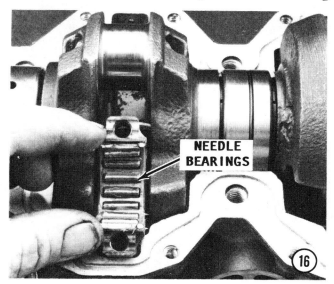

(16)

(18)

CRANKSHAFT INSTALLATION 3-39

of catching. Any step on the outside will mean a step on the inside of the rod and cap. Just a whisker of a lip will cause one of the needle bearings to catch and fail to rotate. The needle will quickly flatten and the rod will begin to "knock". Needle bearings **MUST** rotate or the function of the bearing is lost.

Tighten the rod cap bolts alternately and evenly in three rounds to the torque value given in the Torque Table in the Appendix. Tighten the bolts to 1/2 the torque value on the first round, to 3/4 the torque value on the second round, and to the full torque value on the third and final round. On each round, check with the pick to be sure the cap remains seated squarely.

Install the other rod caps in the same manner.

After the rods have been connected to the crankshaft, rotate the crankshaft until the rings on one cylinder are visible through the exhaust port. Use a screwdriver and push on each ring to be sure it has spring tension, Illustration #21. It will be necessary to move the piston slightly, because all of the rings will not be visible at one time. If there is no spring tension, the ring was broken during installation. The piston must be removed and a new ring installed. Repeat the tension test at the intake port. Check the other cylinders in the same manner.

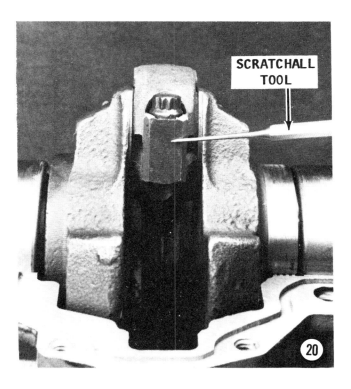

3-19 CRANKCASE COVER INSTALLATION

First, check to be sure the mating surfaces of the crankcase cover and the cylinder block are clean. Pay particular attention to the labyrinth seal grooves in the bearing area, when these seals are used. The mating surfaces and the seal grooves **MUST** be free of any old sealing compound or other foreign material.

V4 and V6 Engines

Check to be sure the sealing rings are all down in place and that the ring openings are

NOT close to the split between the two crankcase halves. Also double check to be sure the pins in the block are still indexed into the holes in the center main bearing on the V4 powerhead and the two center main bearings on the V6. This can be checked by attempting to rotate the bearing. The attempt should **FAIL**.

Three-Cylinder Engines Only

Check to be sure the main bearings are fully seated into the powerhead. The bearings can be checked by attempting to "rock" the bearing in place. The attempt should fail. The bearing should be locked in place with the pin indexed into the hole in the block.

Remove the wooden block from the studs at the upper bearing area. After the wooden block has been removed, **DO NOT** rotate the crankshaft. Rotating the crankshaft will cause it to "lift" and the bearings indexed over the pins in the block will become misaligned.

CRITICAL WORDS

The remainder of the cylinder block installation work should be performed **WITHOUT** interruption. Do not begin the work if a break in the sequence is expected, coffee, tea, lunch, whatever. The sealer will begin to set almost immediately, therefore, the crankcase cover installation, main bearing bolt installation and tightening, and the side bolt installation and tightening, must move along rapidly.

Apply just a small amount of 1000 Sealer into the groove in the cylinder block to hold the seal in place. Install a new "spaghetti" seal into the groove.

SPECIAL NOTE

After the seal on both sides of the cylinder block has been installed, apply a light coating of 1000 Sealer to the outside edge of the "spaghetti" seal, Illustration #22. This "spaghetti" seal is not used on some models. If the seal is not used double check to be sure the two surfaces are clean and smooth. Eject about a 3/8" dab of the seal at very close intervals along the surface of the block. Repeat this procedure on the mating surface of the crankcase cover.

Since 1980, the cylinder and crankcase assemblies do not have the grooves on the mating surfaces for the neoprene "spaghetti" seals. The new sealing method is to use OMC Gel-Seal (P/N 322702). This new seal is to be used on all 2-cylinder through 6-cylinder models. **DO NOT** use similar appearing jel-type sealants, since some of them contain fillers that have a shimming affect. Such shimming could cause improper bearing location, bearing misalignment, and tight armature plate bearings.

BAD NEWS

The Gel-Seal will begin to cure as soon as it is confined between the two surfaces. For this reason, the work should proceed at a good steady pace without interruptions. If the Gel-Seal is allowed to cure before drawing down and tightening the screws to their proper torque value, the seal may act as a shim. This is **BAD NEWS**, because the shimming affect may cause bearing misalignment, improper bearing location, and tight armature plate bearings.

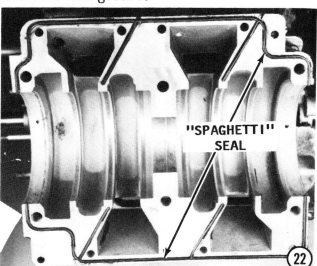

CRANKCASE COVER INSTALLATION 3-41

Lay down a small bead of the Gel-Seal along one flange of the crankcase as shown in Illustration #22A. Take care to apply sealant inside all bolt holes. Keep the sealant at least 1/4" from the labyrinth seals. If the motor is to be operated the same day, the surface opposite the one with the Gel-Seal should be sprayed with OMC LocQuic Primer. Wait several hours before starting the engine. If LocQuic Primer is not used, the assembly should sit overnight before the engine is started.

Next, lower the crankcase cover into place on the cylinder block. Install the two guide centering pins through the cover and into the block, Illustration #23. The centering pins are tapered, therefore, check the crankcase and notice which side has the large hole and which has the small hole. The pin must be inserted into the large hole first. If the pin is installed into the small hole first, the crankcase cover or the cylinder block will break.

3-20 MAIN BEARING BOLT AND CRANKCASE SIDE BOLT INSTALLATION

Apply a coating of 1000 Sealer to the threads of the main bearing bolts. Install and tighten the main bearing bolts finger-tight and then just a bit more, Illustration #24 for V4 engines, Illustration #25 for 3-cylinder engines, and Illustration #26 for V6 engines.

Tighten the main bearing bolts alternately and evenly in three rounds to the torque value given in the Torque Table in the Appendix. Be sure to check the Specifications in the Appendix for the engine being serviced.

Tighten the bolts to 1/2 the total torque value on the first round, to 3/4 the total torque value on the second round, and to the full torque value on the third and final round.

As an example: If the total torque value specified is 200 in lbs, the bolts should be

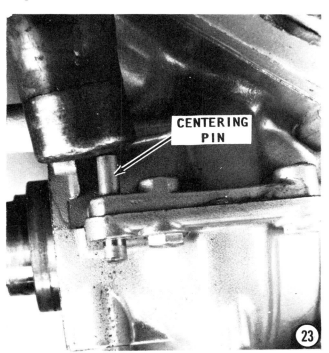

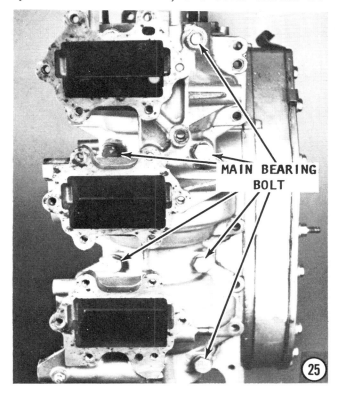

(26)

tightened to 100 in lbs on the first go-around; to 150 in lbs on the second round; and to the full 200 in lbs on the third round.

Install and tighten the crankcase side bolts, Illustration #27, to the torque value given in the Appendix.

Three-Cylinder Engine

Install the other two attaching bolts through the lower bearing cap and tighten them securely.

V4 and V6 Engines

Install the attaching bolts to secure the

(27)

(28)

lower and upper bearing caps. Tighten the bolts **EVENLY** and **SECURELY**.

Install two bolts through the lower bearing cap into the retainer plate. These are special bolts with sealing qualities built in, Illustration #28. Therefore, **NEVER** use the bolts a second time. Discard the old bolts that were removed and install **NEW** bolts. Remove the two 1/4x28 guide pins and install the other two sealing bolts, Illustration #29. Tighten the bolts **EVENLY** and **ALTERNATELY**.

Install the Woodruff key in the crankshaft. Slide the flywheel onto the crankshaft. Rotate the flywheel through several revolutions and check to be sure all moving parts indicate smooth operation without evidence of binding or "rough" spots.

Remove the flywheel.

3-21 EXHAUST COVER AND BYPASS COVER INSTALLATION

Coat both sides of a **NEW** gasket with 1000 Sealer, and then place the gasket in position on the exhaust side of the cylinder block, on the three-cylinder engines and

(29)

HEAD INSTALLATION 3-43

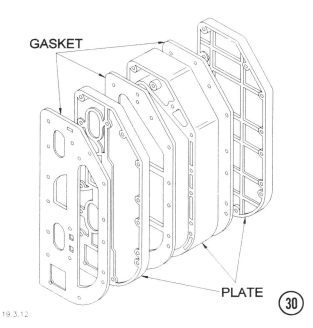

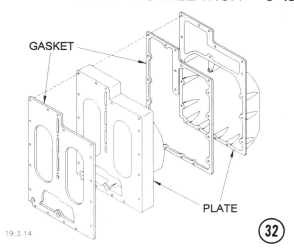

between the heads on the V4. Install the inner plate. Coat both sides of another **NEW** gasket with sealer, and then install the gasket and exhaust cover.

Secure the exhaust cover in place with the attaching hardware.

Coat both sides of a **NEW** gasket with sealer, and then place it in position on the cylinder block, Illustration #30 for 3-cylinder engines, Illustration #31 for the 85 hp and 115 hp -- 1973 thru 1977, Illustration #32 for the 135 hp -- 1973 thru 1977; Illustration #33 for 100 hp, 115 hp, and 140 hp 1978 and on; Illustration #34 for 175 hp 1986 and on; 200 hp 1986 and on.

On the V4 and V6 engines, install the bypass covers in the **SAME** location from which they were removed. Secure the covers in place with the attaching hardware, Illustration #35. If a fuel pump is used, be sure the same bypass cover is installed in the position from which it was removed.

3-22 REED BOX INSTALLATION

Install the reed box and intake manifold onto the cylinder block. A gasket is installed on both sides of the reed box. The reeds and reed stops face inward toward the cylinder, Illustration #36 for 3-cylinder engines, Illustration #37 for V6 engines, and Illustration #38 for V4 engines.

3-23 HEAD INSTALLATION

Place a **NEW** head gasket in place on the cylinder block. Head gaskets for late model powerheads are clearly identified as to which side faces the block and the head. Take an extra minute to be sure the gasket is properly positioned. If the gasket does not have any identification, it may be installed with either side facing the block or head. Check to be sure all bolt holes are

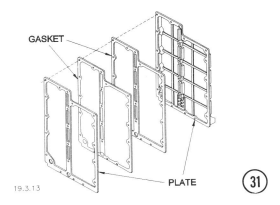

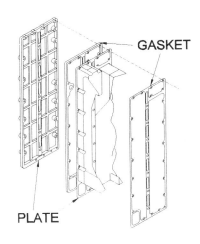

3-44 POWERHEAD

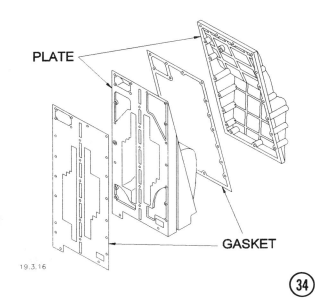

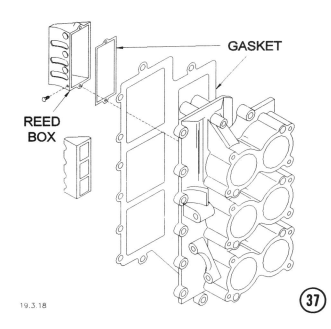

properly aligned, indicating the gasket is positioned properly end-for-end. **NEVER** use automotive type head gasket sealer. The chemicals in the sealer will cause electrolytic action and eat the aluminum faster than you can get to the bank for money to buy a new cylinder block.

CAREFULLY place the head in position. Install the head bolts and tighten them finger tight, then just a bit more, Illustration **#39** and **#40**. Now, tighten the bolts **ALTERNATELY** and **EVENLY** in three rounds to the torque value specified in the Appendix. On the first round tighten the bolts to 1/2 the total torque value, on the second round to 3/4 the total torque value, and to the full torque value on the third and final round. If servicing a V4 or V6 powerhead, repeat the procedure for the other head, Illustration **#41**.

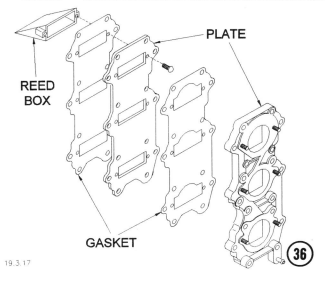

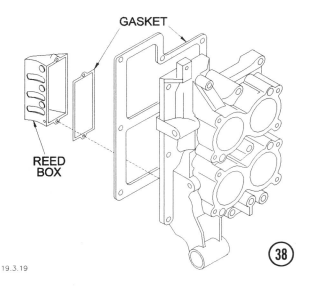

BREAK-IN PROCEDURES 3-45

Install the assembled powerhead to the exhaust housing and tighten the attaching bolts alternately and evenly in three rounds to the torque value specified in the Appendix. Tighten the bolts to 1/2 the torque value on the first round, to 3/4 the total torque value on the second round, and to the full torque value on the third and final round.

Install all powerhead accessories including the flywheel, carburetor, ignition components, starter, etc. If any doubts or difficulties are encountered, follow the procedures outlined in the chapters covering the particular component.

Connect the hoses between the intake manifold and the bypass covers, also the fuel primer hose to the same fittings and routed in the same manner as noted before disassembling. These hoses should have been identified during disassembling under Section 3-3, to ensure proper routing at this time. If they were not identified, see the routing illustration in the Appendix.

The complete outboard unit is now ready to be started and "broke-in" according to the procedures outlined in the next section.

3-24 BREAK-IN PROCEDURES

Mount the engine in a test tank or body of water. If this is not possible, connect a flush attachment and garden hose to the lower unit. **NEVER** operate the engine above idle speed using the flush attachment.

If the engine is operated above an idle speed, the unit must be **IN GEAR**, preferably with a test wheel attached to the propeller shaft. If the engine is operated above idle speed with no load on the propeller, the engine could **RUNAWAY** resulting in serious damage or destruction of the unit.

UNITS WITH VRO SYSTEM

A complete new outboard unit, a new powerhead, or a rebuilt powerhead, must have oil mixed in the fuel tank **IN ADDITION** to the VRO system. The mixture should be 50:1 (1 pint oil to 6 gallons of fuel) and the unit **MUST** be operated with this mixture during the first 10 hours of service.

CRITICAL WORDS

To convince himself the VRO system is working properly, the operator should observe a drop in the oil supply in the VRO oil reservoir during the 10-hour break-in period.

At the end of the 10 hour period, the powerhead mounted fuel filter should be inspected. Remove any dirt or foreign matter collected in the filter..

CAUTION: Water must circulate through the lower unit to the engine any time the engine is run to prevent damage to the water pump in the lower unit. Just five seconds without water will damage the water pump.

a- **ALWAYS** use OMC or BIA certified TC-W oil lubricant.

b- For all 1971-84 engines in this manual use 50:1 oil mixture (1 pint oil per 6 gallons of gasoline).

c- Since 1985 -- without VRO: during break-in and after, use 50:1 oil mixture.

d- **NEVER** use: automotive oils, premixed fuel of unknown oil quantity, or premixed fuel richer than 50:1.

e- When engine starts, **CHECK** water pump operation. Prior to 1977, a water mist should discharge from the exhaust relief holes at rear of drive shaft housing. Since 1977, a stream of water should discharge from the starboard side.

During the first 5 hours of operation, **DO NOT** operate the engine at full throttle (except for **VERY** short periods). Perform the break-in as follows:

1- Operate at 1/2 throttle for first 2 hours.

2- Operate at any speed after first 2 hours **BUT NOT** at sustained full throttle for the next 3 hours.

3- While the engine is operating during the initial period, check the fuel, exhaust, and water systems for leaks.

4- Refer to Chapter 5 for synchronizing procedures.

After the test period, disconnect the fuel line. Remove the engine from the test tank. Install the engine hood.

OMC approved oil for Johnson/Evinrude outboard engines. Only quality grade oil should be used for any engine. The added expense is ridiculously small compared to the cost of the engine or an overhaul.

*Using a flush attachment and garden hose while operating the engine at idle speed. **NEVER** operate the engine in gear or above idle speed with such a device.*

4
FUEL

4-1 INTRODUCTION

The carburetion and ignition principles of two-cycle engine operation **MUST** be understood in order to perform a proper tune-up on an outboard motor.

If you have any doubts concerning your understanding of two-cycle engine operation, it would be best to study the operation theory section in the first portion of Chapter 3, before tackling any work on the fuel system.

4-2 GENERAL CARBURETION INFORMATION

The carburetor is merely a metering device for mixing fuel and air in the proper proportions for efficient engine operation. At idle speed, an outboard engine requires a mixture of about 8 parts air to 1 part fuel. At high speed or under heavy duty service, the mixture may change to as much as 12 parts air to 1 part fuel.

Float Systems

A small chamber in the carburetor serves as a fuel reservoir. A float valve admits fuel into the reservoir to replace the fuel consumed by the engine.

Fuel level in each chamber is extremely critical and must be maintained accurately. Accuracy is obtained through proper adjustment of the float. This adjustment will provide a balanced metering of fuel to each cylinder at all speeds.

Following the fuel through its course, from the fuel tank to the combustion chamber of the cylinder, will provide an appreciation of exactly what is taking place. In order to start the engine, the fuel must be moved from the tank to the carburetor by a squeeze bulb installed in the fuel line.

The fuel systems for engines covered in this manual are equipped with a manually-operated squeeze bulb in the line to transfer fuel from the tank to the engine until the engine starts.

After the engine starts, the fuel passes through the fuel pump to the carburetor. All systems have some type of filter installed somewhere in the line between the tank and the carburetor.

At the carburetor, the fuel passes through the inlet passage to the needle and

Float properly adjusted. Notice how the surface of the float is parallel with the surface of the carburetor.

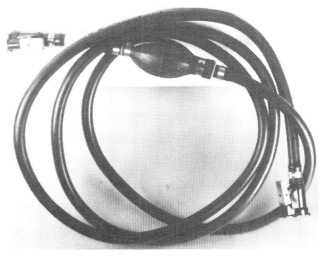

Typical fuel line with squeeze bulb and quick-disconnect fitting at each end. These items may be purchased as an assembled unit.

4-2 FUEL

seat, and then into the float chamber (reservoir). A float in the chamber rides up and down on the surface of the fuel. After fuel enters the chamber and the level rises to a predetermined point, a tang on the float closes the inlet needle and the fuel flow entering the chamber is cutoff. When fuel leaves the chamber as the engine operates, the fuel level drops and the float tang allows the inlet needle to move off its seat and fuel once again enters the chamber. In this manner a constant reservoir of fuel is maintained in the chamber to satisfy the demands of the engine at all speeds.

A fuel chamber vent hole is located near the top of the carburetor body to permit atmospheric pressure to act against the fuel in each chamber. This pressure assures an adequate fuel supply to the various operating systems of the engine.

Air/Fuel Mixture

A suction effect is created each time the piston moves upward in the cylinder. This suction draws air through the throat of the carburetor. A restriction in the throat, called a venturi, controls air velocity and has the effect of reducing air pressure at this point.

The difference in air pressures, at the throat and in the fuel chamber, causes the fuel to be pushed out metering jets extending down into the fuel chamber. When the fuel leaves the jets, it mixes with the air passing through the venturi. This fuel/air mixture should then be in the proper proportion for burning in the cylinder/s for maximum engine performance.

In order to obtain the proper air/fuel mixture for all engine speeds, high- and low-speed orifices are installed. There is no adjustment with the orifice type. The three-cylinder engines have a low-speed orifice and a high-speed orifice. Late model V4 and V6 engines have three orifices for low-speed, intermediate speed, and high-speed.

Engine operation at sea level compared with performance at high altitudes is quite noticeable. A throttle valve controls the volume of air/fuel mixture drawn into the engine. A cold engine requires a richer fuel mixture to start and during the brief period it is warming to normal operating temperature. A choke valve is placed ahead of the metering jets and venturi to provide the extra amount of fuel required for start and while the engine is cold. When this choke valve is closed, a very rich fuel mixture is drawn into the engine.

Late model engines are equipped with what is commonly referred to as a "primer choke system". This system consists of a solenoid valve, distribution lines, and injection nozzles. The nozzles are tapped into the bypass covers. During engine operation, when the choke system is operating, fuel is injected directly into the cylinder, instead of being routed in the usual manner through the crankcase.

The throat of the carburetor is usually

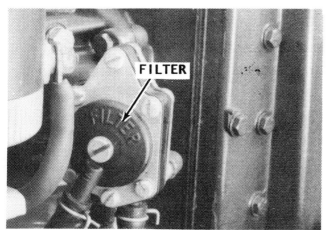

Throw-away fuel pump used on some Johnson-Evinrude engines. The only service possible on this unit is to clean the filter and install a new gasket.

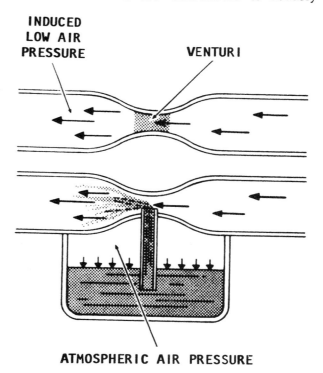

Air flow principle of a modern carburetor.

referred to as the "barrel". Carburetors installed on engines covered in this manual all have a single metering jet with a single throttle and choke plate. Single barrel carburetors are fed by one float and chamber.

4-3 FUEL SYSTEM

The fuel system includes the fuel tank, fuel pump, fuel filters, carburetor, fuel lines with a squeeze bulb, and the associated parts to connect it all together. Regular maintenance of the fuel system to obtain maximum performance, is limited to changing the fuel filter at regular intervals and using **FRESH** fuel. Even with the high price of fuel, removing gasoline that has been standing unused over a long period of time is still the easiest and least expensive preventive maintenance possible.

In most cases this old gas, even with some oil mixed with it, can be used without harmful effects in an automobile using regular gasoline.

If a sudden increase in gas consumption is noticed, or if the engine does not perform

Damaged piston, possibly caused by insufficient oil mixed with the fuel; using too low an octane fuel; or using fuel that had "soured" (stood too long without a preservative additive).

properly, a carburetor overhaul, including boil-out, or attention to the fuel pump, may be required. All engines covered in this manual have a non-serviceable throwaway fuel pump.

4-4 TROUBLESHOOTING

The following paragraphs provide an orderly sequence of tests to pinpoint problems in the system. If an engine has not been

OMC fuel conditioner added to the fuel will keep it fresh for up to one full year.

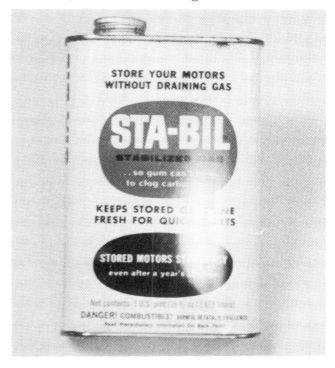

Commercial additives, such as Sta-bil, may be used to keep the fuel in the tank fresh. Under favorable conditions, such additives will prevent the fuel from "souring" for up to twelve months.

used for some time and fuel has remained in the carburetor, it is possible that varnish may have formed. Such a condition could be the cause of hard starting or complete failure of the engine to operate.

Fuel Problems

Many times fuel system troubles are caused by a plugged fuel filter, a defective fuel pump, or by a leak in the line from the fuel tank to the fuel pump. Aged fuel left in the carburetor and the formation of varnish could cause the needle to stick in its seat and prevent fuel flow into the bowl. A defective choke may also cause problems. **WOULD YOU BELIEVE,** a majority of starting troubles, which are traced to the fuel system, are the result of an empty fuel tank or aged fuel.

Fuel will begin to sour in three to four months and will cause engine starting problems. Therefore, leaving the motor sitting idle with fuel in the carburetor, lines, or tank, during the off-season, usually results in very serious problems. A fuel additive such as Sta-Bil or OMC 2+4 Fuel Conditioner may be used to prevent gum from forming during storage or prolonged idle periods.

For many years there has been the widespread belief that simply disconnecting the fuel line at the engine or at the tank, and then running the engine until it stops is the proper procedure before storing the engine for any length of time. Right? **WRONG!**

First, it is **NOT** possible to remove all fuel in the carburetor by operating the engine until it stops. Considerable fuel is trapped in the float chamber, other passages, and in the line leading to the carburetor. The **ONLY** guaranteed method of removing **ALL** fuel, is to take the time to remove the carburetor and drain the fuel. On all engines using high-speed orifice carburetors, the high-speed orifice plug can be removed to drain fuel from the carburetor.

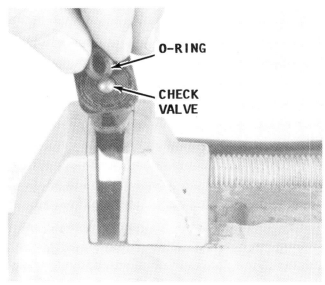

*Fuel connector with the O-ring visible. The O-rings have a relatively short life and **MUST** be replaced at regular intervals, as explained in the text.*

Secondly, if the engine is operated with the fuel supply disconnect at the "quick-disconnect" until it stops, the fuel and oil mixture inside the engine is removed, leaving bearings, pistons, rings, and other parts without any protective lubricant.

Proper procedure involves: disconnecting the fuel line at the tank; operating the engine until it begins to run **ROUGH**; then stopping the engine, which will leave some fuel/oil mixture inside the engine; and finally removing or draining the carburetor. By disconnecting the fuel supply, all **SMALL** passages are cleared of fuel, even though some fuel is left in the carburetor. A light oil should be put in the combustion chamber as instructed in the **Owner's Manual.** On all

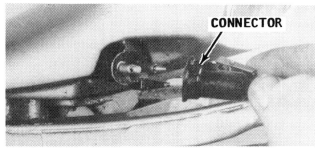

Female fuel line connector ready to be mated with the male portion of the connector.

Fouled spark plug, possibly caused by operator's habit of over-choking or a malfunction holding the choke closed. Either of these conditions will deliver a too-rich fuel mixture to the cylinder.

engines using high-speed orifice carburetors, the high-speed orifice plug can be removed to drain fuel from the carburetor.

Choke Problems

When the engine is hot, the fuel system can cause starting problems. After a hot engine is shut down, the temperature inside the fuel bowl may rise to 200°F and cause the fuel to actually boil. All carburetors are vented to allow this pressure to escape to the atmosphere. However, some of the fuel may percolate over the high-speed nozzle.

If the choke should stick in the open position, the engine will be hard to start. If the choke should stick in the closed position, the engine will flood making it **VERY** difficult to start.

In order for this raw fuel to vaporize enough to burn, considerable air must be added to lean out the mixture. Therefore, the only remedy is to remove the spark plugs; ground the leads; crank the engine through about 10 revolutions; clean the plugs; install the plugs again; and start the engine.

If the needle valve and seat assembly is leaking, an excessive amount of fuel may enter the intake manifold in the following manner: after the engine is shut down, the pressure left in the fuel line will force fuel past the leaking needle valve. This extra fuel will raise the level in the fuel bowl and cause fuel to overflow into the intake manifold.

A continuous overflow of fuel into the intake manifold may be due to a sticking inlet needle or to a defective float which would cause an extra high level of fuel in the bowl and overflow into the intake manifold.

Procedures to troubleshoot the "primer choke system" are given at the end of this chaper.

FUEL PUMP TESTS

CAUTION: Gasoline will be flowing in the engine area during this test. Therefore, guard against fire by grounding the high-tension wire to prevent it from sparking.

Testing System with Squeeze Bulb

An adequate safety method is to ground each spark plug lead. Disconnect the fuel line at the "quick-disconnect" at the engine. Place a suitable container over the end of the fuel line to catch the fuel discharged. Insert a small screwdriver into the end of the line to open the check valve, and then squeeze the primer bulb and observe if there is satisfactory fuel flow from the line.

If there is no fuel discharged from the line, the check valve in the squeeze bulb may be defective, or there may be a break or obstruction in the fuel line.

If there is a good fuel flow, remove the fuel lines at the carburetors and connect the "quick-disconnect" at the engine. Crank the engine. If the fuel pump is operating properly, a healthy stream of fuel should pulse out of the line.

Continue cranking the engine and catching the fuel for about 15 pulses to determine if the amount of fuel decreases with each pulse or maintains a constant amount. A decrease in the discharge indicates a restriction in the line. If the fuel line is plugged, the fuel stream may stop. If there is fuel in the fuel tank but no fuel flows out

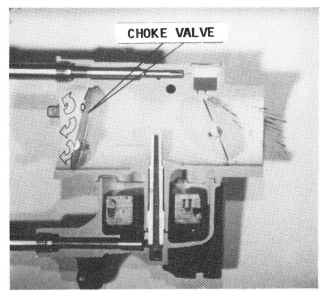

The choke plays a most important role during engine start and in controlling the amount of air entering the carburetor, under various load conditions.

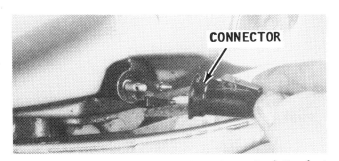

Insert a small screwdriver into the end of the fuel line to open the check valve, as described in the text.

the fuel line while the engine is being cranked, the problem may be in one of several areas:

1- Plugged fuel line from the fuel pump to the carburetor.

2- Defective O-ring in fuel line connector into the fuel tank.

3- Defective O-ring in fuel line connector into the engine.

4- Defective fuel pump.

5- The line from the fuel tank to the fuel pump may be plugged; the line may be leaking air; or the squeeze bulb may be defective.

6- Defective fuel tank.

7- If the engine does not start even though there is adequate fuel flow from the fuel line, the fuel inlet needle valve and the seat may be gummed together and prevent adequate fuel flow.

FUEL LINE TEST

On all factory installations covered in this manual, the fuel line is provided with quick-disconnect fittings at the tank and at the engine, as produced by the manufacturer. Owners may install a built-in tank with a permanent-type fuel line to the engine. If there is reason to believe the problem is at the quick-disconnects, the hose ends can be replaced as an assembly, or new O-rings may be installed. A supply of new O-rings should be carried on board for use in isolated areas where a marine store is not available. Replacement procedures are presented in Section 4-11.

For a small additional expense, the entire fuel line can be replaced eliminating this entire area as a problem source for many future seasons.

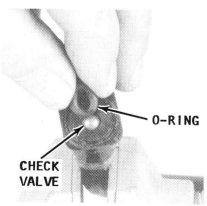

Fuel connector with the O-rings visible. These O-rings have a relatively short life and may be the source of fuel problems. The O-rings MUST be replaced on a regular basis.

A typical squeeze bulb with the directional arrow clearly visible. The squeeze bulb must be installed with the arrow pointing in the direction of fuel flow, toward the engine.

The V6 engine requires a larger fuel line than the 3-cylinder or V4 engine. The line is connected directly to the fuel tank without a quick-disconnect fitting. If the fuel line is replaced, check to be sure the larger diameter line is installed.

The primer squeeze bulb can be replaced in a short time. A squeeze bulb assembly kit, complete with the check valves installed, may be obtained from the local Johnson/Evinrude dealer. The replacement kit will also include two tie straps to secure the bulb properly in the line.

An arrow is clearly visible on the squeeze bulb to indicate the direction of fuel flow. The sqeeze bulb **MUST** be installed correctly in the line because the check valves in each end of the bulb will allow fuel to flow in **ONLY** one direction. Therefore, if the squeeze bulb should be installed backwards, in a moment of haste to get the job done, fuel will not reach the carburetor. To replace the bulb, see Section 4-11.

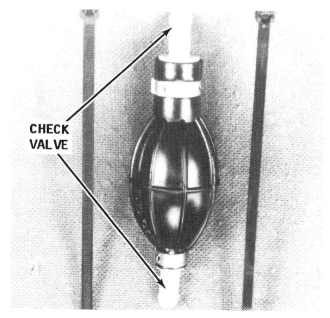

Parts in a squeeze bulb replacement kit include the squeeze bulb, two check valves, and the two tie straps to secure the bulb in the line.

Boats equipped with larger horsepower engines usually have built-in fuel tanks. The fuel system is provided with an anti-siphon valve to prevent fuel from being siphoned from the tank in the event the fuel line is broken or disconnected. The valve is mounted on the tank top. It should be removed periodically and checked to verify an adequate fuel flow under normal conditions and no fuel flow when the valve is closed.

ROUGH ENGINE IDLE

If an engine does not idle smoothly, the most reasonable approach to the problem is to perform a tune-up to eliminate such areas as: defective ignition parts; faulty spark plugs; and synchronization out of adjustment.

Other problems that can prevent an engine from running smoothly include: an air leak in the intake manifold; uneven compression between the cylinders; and sticky or broken reeds.

Of course any problem in the carburetor affecting the air/fuel mixture will also prevent the engine from operating smoothly at idle speed. These problems usually include: too high a fuel level in the bowl; a heavy float; leaking needle valve and seat; defective automatic choke; or an improper orifice installed.

"Sour" fuel (fuel left in a tank without a preservative additive) will cause an engine to run rough and idle with great difficulty.

If the fuel/oil mixture is too strong on the oil side, the engine will run rough. The only solution to this problem is to drain the fuel and fill the tank with a **FRESH ACCURATE** mixture.

Drain and Primer Testing

A check valve is installed in the cylinder drain. If this valve is not operating correctly, the engine will operate roughly, will fail to idle properly, and will not achieve maximum rpm at full throttle.

A syringe, a 1/8" ID piece of clear plastic hose, a short piece of 3/32" OD brass or copper tubing, and some Isopropyl Alcohol are all that is required to make these simple tests. This equipment may also be used to test idle and air bleed passages in carburetors.

Install the plastic tubing onto the syringe and fill them both with alcohol. Remove the primer hose from the nipple on the bypass cover. Install the end of the syringe hose to the fitting. Now, push on the syringe plunger slowly. If bubbles or liquid moves forward, the orifice is open or clean. If the liquid does not move, the orifice is plugged.

Disconnect the hose from the intake bypass nipple. Install the syringe tubing to the drain hose. Now, push in slightly on the syringe plunger. If the fluid fails to move,

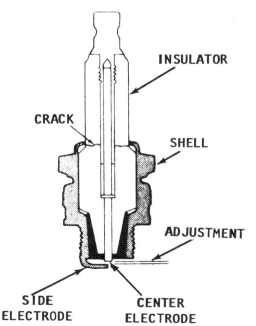

Cross-sectional view of a spark plug showing the principle parts with important comments for satisfactory service.

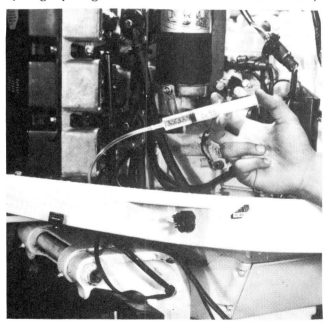

Pushing on the syringe to test the check valve or primer orifice as described in the text.

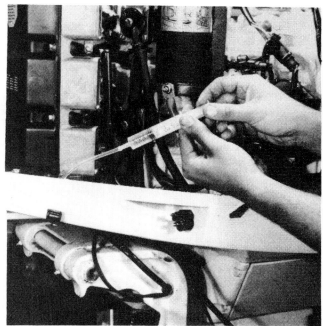

Pulling on the syringe to test the check valve, as described in the text.

the check valve is closed and operating properly. If fluid flows in the hose, the check valve is stuck open and should be replaced.

Pull out on the syringe plunger. Some dark fluid should be pulled through the hose. This action further verifies the check valve is operating properly.

If the valve fails either test, it **MUST** be replaced.

After all orifices and check valves have been tested, reconnect the primer hoses.

EXCESSIVE FUEL CONSUMPTION

Excessive fuel consumption can result from one of three conditions, or a combination of all three.

1- Inefficient engine operation.
2- Damaged condition of the hull, including excessive marine growth.
3- Poor boating habits of the operator.

If the fuel consumption suddenly increases over what could be considered normal, then the cause can probably be attributed to the engine or boat and not the operator.

Marine growth on the hull can have a very marked effect on boat performance. This is why sail boats always try to have a haul-out as close to race time as possible. While you are checking the bottom take note of the propeller condition. A bent blade or other damage will definitely cause poor boat performance.

If the hull and propeller are in good shape, then check the fuel system for possible leaks. Check the line between the fuel pump and the carburetor while the engine is running and the line between the fuel tank and the pump when the engine is not running. A leak between the tank and the pump many times will not appear when the engine is operating, because the suction created by the pump drawing fuel will not allow the fuel to leak. Once the engine is turned off and the suction no longer exists, fuel may begin to leak.

If a minor tune-up has been performed and the spark plugs, ignition parts, and synchronization are properly adjusted, then the problem most likely is in the carburetor, indicating an overhaul is in order. Check for leaks at the needle valve and seat. Use extra care when making any adjustments affecting the fuel consumption, such as the float level or automatic choke.

ENGINE SURGE

If the engine operates as if the load on the boat is being constantly increased and decreased, even though an attempt is being made to hold a constant engine speed, the problem can most likely be attributed to the fuel pump. The only corrective solution is to replace the pump.

4-5 JOHNSON/EVINRUDE CARBURETORS

This section provides complete detailed procedures for: removal; disassembly; cleaning and inspecting; assembling including bench adjustments; installation; and operating adjustments for the four carburetors installed on powerheads covered in this manual.

The Type III carburetor is a frontdraft carburetor, almost the same as the Type II except with a low-speed orifice, an intermediate-speed orifice, and a high-speed orifice.

These carburetors are equipped with either a manual choke, an electric choke, or the "primer choke system".

The manual backup permits the operator to operate the choke in the event the battery is dead. The following table lists the types of carburetors installed on the various horsepower and model years for engines covered in this manual.

CARBURETOR INSTALLATIONS

Type I
Three carburetors per powerhead.
Frontdraft, single barrel with
fixed low-speed orifice
and high-speed orifice
65 hp -- 1972 and 1973
70 hp -- 1974 thru 1982
75 hp -- 1974 thru 1982

Type II
Two or three carburetors per powerhead.
Frontdraft, double barrel
with low-speed orifice
and high-speed orifice.

All V4 and V6 powerheads
through 1978 **EXCEPT** 85 hp, 115 hp
and 140 hp powerheads thru 1979

Type III
same as Type II except
with low-speed orifice,
intermediate-speed orifice,
and high-speed orifice.

All V4 and V6 powerheads
1979 and on **EXCEPT** the following:
85 hp 1980 only
115 hp, 120hp, and 140 hp - 1980 and on.
200hp and 225hp - 1986 and on

TYPE IV
Four or six carburetors per powerhead.
Unique design from Type III
with low-speed orifice,
intermediate-speed orifice,
and high-speed orifice.

V4 powerheads - 120hp and 140hp
1985 and on.
V6 powerheads - 200hp and 225 hp
1986 and on.

4-6 TYPE I CARBURETOR
65 HP -- 1972 AND 1973
70 HP -- 1974 THRU 1982
75 HP -- 1974 THRU 1982

Three carburetors are installed on each engine, one for each cylinder.
This carburetor is the single barrel, front draft. It has a fixed low-speed orifice and a high-speed orifice in the float bowl.

All of the Type I carburetors utilize an electric choke mounted on the side of the engine. On most models, the electric choke does not have to be removed in order to remove and service the carburetor.

REMOVAL

Preliminary Tasks

1- Disconnect the battery cables at the battery as a precaution against an accidental spark igniting the fuel or fuel fumes present during the service work. Disconnect the quick-disconnect on the fuel line to the engine. Remove the hood. Disconnect the fuel line from the junction on the port side of the engine. This is the line from the fuel pump to the carburetors. Remove the low-speed adjustment lever at the bottom of the air silencer.

2- Remove the air silencer cover, and then the air silencer. Disconnect the hose at the bottom of the air silencer. (This hose is connected to a fitting at the bottom of the crankcase.)

4-10 FUEL

3- Disconnect the choke and throttle linkage between the three carburetors on the starboard side of the engine. The linkage will snap out of a retainer on the carburetor cams. The nylon retainers **MUST** be removed before the carburetor is immersed in any type of cleaning solution. Disconnect the choke wire from the electric choke, by first removing the O-ring and coil spring.

4- Remove the six nuts securing the carburetor to the intake manifold. Identify all of the carburetors as an aid to installation. The carburetors **MUST** be installed in their original positions. The fuel connections and the choke arrangement is different for each carburetor. The carburetors can now be removed as an assembly.

SPECIAL WORDS

Only one carburetor will be rebuilt in the following procedures. Service of the other two units is to be performed in the same manner.

The carburetor parts should be kept with the individual unit to ensure all items are installed back in their original positions.

DISASSEMBLING

5- Disconnect the fuel hoses between the three carburetors. This is accomplished

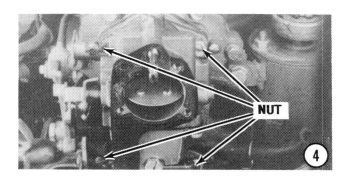

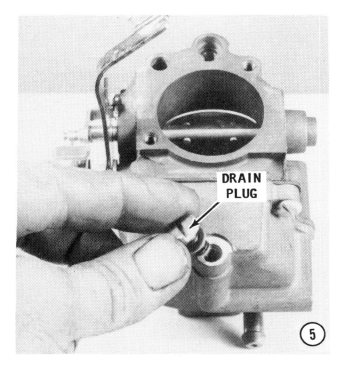

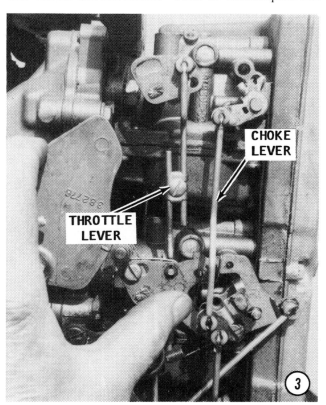

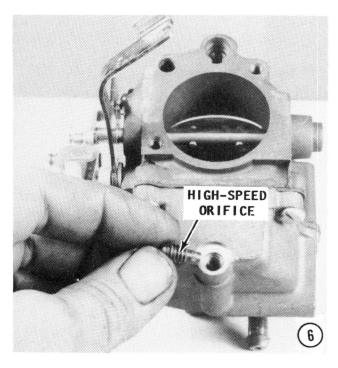

TYPE I CARBURETOR   4-11

by simply cutting the tie strap or working the clip securing the hose to the carburetor fitting. Remove the drain plug or the high-speed orifice plug from the bottom of the carburetor bowl.

6- Remove the orifice from the float bowl, using the proper size screwdriver or special OMC tool No. 317002.

7- Remove the low-speed plug from the carburetor. Using the special orifice tool, OMC No. 317002, remove the low-speed orifice.

8- Turn the carburetor upside-down and remove the screw securing the float bowl to the carburetor body. Lift the bowl free of the carburetor.

9- Remove and **DISCARD** the bowl gasket and the small gasket around the high-speed nozzle. The high-speed nozzle is **NOT** removable.

10- Remove the float hinge pin, and then remove the float.

11- Remove the inlet needle and seat from the carburetor body.

CLEANING AND INSPECTING

NEVER dip rubber parts, plastic parts, or nylon parts, in carburetor cleaner. These parts should be cleaned **ONLY** in solvent, and then blown dry with compressed air.

Place all metal parts in a screen-type tray and dip them in carburetor cleaner until they appear completely clean, then wash with solvent or clean water, and blow dry with compressed air.

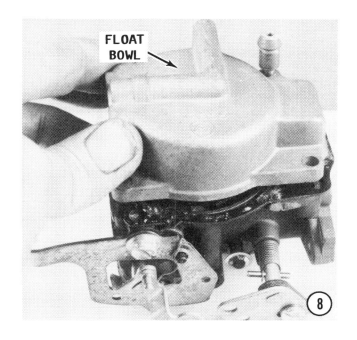

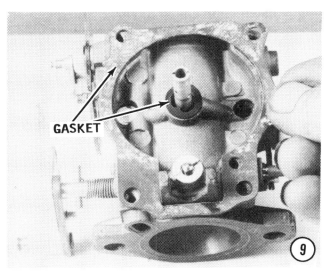

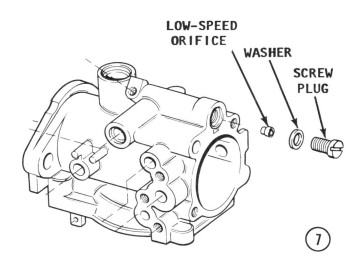

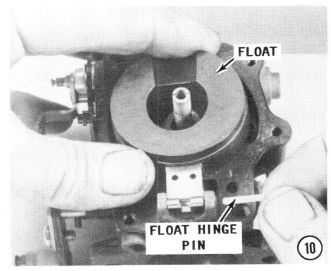

4-12 FUEL

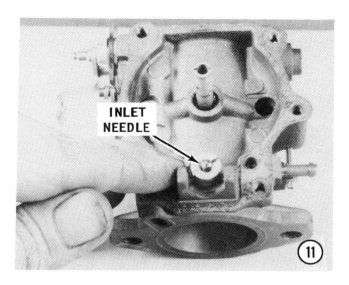

Blow out all passages in the castings with compressed air. Check all parts and passages to be sure they are not clogged or contain any deposits. **NEVER** use a piece of wire or any type of pointed instrument to clean drilled passages or calibrated holes in a carburetor.

Move the throttle shaft back-and-forth to check for wear. If the shaft appears to be too loose, replace the complete throttle body because individual replacement parts are **NOT** available.

Inspect the main body, airhorn, and venturi cluster gasket surfaces, for cracks and burrs which might cause a leak. Check the float for deterioration. Check to be sure the float spring has not been stretched. If

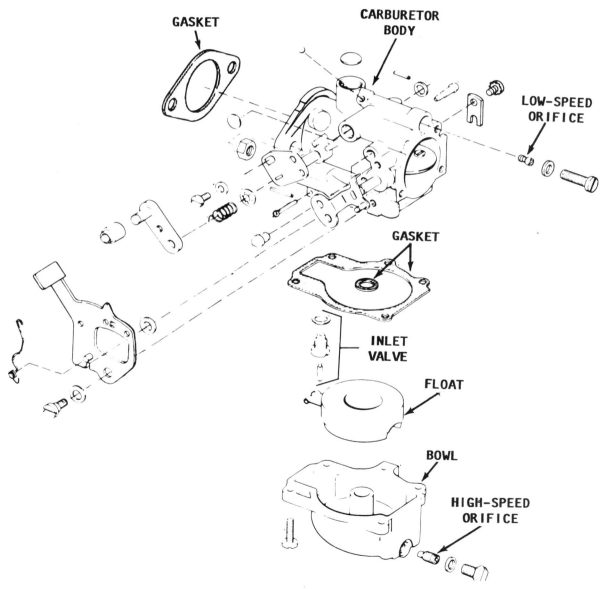

Exploded drawing of the Type I carburetor, used on all 3-cylinder engines.

TYPE I CARBURETOR 4-13

any part of the float is damaged, the unit must be replaced. Check the float arm needle contacting surface and replace the float if this surface has a groove worn in it.

Check the orifices for cleanliness. The orifice has a stamped number. This number represents a drill size. Check the orifice with the shank of the proper size drill to verify the proper orifice is used. The local OMC dealer will be able to provide the correct size orifice for the engine and carburetor being serviced.

Most of the parts that should be replaced during a carburetor overhaul are included in overhaul kits available from your local marine dealer. One of these kits will contain a matched fuel inlet needle and seat. This combination should be replaced each time the carburetor is disassembled as a precaution against leakage.

ASSEMBLING TYPE I CARBURETOR

1- Install the inlet seat using the proper size screwdriver. Inject just a drop of lightweight oil into the seat. Thread the inlet needle valve into the seat, and tighten it until it barely seats. **DO NOT** overtighten or the needle valve may be damaged.

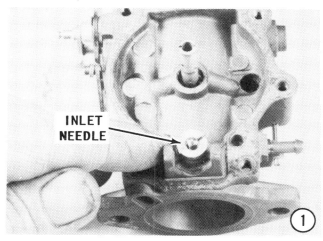

2- Position **NEW** gaskets in place on the carburetor body and on the nozzle.

3- Lower the float down over the nozzle, and then slide the hinge pin into place.

4- Hold the carburetor body in a horizontal position and check to be sure the float is parallel to the carburetor surface, as shown. If the float is not parallel,

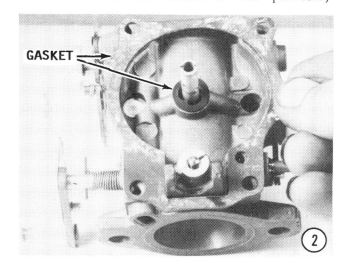

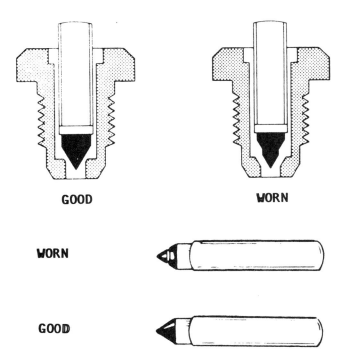

Cross-sectional drawing to allow comparison of a new needle and seat with one badly worn. Notice how the edges of the worn valve and set have become beveled.

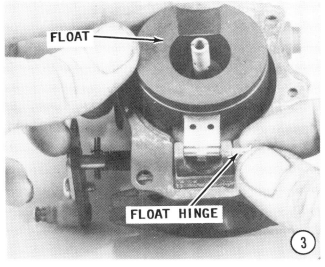

CAREFULLY bend the float tang until the float is in a parallel position and both sides are equal distance from the carburetor surface.

5- Place the float bowl in position, and then secure it in place with the attaching screws.

6- Using special tool, OMC No. 317002, install the low-speed orifice. Install the orifice plug.

WORDS OF CAUTION

Take time to use the proper size screwdriver, or special tool OMC No. 317002, to install the orifice. If the orifice is damaged and the edge of the opening burred, because the wrong size screwdriver was used, the flow of fuel will be restricted and the orifice will not function properly. Damage to the orifice will also make it very difficult to remove the orifice during the next carburetor service.

7- Install the high-speed orifice into the carburetor bowl, using the proper size screwdriver to prevent damaging the orifice.

8- Thread the plug, with a new gasket, into place and tighten it snugly.

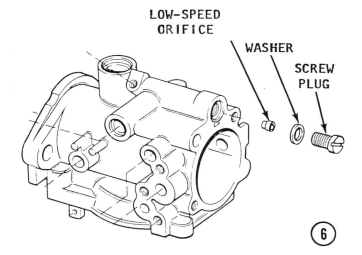

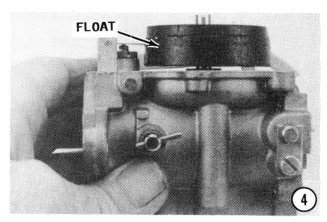

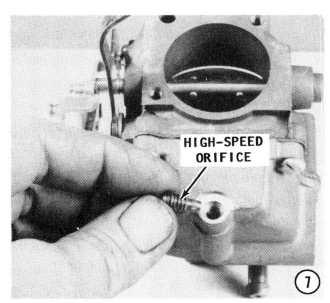

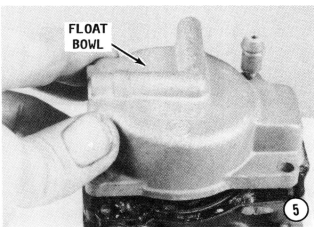

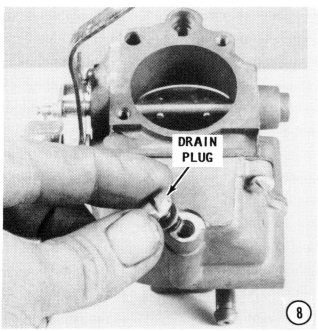

TYPE I CARBURETOR 4-15

SECOND AND THIRD CARBURETORS

Perform Steps 1 thru 9 to assemble the second and third carburetors.

INSTALLATION

9- Connect the fuel line between the carburetors. Clean the mating surface of the intake manifold. Check to be sure all old gasket material has been removed. Slide a **NEW** gasket down over the studs into place on the manifold. Install the three carburetors at the same time onto the studs and secure them in place with the nuts. Tighten the nuts **ALTERNATELY** and **EVENLY**. Connect the fuel line between the fuel pump and the carburetors.

10- Position a **NEW** air silencer gasket in place on the front of the engine. Connect the hose from the crankcase to the back of the air silencer. Coat the threads of the air silencer retaining screws with Loctite and then install the air silencer. Install the air silencer cover and at the same time slide the plastic adjusting knob onto the low-speed needle for each carburetor. **DO NOT** push the knobs all the way home onto the needle at this time. You may seriously consider **NOT** to install and use the piece of linkage connecting the low-speed needle valve of each carburetor.

11- On the starboard side of the engine, install the choke and throttle retainers. Connect the choke and throttle linkage between the carburetors. Connect the choke coil solenoid wire onto the linkage stud, and

then slide the O-ring and coil spring onto the stud to secure the wire in place. Adjust the choke butterflies by loosening the screw between the top and bottom carburetor linkage. The center carburetor is the base unit. Adjustments are made from the center carburetor to the top and bottom carburetors. Make the adjustment to close the upper and lower carburetor choke butterflies when the center butterfly is closed. After the adjustment is satisfactory, tighten the screw.

Adjust the throttle butterflies in the same manner. Loosen the screw between the top and center and the bottom and

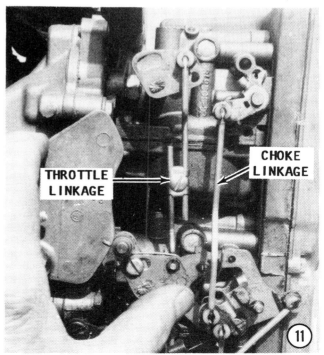

center carburetor linkage. Make the adjustment to close upper and lower throttle butterflys when the center butterfly is closed. After the adjustment is correct, tighten the screw.

EXPLANATION

This linkage permits adjustment of the three carburetors simultaneously, while using only one knob. Sounds great! But when the one carburetor is adjusted, the others are also changed. A great many professional mechanics have discovered that the linkage is not required. They feel it is far more efficient to adjust each carburetor individually.

GOOD WORDS

It is best to synchronize the fuel and ignition systems at this time. See Chapter 5. After the synchronization has been completed, proceed with the following work.

12- Mount the engine in a test tank or body of water. If this is not possible, connect a flush attachment and garden hose to the lower unit. **NEVER** operate the engine above idle speed using the flush attachment. If the engine is operated above idle speed with no load on the propeller, the engine could **RUNAWAY** resulting in serious damage or destruction of the unit.

CAUTION: Water must circulate through the lower unit to the engine any time the engine is run to prevent damage to the water pump in the lower unit. Just five seconds without water will damage the water pump.

Start the engine and allow it to warm to operating temperature. Pop the three rubber caps, one for each carburetor, out of the front cover of the air silencer. Adjust the low-speed idle by turning the low-speed needle valve **CLOCKWISE** until the engine begins to misfire or the rpm drops noticeably. From this point, rotate the needle valve **COUNTERCLOCKWISE** until the engine is operating at the highest rpm. If the engine coughs and operates as if the fuel is too lean, but the idle adjustments have been correctly made, then recheck the synchronization between the fuel and ignition systems.

After the idle has been adjusted, push the idle knob retainers onto the needle until it engages the spline on the low-speed needle. Repeat the procedure for the other two carburetors. Replace the rubber caps into the front cover of the air silencer.

4-7 TYPE II CARBURETOR WITH LOW-SPEED ORIFICE AND HIGH-SPEED ORIFICE

All V4 and V6 powerheads
thru 1978
EXCEPT
85 hp, 115 hp, and 140 hp
thru 1979

TYPE III CARBURETOR WITH LOW-SPEED ORIFICE INTERMEDIATE ORIFICE AND HIGH-SPEED ORIFICE

All V4 and V6 powerheads
1979 and on
EXCEPT
85 hp 1980 only; also
115 hp, 120hp, and 140 hp - 1980 and on.
200hp and 225hp - 1986 and on.

DESCRIPTION

The Type II and Type III carburetors are frontdraft type, with two throats ("double barrel"). The Type II has a low-speed orifice and a high-speed orifice. The Type III has a fixed low-speed orifice, an intermediate orifice, and a high-speed orifice. Both models are covered in this section.

Two of the Type II or Type III carburetors are installed on the V4 model engines and three are installed on the V6 engines.

TYPE II AND TYPE III CARBURETOR 4-17

This arrangement provides a separate "barrel" for each cylinder. The float in each carburetor serves two cylinders.

REMOVAL

1- Disconnect the battery cables at the battery as a precaution against an accidental spark igniting the fuel or fuel fumes present during the service work. Disconnect the quick-disconnect on the fuel line to the engine. Remove the hood. Remove the retaining screws from the front of the air silencer. Remove the drain hose from the bottom of the air silencer. Remove the adjusting arms to the low-speed needle valves, if used. Remove the four screws securing the retainer base to the carburetors. Lift the retainer base upward and out of the way.

Disconnect the choke linkage. Slip the retaining ring and the choke solenoid spring back from the upper choke arm. Remove the screws securing the choke yoke to the air silencer. Unclamp the wires to the choke solenoid. Set the choke solenoid aside.

2- Remove the fuel pump retaining screws. Move the fuel pump aside to allow the carburetors to be removed. It is not necessary to disconnect the lines to and from the fuel pump. Disconnect the fuel line hoses between the two carburetors.

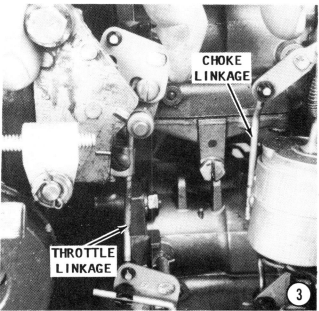

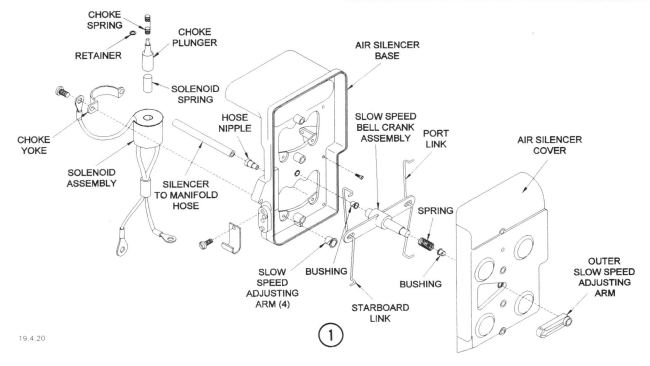

4-18 FUEL

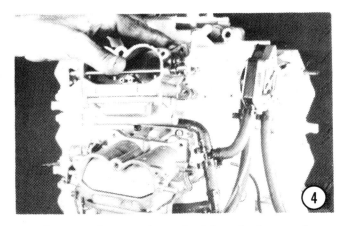

3- On the starboard side of the engine, notice the arms between the carburetors for throttle and choke action. Loosen the linkage fastener, slide the fastener upward and the linkage arms will then come free.

4- Remove the carburetor retaining nuts and lockwashers. Lift the carburetors free of the intake manifold.

SPECIAL WORDS

Only one carburetor will be rebuilt in the following procedures. Service of the other unit is to be performed in the same manner. The basic carburetor is the unit with the orifices. Differences for units with the low-speed adjustments will be noted.

The carburetor parts should be kept with the individual unit to ensure all items are installed back in their original positions.

DISASSEMBLING

5- Lay the carburetors on the bench on the throat openings. Disconnect and remove the hoses between the carburetors.

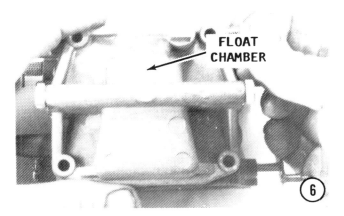

6- Turn the carburetor body upsidedown, and remove the four float chamber screws.

7- Lift the bowl assembly free of the carburetor body. Remove and discard the gasket.

8- Remove the hinge pin and then the float. Remove the needle from the seat, then the seat, and finally the gasket.

9- Remove the two low-speed orifice screws. These screws are located just above the high-speed orifices on both sides of the carburetor body. Use the proper size screwdriver, or special tool OMC No. 317002, and remove the two low-speed orifices. If working on a Type III carburetor, remove the intermediate orifice plug and then the intermediate orifice.

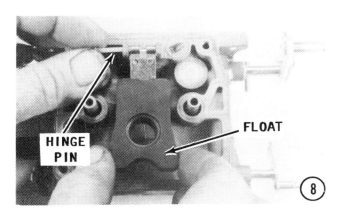

TYPE II AND TYPE III CARBURETOR 4-19

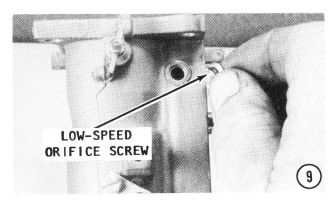

9

SPECIAL OMC WORDS

Since 1983, OMC dictates carburetor parts should **NOT** be submerged in carburetor cleaner, as has been the practice since carburetors were invented. Their approved procedure is to place the parts in a shallow tray and then spray them with an aerosol carubretor cleaner.

A syringe, short section of clear plastic hose, and Isopropyl Alchohol should be used to clear passages and jets.

Blow out all passages in the castings with compressed air. Check all parts and passages to be sure they are not clogged or contain any deposits. **NEVER** use a piece of wire or any type of pointed instrument to clean drilled passages or calibrated holes in a carburetor.

Move the throttle shaft back-and-forth to check for wear. If the shaft appears to be too loose, replace the complete throttle body because individual replacement parts are **NOT** available.

Inspect the main body, airhorn, and venturi cluster gasket surfaces for cracks and

10- Remove the two drain plugs, one on each side of the carburetor body.

11- Use the proper size screwdriver, or special tool OMC No. 317002, and remove the two high-speed orifices.

CLEANING AND INSPECTING

NEVER dip rubber parts, plastic parts, or nylon parts, in carburetor cleaner. These parts should be cleaned **ONLY** in solvent, and then blown dry with compressed air.

Place all metal parts in a screen-type tray and dip them in carburetor cleaner until they appear completely clean, then wash with solvent or clean water, and blow dry with compressed air.

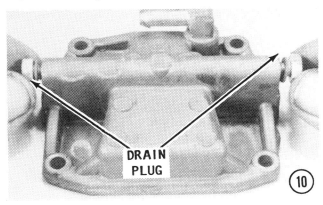

10

11

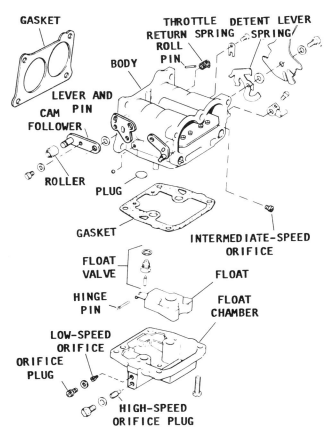

Exploded drawing of the Type III carburetor with low speed, intermediate speed, and high speed orifices.

burrs which might cause a leak. Check the float for deterioration. Check to be sure the float spring has not been stretched. If any part of the float is damaged, the unit must be replaced. Check the float arm needle contacting surface and replace the float if this surface has a groove worn in it.

Check the orifices for cleanliness. The orifice has a stamped number. This number represents a drill size. Check the orifice with the shank of the proper size drill to verify the proper orifice is used. The local OMC dealer will be able to provide the correct size orifice for the engine and carburetor being serviced.

Most of the parts that should be replaced during a carburetor overhaul are included in overhaul kits available from your local marine dealer. One of these kits will contain a matched fuel inlet needle and seat. This combination should be replaced each time the carburetor is disassembled as a precaution against leakage.

ASSEMBLING

GOOD WORDS

Make every effort to keep all parts as clean as possible. Do not lay parts on material which may contain dust, dirt, or lint.

Replace all gaskets and O-rings. Never attempt to use the original parts a second time.

1- Install the inlet seat using the proper size screwdriver. Inject just a drop of lightweight oil into the seat. Thread the inlet needle valve into the seat, and tighten it until it barely seats. **DO NOT** overtighten or the needle valve may be damaged.

2- Position **NEW** gaskets in place on the carburetor body and on the nozzle.

3- Lower the float down over the nozzle, and then slide the hinge pin into place. Hold the carburetor body in such a manner that allows the float to hang free. The float should be parallel with the carburetor body. If it is not, bend the tab as close to the float as possible until the float is parallel with the body. Take care to bend the tab on both sides of the float and to bend it squarely.

WORDS OF CAUTION

Take time to use the proper size screwdriver, or special tool OMC No 317002, to install an orifice. If the orifice is damaged and the edge of the opening burred, because the wrong size screwdriver was used, the flow of fuel will be restricted and the orifice will not function properly. Damage to the orifice will also make removal very difficult during the next carburetor service.

4- Use special tool OMC No. 317002 and install the high-speed orifices, one on each side of the carburetor body.

5- Install the drain plugs with **NEW** gaskets.

6- Install the low-speed orifices into the carburetor body just above the high-speed orifices.

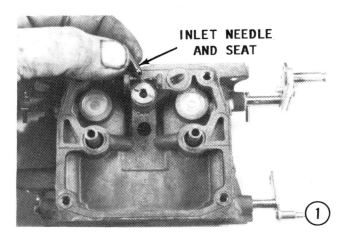

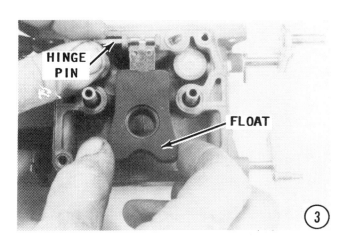

TYPE II AND TYPE III CARBURETOR 4-21

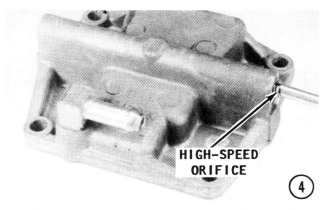

7- Install the screws, with **NEW** washers behind the orifices. If working on a Type IIB carburetor, install the intermediate orifice, and then the plug.

8- Install the float chamber and secure it with the four screws. Tighten the screws **ALTERNATELY** and **EVENLY**.

INSTALLATION

9- Install and connect the fuel hoses between the carburetors.

10- Place a **NEW** gasket in position on the intake manifold. Install the lower carburetor first, and secure it in place with the four attaching screws. Install the upper carburetor on the V4 engines or the middle and upper carburetors on the V6 engines.

11- Connect the fuel hoses between the carburetors.

Install a **NEW** gasket to the face of the carburetor, and then the air silencer base. See illustration, Page 4-17. Install the choke to the air silencer base. Connect the linkage to the carburetor choke arm. Secure the linkage in place with the small O-ring. Sometimes an E-clip is used.

12- Connect the linkage between the carburetors on the starboard side of the engine. Make sure the throttle butterflies are fully closed when the connections are

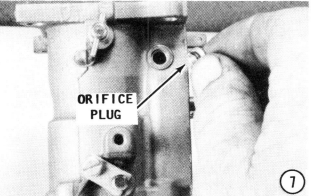

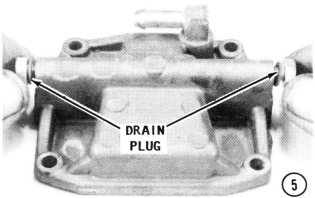

made. If the carburetor has the "primer choke system" the throttle butterflies are not used. Tighten the retainer securely. Connect the choke linkage between the carburetors. Make sure the choke butterflies are wide open when the connections are made. Tighten the retainer securely.

Check to be sure the choke butterfly in each carburetor seats fully closed. Check the throttle butterflies to be sure both are fully closed. Now, move the linkage and check to be sure they both open fully and to the same degree. If an adjustment is required, the retainer on the linkage can be moved.

Synchronization

To synchronize the fuel system with the ignition system, see the appropriate section in Chapter 5.

Closing Tasks

Mount the engine in a test tank or body of water. If this is not possible, connect a flush attachment and garden hose to the lower unit. **NEVER** operate the engine above idle speed using the flush attachment. If the engine is operated above idle speed with no load on the propeller, the engine could **RUNAWAY** resulting in serious damage or destruction of the unit.

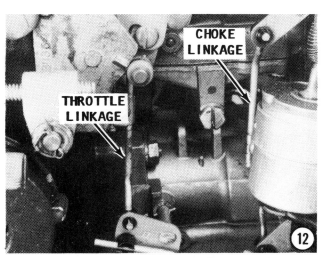

CAUTION: Water must circulate through the lower unit to the engine any time the engine is run to prevent damage to the water pump in the lower unit. Just five seconds without water will damage the water pump.

Start the engine and allow it to warm to operating temperature. Adjust the low-speed idle by turning the low-speed needle valve **CLOCKWISE** until the engine begins to misfire or the rpm drops noticeably. From this point, rotate the needle valve **COUNTERCLOCKWISE** until the engine is operating at the highest rpm. If the engine coughs and operates as if the fuel is too lean, but the idle adjustments have been correctly made, then recheck the synchronization between the fuel and ignition systems.

**4-8 TYPE IV CARBURETOR
UNIQUE DESIGN
WITH LOW-SPEED ORIFICE
INTERMEDIATE-SPEED ORIFICE
AND HIGH-SPEED ORIFICE**

V4 powerheads - 120hp and 140hp
1985 and on.
V6 powerheads - 200hp and 225 hp
1986 and on.

REMOVAL

1- Disconnect the battery cables at the battery as a precaution against an accidental spark igniting the fuel or fuel fumes present during the service work. Disconnect the quick-disconnect on the fuel line to the powerhead. Remove the hood.

GOOD WORDS

Perform the following step for each carburetor to be removed.

TYPE II AND TYPE III CARBURETOR 4-23

2- Remove the retaining screws from the front of the air silencer, and then remove the silencer. Disconnect the main fuel hose to the carburetor. Loosen the throttle shaft link. Cut any remaining fuel line "tie wraps"; remove the attaching hardware; and then lift the carburetor free of the powerhead.

DISASSEMBLING

1- Remove the four bolts securing the black plastic main carburetor body to the aluminum throttle body. Separate the two parts. Remove and discard the O-ring. Observe the two air bleed orifices located on the front face of the carburetor. The upper orifice is the intermediate orifice and lower orifice is the idle orifice. Make a note of the number embossed on each orifice and its location, upper or lower, to ensure the proper orifice is installed back in the correct location. Using the proper size screwdriver, remove both orifices.

2- Remove the two screws securing the side cover to the main carburetor body. Remove the side cover, and then remove and discard the gasket.

3- Remove the five Phillips head screws securing the float bowl to the main carburetor body. Lift off the float bowl. Remove and discard the float bowl gasket. Do **NOT** disturb the two tubes on the side of the float chamber. It is not necessary to remove these tubes during a standard overhaul.

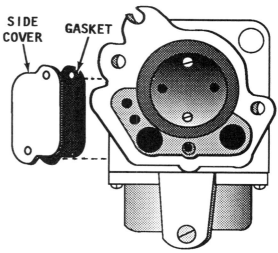

Line drawing depicting the side cover and gasket removed.

4- Loosen and remove the float pin securing screw. Lift out the float, pin, and needle valve assembly. The needle valve assembly is hooked onto the float tab with a very small wire clip.

5- Remove the plug from the float bowl. Using the proper size screwdriver, reach into the hole uncovered by removal of the plug and remove the high speed fuel orifice.

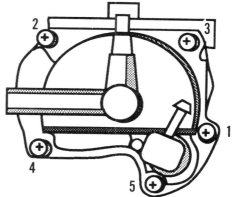

Line drawing to indicate float bowl tightening sequence.

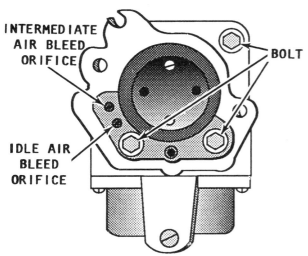

Simple cross-section drawing showing location of the bolts securing the two bodies together and the intermediate orifice and the idle air bleed orifice.

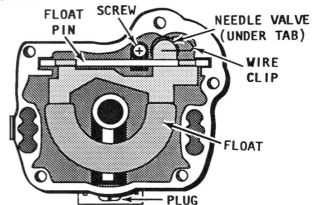

Line drawing of the float bowl interior to indicate the correct positioning of the needle valve wire clip.

Do **NOT** confuse this orifice with the two previously removed in Step No. 1. Remove the needle valve seat from the main carburetor body.

CLEANING AND INSPECTING

NEVER dip rubber parts, plastic parts, or nylon parts, in carburetor cleaner. These parts should be placed in a shallow tray and then sprayed with an aerosol carubretor cleaner.

A syringe, short section of clear plastic hose, and Isopropyl Alchohol should be used to clear passages and jets.

Place all metal parts in a screen-type tray and dip them in carburetor cleaner until they appear completely clean, then wash with solvent or clean water, and blow dry with compressed air.

Blow out all passages in the castings with compressed air. Check all parts and passages to be sure they are not clogged or contain any deposits. **NEVER** use a piece of wire or any type of pointed instrument to clean drilled passages or calibrated holes in a carburetor.

Inspect the main body, airhorn, and venturi cluster gasket surfaces for cracks and burrs which might cause a leak. Check the float for deterioration. Check to be sure the float spring has not been stretched. If any part of the float is damaged, the unit must be replaced. Check the float arm needle contacting surface and replace the float if this surface has a groove worn in it.

Check the orifices for cleanliness. The orifice has a stamped number. This number represents a drill size. Check the orifice with the shank of the proper size drill to verify the proper orifice is used. The local OMC dealer will be able to provide the correct size orifice for the engine and carburetor being serviced.

Most of the parts that should be replaced during a carburetor overhaul are included in overhaul kits available from your local marine dealer. One of these kits will contain a matched fuel inlet needle and seat. This combination should be replaced each time the carburetor is disassembled as a precaution against leakage.

ASSEMBLING

1- Install and tighten the needle valve seat. Using the proper size screwdriver, install the high speed fuel orifice into the side opening in the main carburetor body. Tighten this brass orifice just snug. Install the float chamber plug and tighten the plug to a torque value of 32 in lb (3.5Nm).

2- Hook the needle valve over the float tab with the wire clip portside. Lower the float assembly into place inside the float chamber with the needle valve entering the needle vavle seat and the ends of the float pin indexing into the two notches of the float chamber. Secure the float pin with the Phillips head screw. Tighten this screw securely. Check to be sure the float is free to move up and down and the wire clip is positioned on top of the float tab, as shown in the accompanying illustration.

Float Adjustment

3- Place a new float bowl gasket over the installed float. Invert the float bowl and allow the float to drop freely. The float should be level with the gasket surface $\pm$ 1/32" (0.8mm). If an adjustment is required, remove the float assembly and gently bend the float tab, as necessary.

4- Install the float bowl. Secure the bowl to the main carburetor body with the

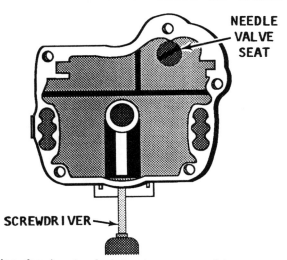

Line drawing to depict using a screwdriver to remove/install the high speed orifice.

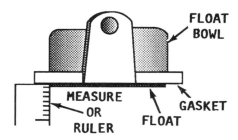

Line drawing to depict the correct method of making the float drop measurement.

five attaching screws. Tighten the screws to a torque value of 20 in lb (2.5Nm) in the sequence indicated in the accompanying illustration.

5- Position the side cover and a new gasket in place. Install and tighten the two attaching screws, alternately and evenly to a torque value of 20 in lb (2.5Nm).

6- Place a new O-ring into the groove of the main carburetor body and align the body to the carburetor throttle body. Install and tighten the four attaching bolts alternately and evenly to a torque value of 42 in lb (5.5Nm).

INSTALLATION

Position new carburetor mounting gaskets on the intake manifold. The manufacturer recommends **NO** sealer be used at this location. Install each carburetor in the same location from which it was removed. Tighten the two attaching bolts alternately and evenly to a torque value of 10 ft lb (13.6Nm). Install the fuel hoses and secure each hose with a new tie strap. Connect the throttle linkage and the primer hoses.

Install the air box and gasket. Tighten the attaching hardware to a torque value of 6 ft lb (8Nm).

4-9 PRIMER CHOKE SYSTEM

The "primer choke system" consists of a solenoid valve, distribution lines, and injection nozzles. The nozzles are tapped into the bypass covers. During engine cranking when the choke system is operating, fuel is injected through metered holes in the nozzles directly into the cylinders, instead of being routed in the usual manner through the crankcase.

During engine operation, from the fuel tank, the fuel passes through the fuel line, to the fuel pump, and into the carburetor. From the carburetor the fuel and air mixture passes through the crankcase and into the cylinder.

The "primer choke system" injects fuel directly into the cylinder. The system is controlled by the "push-in" type key switch. As the key is pushed in, the solenoid is activated, moving a small plunger which acts as a pump, injecting fuel through the nozzles directly into the cylinder to assist engine start.

If the battery is dead and the choking effect is desired, a lever on the solenoid may be moved to the **MANUAL** position opening the seat in the valve. When the squeeze bulb is activated, fuel will pass through the nozzles directly into the cylinders The lever is then returned to the **RUN** position during actual cranking of the engine.

WORD OF CAUTION

If the fuel tank has been exposed to direct sunlight, pressure may have developed inside the tank. Therefore, when the solenoid lever is moved to the **MANUAL** position, an excessive amount of fuel may be forced into the cylinders. As a safety precaution, under possible fuel tank pressure conditions, the fuel tank cap should be opened slightly to allow the pressure to escape before attempting to start the engine.

SOLENOID TESTING

Connect an ohmmeter to the solenoid between the blue/white stripe lead and the black (ground) wire. Observe the reading. The ohmmeter should indicate 5.5 ± 1.5 ohms. If the reading is not within the prescribed range, the solenoid is defective and must be replaced.

The accompanying illustration will be helpful in ordering and replacing parts of the "primer choke system".

The fuel hoses should be checked to ensure they remain flexible and are clear to permit an adequate fuel supply to pass through. Pay particular attention to any evidence of a crack in a fuel line which may permit fuel to escape and cause a very hazardous condition. The diagrams in the Appendix may be very helpful during replacement of fuel lines to ensure correct routing.

OMC tool No. 326623, is available to clean the metered holes in the nozzles.

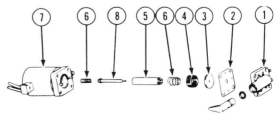

1. Cover
2. Gasket
3. Seal
4. Filler
5. Plunger
6. Spring
7. Solenoid body
8. Plunger valve

Exploded drawing of a primer solenoid valve.

4-10 VARIABLE RATIO OIL SYSTEM

DESCRIPTION

Since 1983, V4, V6, and V8 model units have been equipped with a Variable Ratio Oil System, commonly referred to as simply VRO. Since late 1987, the 3-cylinder in-line powerhead has been equipped with the VRO system. Purpose of the system is to mix oil with the fuel in the proper ratio at all engine speeds to ensure adequate lubrication. The system replaces the age-old method of manually adding a quantity of oil to the fuel tank.

The VRO system consists of an oil reservoir (tank), a VRO oil line primer, a pump to move the oil from the tank to the powerhead, a warning horn, a spark arrestor in the pulse hose to the VRO pump, an oil inlet filter, a vacuum switch in the fuel line (only on the V6 and V8 models) and the necessary hoses and fittings to connect the various items for efficient operation. All connections in the system **MUST** be airtight to prevent serious damage to the powerhead.

As the name implies, the VRO pump moves the oil from the oil reservoir to the powerhead. However, it is a **dual** pump and also pumps fuel. Pumping action of the pump stops automatically if fuel is not available at the pump for any reason. This automatic shut down of the pump prevents the carburetors from filling with oil.

On L3, V4, V6, and V8 powerheads equipped with the VRO system, the warning horn, located in the control box or behind the console, serves two functions.

First, the horn will sound for 1/2 second every 20 seconds if the oil tank level reaches 1/4 of the tank's capacity. The low oil warning circuit consists of a sending unit in the oil reservoir, a ground wire to the engine and a wire to the warning horn through the key switch.

Secondly, the warning horn is connected to an overheat circuit and two sending units mounted in the cylinder heads and wired through the key switch. The horn will sound continuously if the temperature of the powerhead exceeds $99^{\circ}C$ ($211^{\circ}F$). The horn is deactivated when the powerhead is allowed to cool to $79^{\circ}C$ ($175^{\circ}F$).

On V6 and V8, powerheads prior to 1988, the warning horn serves three purposes. In addition to the two functions of low oil and overheat conditions just described, the horn will also sound if there is a restriction in the fuel line and the vacuum in the **IN** hose falls below 17.7 cm (7") of mercury. A vacuum switch is mounted on the port side of the powerhead with a ground connection and a wire connected to the warning horn through the key switch. With this arrangement, the horn will sound if there is a restriction in the fuel **IN** line causing insufficient fuel delivery to the engine.

On V6 and V8 powerheads since 1988, the function of the "low oil" warning horn was changed to a "no oil" horn. Since 1988 L3 and V4 powerheads have been equipped with the "no oil" horn.

The spark arrestor on all models is installed in the pulse hose to the VRO pump. This flame arrestor prevents a backfire flame from entering the VRO pump. A clamp positioned on the hose prevents the spark arrestor from migrating up the hose to the pump.

GOOD WORDS

Anytime the pulse hose is disconnected at the powerhead, **TAKE CARE** to be sure the spark arrestor remains in the hose. The spark arrestor **MUST** be properly positioned in the pulse hose to prevent serious and permanent damage to the VRO pump.

The oil inlet filter, located in the VRO reservoir oil pickup line, prevents any dirt or foreign material from entering the pump. If the filter should need cleaning, the hose assembly should be removed and reverse flushed using clean solvent. **DO NOT** at-

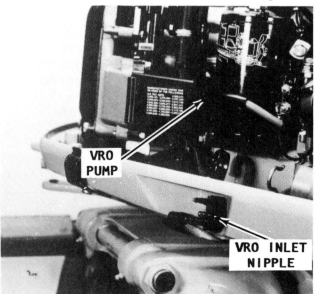

Typical variable ratio oil system (VRO) installation with principle parts identified.

tempt to remove the filter because the filter and hose are serviced and replaced as an assembly.

NEW POWERHEAD BREAK-IN PROCEDURE

A complete new outboard unit, a new powerhead, or a rebuilt powerhead, must have oil mixed in the fuel tank **IN ADDITION** to the VRO system. The mixture should be 50:1 (1 pint oil to 6 gallons of fuel) and the unit **MUST** be operated with this mixture during the first 10 hours of service.

CRITICAL WORDS

To convince himself the VRO system is working properly, the operator should observe a drop in the oil supply in the VRO oil reservoir during the 10-hour break-in period.

At the end of the 10 hour period, the powerhead mounted fuel filter should be inspected. Remove any dirt or foreign matter collected in the filter.

TROUBLESHOOTING VRO SYSTEM

This short section list a few of the probable problems that might occur in the system with suggested corrective action.

The next section -- **SERVICING** outlines in detail how the tests and service work is to be performed.

Low Oil Level Warning Horn Sounds
All Models

Warning horn sounds for 1/2 second every 20 seconds:

a- Oil level in the oil reservoir is below 1/4 full. Add oil to the reservoir.

b- Disc on the pickup unit may not be positioned properly. Remove the pickup unit from the reservoir and correct as required.

L3 and V4 Models
Prior to 1988
Horn Sounds Continuously

a- Powerhead has overheated. Back off the throttle until the horn stops. If horn continues to sound, stop the engine immediately and check the water intake. Start the engine and run it at a fast idle in neutral, or shift to forward and proceed slowly. Observe the water pump indicator which should discharge a steady stream of water. After operating the engine for about two minutes, the powerhead should cool and the horn will be deactivated at 79°C (175°F).

b- If horn continues to sound, shut down the engine to prevent serious powerhead damage. Check the horn circuit for the cause of the overheat condition.

V6 and V8 Models Prior to 1988
And All Models Since 1988
Warning Horn Sounds Continuously

a- Horn may sound if there is a restriction in the fuel system causing insufficient delivery of fuel to the powerhead. Back off the throttle. If the horn stops when the throttle is reduced, there is probably a restriction in the fuel system.

b- If a horn continues to sound when the throttle is reduced the powerhead may be overheated. Shut down the engine and check the cause of the overheat condition.

SERVICING VRO SYSTEM

Servicing consists of making simple tests and checks. A vacuum gauge, pressure gauge, a "T" fitting, a short section of clear plastic hose, a source of low-pressure compressed air, and a couple of normal shop tools are all that is required to service the VRO system.

Check Fuel and Oil Circuits

1- Verify there is more than 1/4 tank of oil in the reservoir. Disconnect the oil inlet hose from the VRO pump. Be prepared to catch oil as it is ejected from the end of the hose. Squeeze the bulb and verify oil is ejected from the open end of the hose. The presence of oil verifies a clear line from the tank to this point. If no oil is ejected, clean the line from the tank and repeat the test.

2- Disconnect the fuel mixed outlet hose from the VRO pump.

3- Insert a "T" fitting into the end of the hose. Insert a drop of oil into each end of a short section of clear plastic hose. Connect one end of the clear piece of plastic hose to one arm of the "T" and the other end of the hose to the VRO pump. Connect a 0-15 psi pressure gauge to the leg of the "T". Secure the connections with tie straps or hose clamps. Mount the engine in an adequate size test tank or move the boat to a body of water.

NEVER operate the engine using a flush attachment for this test. If the engine is operated above idle speed with no load on

the propeller, the engine could **RUNAWAY** resulting in serious damage or destruction of the unit.

CAUTION: Water must circulate through the lower unit to the engine any time the engine is run to prevent damage to the water pump in the lower unit. Just five seconds without water will damage the water pump.

Start the engine and shift the unit into gear. Advance the throttle to the near wide open position. Check the pressure gauge. The gauge should indicate 34 kPa (5 psi) to 103 kPa (15 psi) pressure at near full throttle. Each time the pump pulses a small squirt of oil, in addition to the fuel passing through, should be observed in the clear plastic hose discharging from the pump.

Also, the fuel pressure will drop approximately 6.8 kPa (1 psi) to 13.7 kPa (2 psi) and a "click" sound may be heard each time the pump pulses and discharges oil.

Results

Fuel pressure satisfactory and oil is observed discharging from the pump through the clear plastic hose: Fuel and oil systems are verified satisfactory.

No fuel pressure: Check quantity of fuel in the tank to be sure fuel level reaches the pickup. Add fuel if required. Check fuel line to be sure it is not pinched or kinked restricting fuel flow. Check engine pulse hose to be sure it is not pinched, leaking or disconnected. If all above conditions are satisfactory, the VRO pump is defective and **MUST** be replaced as a unit.

Low fuel pressure: Check for restricted fuel filter at the engine. Check the engine pulse hose to be sure it is not pinched, kinked, leaking, or restricting fuel flow. Squeeze the fuel primer bulb a few times to force a possible fuel vapor-lock out of the system.

Warning Horn Check

4- Slide the "boot" down the wire to clear the knife disconnect between the temperature switch and the horn lead. This is not an easy task, but with a pair of needle nose pliers and some patience, it can be done. After the "boot" is clear, disconnect the fitting. Turn the key switch to the **ON** position. Now, make contact with the lead onto a "clean" place on the powerhead. The horn should sound. If the horn does not

VRO SYSTEM 4-29

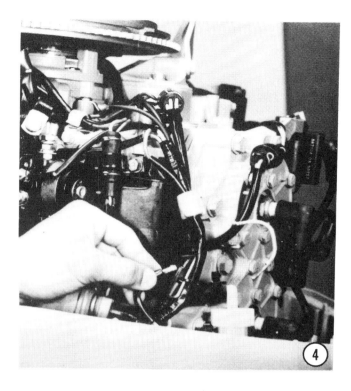

sound, there is a problem between the knife disconnect and the dashboard or wherever the horn is installed.

Clean Vacuum Pulse Hose

5- Remove the vacuum pulse hose. The hose may be cleaned by back-flushing with clean solvent. **TAKE CARE** to ensure the spark arrestor installed in the hose stays in the hose. A clamp is positioned on the hose to prevent the spark arrestor from migrating up the hose to the pump. If the hose is damaged and requires replacement, the hose and flame arrestor are purchased as an assembly. The flame arrestor cannot be purchased separately.

VRO Pump

If troubleshooting and service work indicates the VRO pump to be defective, it must be replaced. The pump cannot be serviced or repaired. The pump is removed by first disconnecting the hoses and then removing the three mounting bolts. Lift the VRO pump free.

Engine Mounted Fuel Filter

6- If a fuel inlet filter is mounted at the engine, this filter can be separated and inspected without removing the hoses. The filter should be inspected at the end of the 10-hour break-in period and at regular intervals as part of normal engine maintenance and service.

GOOD WORDS

Use **ONLY** OMC approved clamps on the connections on the inlet side of the VRO pump. A screw type hose clamp will very likely pinch or break the hose causing a suction leak. A tie-wrap will not tighten down to the degree required to prevent a vacuum leak.

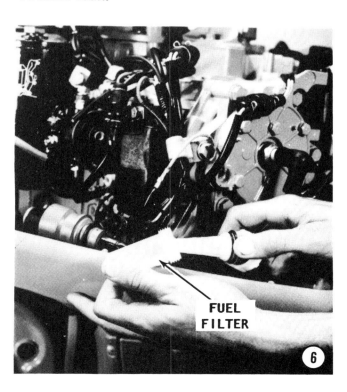

4-30 FUEL

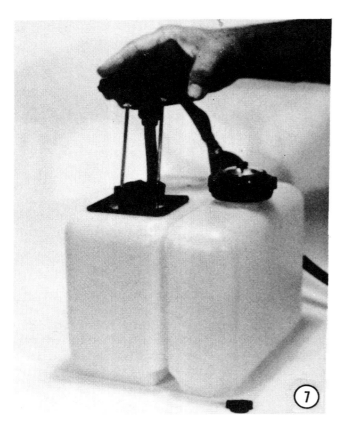

Vacuum Hose Check

This check will verify the system in good condition from the oil pickup to the end of the hose.

7- Remove the VRO pickup unit from the oil reservoir by first removing the four mounting screws and then lifting the pickup straight up and out of the reservoir. Inspect the oil pickup filter, and clean it, if necessary.

8- Purge the system of oil and any small particles of foreign matter by using low-pressure compressed air through the VRO pickup end.

9- Connect a vacuum gauge to the end of the hose. Insert the plug-type nipple that is shipped with the engine (snapped to the

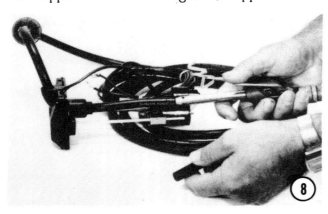

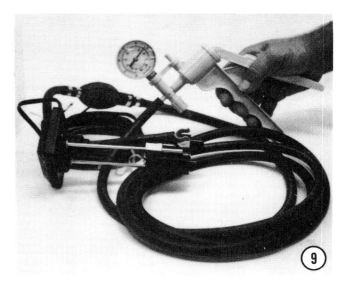

fuel hose at the engine) into the VRO pickup. If the plug has been lost, one can be easily made out of suitable material. Secure the plug with a clamp. Pump the gauge until 17.7 cm (7") of mercury is indicated. The system should hold the vacuum reading.

10- If the system fails to hold the required vacuum reading, check each connection by applying a small amount of oil at each fitting. The oil will **MOMENTARILY** stop the leak and the vacuum reading will stop dropping. Carefully inspect the hose for damage.

GOOD WORDS

OMC **STRONGLY** recommends that the hose from the primer bulb to the VRO pump be one continuous hose with no fittings between. Therefore, if the hose is damaged, the entire length should be replaced. Also,

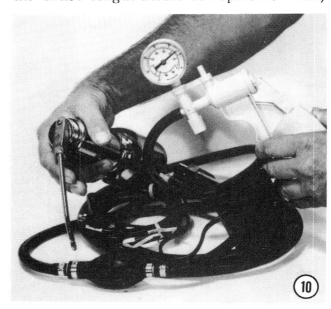

TANK SERVICE 4-31

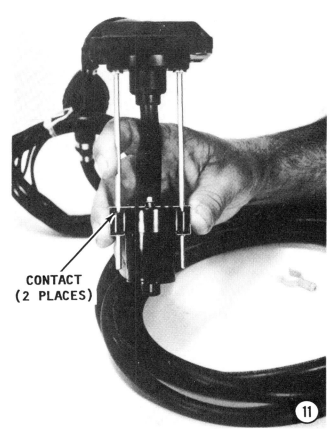

CONTACT (2 PLACES)

11- If the horn sounds indicating low oil, the contacts on the pickup may not be positioned properly. A disc rises slightly when oil is added into the reservoir and lowers slightly as the oil level drops. Check to be sure the contact surface on both sides of the disc is riding equally with the other side to prevent sounding the horn prematurely. If the disc fails to rise, the horn will sound continuously. Clean the float chamber in solvent and the disc should then rise clear of the contacts.

Fuel Line Vacuum Test

12- Connect a "T" fitting and vacuum gauge to the outlet hose from the fuel tank, as shown. Check to be sure all fittings and connections are tight.

NEVER operate the engine above idle speed using a flush attachment for this test. If the engine is operated above idle speed with no load on the propeller, the engine could **RUNAWAY** resulting in serious damage or destruction of the unit.

CAUTION: Water must circulate through the lower unit to the engine any time the engine is run to prevent damage to the water pump in the lower unit. Just five seconds without water will damage the water pump.

Start the engine and operate it at idle speed. Observe the vacuum gauge. The gauge should indicate no more than 17.7 cm (7") of mercury.

if a dual engine installation is used, **DO NOT** "T" into the line from the primer bulb. Use a separate oil reservoir, with separate primer bulb and hose to the VRO pump on the second engine -- a completely separate system for each engine.

4-11 FUEL TANK SERVICE

A squeeze bulb is used to move fuel from the tank to the carburetor until the engine is operating. Once the engine starts, the fuel pump, mounted on the engine, transfers fuel from the tank to the carburetor. The pickup unit in the tank is sold as a complete unit, but without the gauge and float.

On V6 engine installations, the larger diameter fuel line is connected directly to the fuel tank. The quick-disconnect fitting is not used.

1- To replace the pickup unit, first remove the four screws securing the unit in the tank. Next, lift the pickup unit up out of the tank.

2- Remove the two Phillips screws securing the fuel gauge to the bottom of the pickup unit and set the gauge aside for installation onto the new pickup unit.

If the pickup unit is not to be replaced, clean and check the screen for damage. It

Fuel pump pickup assembly. The fuel gauge assembly is not sold as a part of the pickup unit.

is possible to bend a new piece of screen material around the pickup and solder it in place without purchasing a complete new unit. Attach the fuel gauge to the new pickup unit and secure it in place with the two Phillips screws.

3- Clean the old gasket material from the fuel tank and old pickup unit (if the old pickup unit is to be installed for further service). Work the float arm down through the fuel tank opening, and at the same time insert the fuel pickup tube into the tank. It will probably be necessary to exert a little force on the float arm in order to feed it all into the hole. The fuel pickup arm should spring into place once it is through the hole. Secure the pickup and float unit in place with the four attaching screws.

4- The primer squeeze bulb can be replaced in a short time. A squeeze bulb assembly, complete with the check valves installed, may be obtained from the local OMC dealer.

An arrow is clearly visible on the squeeze bulb to indicate the direction of fuel flow. The squeeze bulb **MUST** be installed correctly in the line because the check valves in each end of the bulb will allow fuel to flow in **ONLY** one direction. Therefore, if the squeeze bulb should be installed backwards (in a moment of haste to get the job done), fuel will not reach the carburetor.

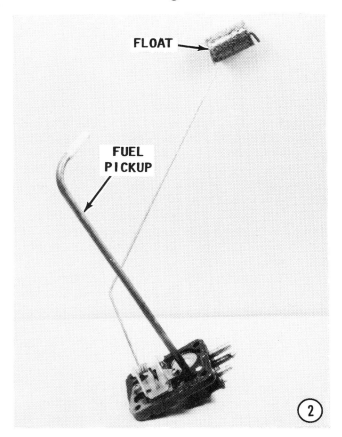

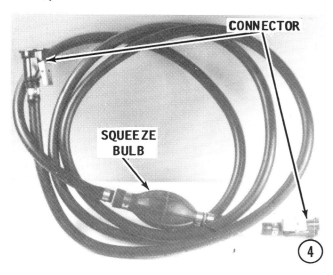

TANK SERVICE 4-33

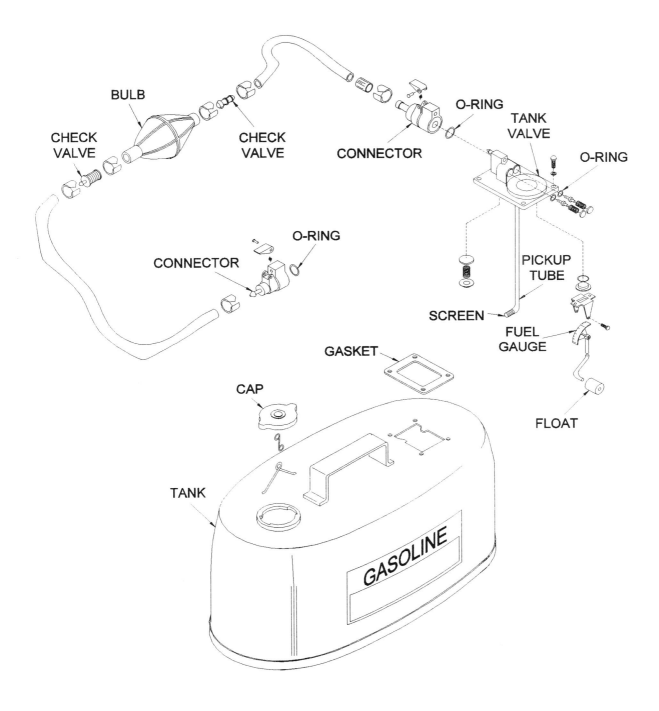

Exploded drawing of a modern non-pressurized fuel tank and fuel line.

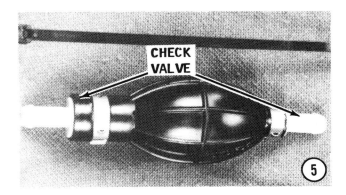

5- To replace the bulb, first unsnap the clamps on the hose at each end of the bulb. Next, pull the hose out of the check valves at each end of the bulb. New clamps are included with a new squeeze bulb. If the fuel line has been exposed to considerable sunlight, it may have become hardened, causing difficulty in working it over the check valve. To remedy this situation, simply immerse the ends of the hose in boiling water for a few minutes to soften the rubber and the hose will then slip onto the check valve without further problems. After the lines on both sides have been installed, snap the clamps in place to secure the line. Check a second time to be sure the arrow is pointing in the fuel flow direction, **TOWARDS** the engine.

6- Use two ice picks or similar tools, and push down the center plunger of the connector and work the O-ring out of the hole.

7- Apply just a drop of oil into the hole of the connector. Apply a thin coating of

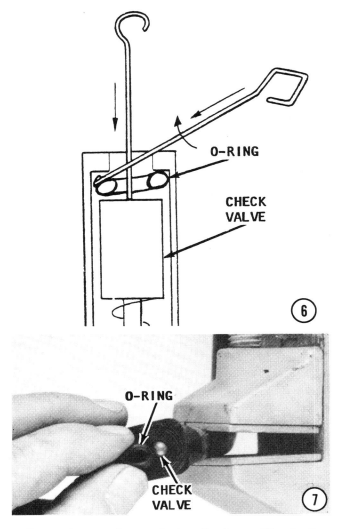

oil to the surface of the O-ring. Pinch the O-ring together and work it into the hole while simultaneously using a punch to depress the plunger inside the connector.

5
IGNITION

5-1 INTRODUCTION

The less an outboard engine is operated, the more care it needs. Allowing an outboard engine to remain idle will do more harm than if it is used regularly. To maintain the engine in top shape and always ready for efficient operation at any time, the engine should be operating every 3 to 4 weeks throughout the year.

The carburetion and ignition principles of two-cycle engine operation **MUST** be understood in order to perform a proper tune-up on an outboard motor.

If you have any doubts concerning your understanding of two-cyle engine operation, it would be best to study the operation theory section in the first portion of Chapter 3, before tackling any work on the ignition system.

CHAPTER COVERAGE

The first five sections of this section apply to **ALL** engines and ignition systems covered in this manual: 5-1, this Introduction; 5-2, Spark Plug Evaluation; 5-3, Polarity Check; 5-4, Wiring Harness, and 5-5, CD Ignition System Description.

All engines covered in this manual have a magneto capacitor discharge (CD) ignition system. However, because the 3-cylinder, V4, and V6 engines are included **AND** because of powerpack changes, the troubleshooting and test procedures vary. To simplify the coverage and as an assist to finding the proper section for the unit being serviced, the CD systems are divided into two major catagories labeled CD No. 1 and CD No. 2.

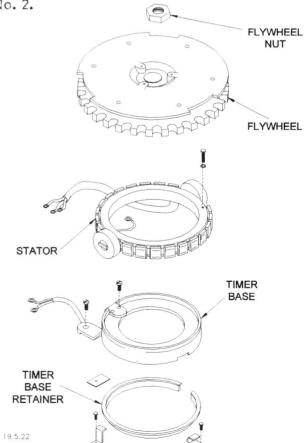

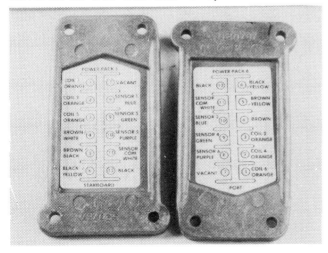

Two Power Packs used with the Type I Capacitor Discharge (CD) flywheel ignition system covered in this Chapter.

Exploded drawing of a typical flywheel, stator, timer base, and timer base retainer arrangement.

5-2 IGNITION

The major difference between the two systems is the type of powerpack installed. The powerpack of CD No. 1 had external terminals with the various leads connected and disconnected from these terminals. The CD No. 2 powerpack is a sealed unit with all wiring connections accomplished inside the unit. This powerpack has a wiring harness with a multiple pin connector.

Each of these systems will have a separate sub-section for the 3-cylinder, V4, and V6 engines.

Towards the end of the chapter, Section 5-14, covers ignition component replacement.

The engine size and model for each section is clearly identified at the head of the section. Therefore, it is an easy matter to find the section in the Table of Contents for the unit being serviced and then to follow **ONLY** those procedures to accomplish the work.

To synchronize the ignition system with the fuel, see Section 5-15.

For detailed timing procedures, see Section 5-17.

5-2 SPARK PLUG EVALUATION

Removal: Remove the spark plug wires by pulling and twisting on only the molded cap. **NEVER** pull on the wire or the connection inside the cap may become separated or the boot damaged. Remove the spark plugs and keep them in order. **TAKE CARE** not to tilt the socket as you remove the plug or the insulator may be cracked.

Examine: Line the plugs in order of removal and carefully examine them to determine the firing conditions in each cylinder. If the side electrode is bent down onto

*Damaged spark plugs. Notice the broken electrode on the left plug. The broken part **MUST** be found and removed before returning the powerhead to service.*

the center electrode, the piston is traveling too far upward in the cylinder and striking the spark plug. Such damage indicates the wrist pin or the rod bearing is worn excessively. In all cases, an engine overhaul is required to correct the condition. To verify the cause of the problem, crank the engine by hand. As the piston moves to the full up position, push on the piston crown with a screwdriver inserted through the spark plug hole, and at the same time rock the flywheel back-and-forth. If any play in the piston is detected, the engine must be rebuilt.

Correct Color: A proper firing plug should be dry and powdery. Hard deposits inside the shell indicate too much oil is being mixed with the fuel. The most important evidence is the light gray color of the porcelain, which is an indication this plug has been running at the correct temperature. This means the plug is one with the correct heat range and also that the air-fuel mixture is correct.

Rich Mixture: A black, sooty condition on both the spark plug shell and the porcelain is caused by an excessively rich air-fuel mixture, both at low and high speeds. The rich mixture lowers the combustion temper-

This spark plug is foul from operating with an overrich condition, possibly an improper carburetor adjustment.

This spark plug has been operating too-cool, because it is rated with a too-low heat range for the engine.

ature so the spark plug does not run hot enough to burn off the deposits.

Deposits formed only on the shell is an indication the low-speed air-fuel mixture is too rich. At high speeds with the correct mixture, the temperature in the combustion chamber is high enough to burn off the deposits on the insulator.

Too Cool: A dark insulator, with very few deposits, indicates the plug is running too cool. This condition can be caused by low compression or by using a spark plug of an incorrect heat range. If this condition shows on only one plug it is most usually caused by low compression in that cylinder. If all of the plugs have this appearance, then it is probably due to the plugs having a too-low heat range.

Fouled: A fouled spark plug may be caused by the wet oily deposits on the insulator shorting the high-tension current to ground inside the shell. The condition may also be caused by ignition problems which prevent a high-tension pulse being delivered to the spark plug.

Carbon Deposits: Heavy carbon-like deposits are an indication of excessive oil in the fuel. This condition may be the result of poor oil grade, (automotive-type instead of a marine-type); improper oil-fuel mixture in the fuel tank; or by worn piston rings.

Overheating: A dead white or gray insulator, which is generally blistered, is an indication of overheating and pre-ignition. The electrode gap wear rate will be more than normal and in the case of pre-ignition, will actually cause the electrodes to melt. Overheating and pre-ignition are usually caused by improper point gap adjustment; detonation from using too-low an octane rating fuel; an excessively lean air-fuel mixture; or problems in the cooling system.

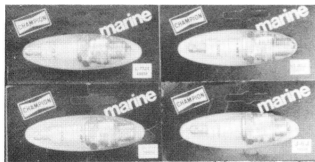

Today, numerous type spark plugs are available for service. ALWAYS check with your local marine dealer to be sure you are purchasing the proper plug for the engine being serviced.

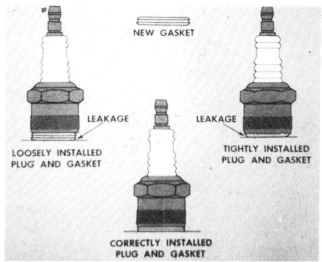

Drawing to illustrate a spark plug properly installed, center, and other plugs, left and right, improperly installed.

Electrode Wear: Electrode wear results in a wide gap and if the electrode becomes carbonized it will form a high-resistance path for the spark to jump across. Such a condition will cause the engine to misfire during acceleration. If all plugs are in this condition, it can cause an increase in fuel consumption and very poor performance during high-speed operation. The solution is to replace the spark plugs with a rating in the proper heat range and gapped to specification.

Red rust-colored deposits on the entire firing end of a spark plug can be caused by water in the cylinder combustion chamber. This can be the first evidence of water

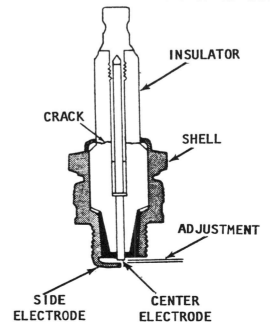

Cut-a-way drawing showing major spark plug parts.

5-4 IGNITION

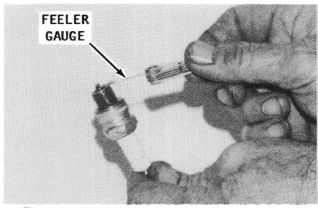

The spark plug gap should always be checked before installing new or used spark plugs.

entering the cylinders through the exhaust manifold because of scale accumulation. This condition **MUST** be corrected at the first opportunity. Refer to Chapter 3, Powerhead.

5-3 POLARITY CHECK

Coil polarity is extremely important for proper battery ignition system operation. If a coil is connected with reverse polarity, the spark plugs may demand from 30 to 40 percent more voltage to fire. Under such demand conditions, in a very short time the coil would be unable to supply enough voltage to fire the plugs. Any one of the following three methods may be used to quickly determine coil polarity.

1- The polarity of the coil can be checked using an ordinary D.C. voltmeter. Connect the positive lead to a good ground. With the engine running, momentarily touch the negative lead to a spark plug terminal. The needle should swing upscale. If the needle swings downscale, the polarity is reversed.

2- If a voltmeter is not available, a pencil may be used in the following manner: Disconnect a spark plug wire and hold the metal connector at the end of the cable about 1/4" from the spark plug terminal.

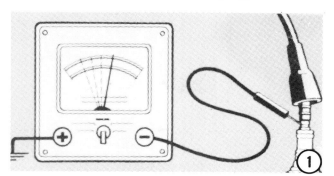

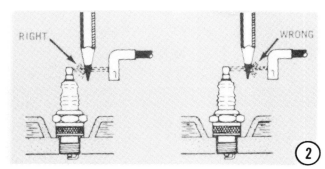

Now, insert an ordinary pencil tip between the terminal and the connector. Crank the engine with the ignition switch ON. If the spark feathers on the plug side and has a slight orange tinge, the polarity is correct. If the spark feathers on the cable connector side, the polarity is reversed.

3- The firing end of a used spark plug can give a clue to coil polarity. If the ground electrode is "dished", it may mean polarity is reversed.

5-4 WIRING HARNESS

CRITICAL WORDS: These next two paragraphs may well be the most important words in this chapter. Misuse of the wiring harness is the most single cause of electrical problems with outboard power plants.

A wiring harness is used between the key switch and the engine. This harness seldom contains wire of sufficient size to allow connecting accessories. Therefore, anytime a new accessory is installed, **NEW** wiring should be used between the battery and the accessory. A separate fuse panel **MUST** be installed on the dash. To connect the fuse panel, use one red and one black No. 10 gauge wire from the battery. If a small amount of 12-volt current should be accidently attached to the magneto system, the coil may be damaged or **DESTROYED**. Such a mistake in wiring can easily happen

if the source for the 12-volt accessory is taken from the key switch. Therefore, again let it be said, **NEVER** connect accessories through the key switch.

5-5 GENERAL TROUBLESHOOTING ALL ENGINES — ALL IGNITIONS

Always attempt to proceed with the troubleshooting in an orderly manner. The shotgun approach will only result in wasted time, incorrect diagnosis, replacement of unnecessary parts, and frustration.

Begin the ignition system troubleshooting with the spark plug/s and continue through the system until the source of trouble is located.

The following test equipment is a **MUST** when troubleshooting this system. Stating it another way, "There is no way on this green earth to properly and accurately test the complete system or individual components without the special items listed."

Preliminary Test

The first area to check on a CD flywheel magneto ignition system is the system ground.

Also check the battery charge and to verify the battery cables are connected properly for correct polarity. If the battery cables are connected backwards, testing will indicate **NO** spark.

If the battery is below a full charge, it will not be possible to obtain full cranking speed during the tests.

Compression Check

A compression check is extremely important, because an engine with low or uneven compression between cylinders **CANNOT** be tuned to operate satisfactorily. Therefore, it is essential that any compression problem be corrected before proceeding with the tune-up procedure. See Chapter 3.

If the powerhead shows any indication of overheating, such as discolored or scorched paint, especially in the area of the top (No. 1) cylinder, inspect the cylinders visually thru the transfer ports for possible scoring. A more thorough inspection can be made if the head is removed. It is possible for a cylinder with satisfactory compression to be scored slightly. Also, check the water pump. The overheating condition may be caused by a faulty water pump.

An overheating condition may also be caused by running the engine out of the water. For unknown reasons, many operators have formed a bad habit of running a small engine without the lower unit being submerged. Such a practice will result in an overheated condition in a matter of seconds. It is interesting to note, the same operator would never operate or allow anyone else to run a large horsepower engine without water circulating through the lower unit for cooling. Bear in mind, the laws governing operation and damage to a large unit **ALL** apply equally as well to the small engine.

Checking Compression

1- Remove the spark plug wires. **ALWAYS** grasp the molded cap and pull it loose with a twisting motion to prevent damage to the connection. Remove the spark plugs and keep them in **ORDER** by cylinder for evaluation later. Ground the spark plug leads to the engine to render the ignition system inoperative while performing the compression check.

2- Insert a compression gauge into the No. 1, top, spark plug opening. Crank the engine with the starter, or pull on the starter cord, through at least 4 complete piston strokes with the throttle at the wide-open position, or until the highest possible reading is observed on the gauge. Record the reading.

Repeat the test and record the compression for each cylinder. A variation between cylinders is far more important than the actual readings. A variation of more than 5 psi between cylinders indicates the lower compression cylinder may be defective. The problem may be worn, broken, or sticking piston rings, scored pistons or worn cylinders. These problems may only be determined after the head has been removed. Removing the head on an outboard engine is not that big a deal, and may save many hours of frustration and the cost of purchasing unnecessary parts to correct a faulty condition.

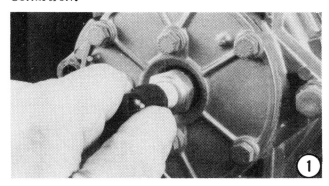

5-6 IGNITION

Spark Plugs

3- Check the plug wires to be sure they are properly connected. Check the entire length of the wire/s from the plug/s to the coils. If the wire is to be removed from the spark plug, **ALWAYS** use a pulling and twisting motion as a precaution against damaging the connection.

4- Attempt to remove the spark plug/s by hand. This is a rough test to determine if the plug is tightened properly. You should not be able to remove the plug without using the proper socket size tool. Remove the spark plug/s and keep them in order. Examine each plug and evaluate its condition as described in Section 5-2.

If the spark plugs have been removed and the problem cannot be determined, but the plug appears to be in satisfactory condition, electrodes, etc., then replace the plugs in the spark plug openings.

A conclusive spark plug test should always be performed with the spark plugs installed. A plug may indicate satisfactory spark when it is removed and tested, but under a compression condition may fail. An example would be the possibility of a person being able to jump a given distance on the ground, but if a strong wind is blowing, his distance may be reduced by half. The same is true with the spark plug. Under good compression in the cylinder, the spark may be too weak to ignite the fuel properly.

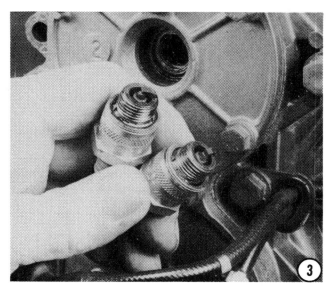

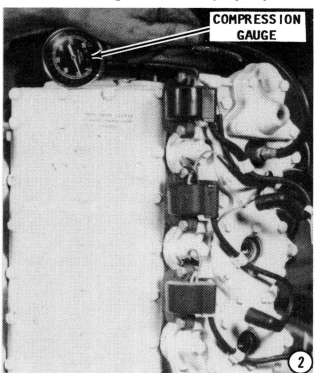

Therefore, to test the spark plug under compression, replace it in the engine and tighten it to the proper torque value. Another reason for testing for spark with the plugs installed is to duplicate actual operating conditions regarding flywheel speed. If the flywheel is rotated with the pull cord with the plugs removed, the flywheel will rotate much faster because of the no-compression condition in the cylinder, giving the **FALSE** indication of satisfactory spark.

A spark tester capable of testing for spark while cranking and also while the engine is operating, can be purchased from your local marine dealer or automotive parts house. An inexpensive tester will give the same information as a more costly unit.

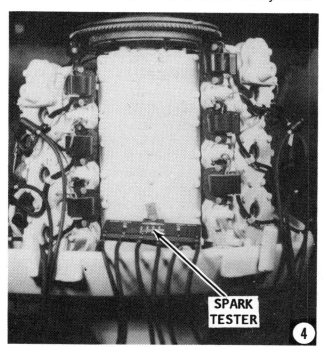

5-6
TYPE I CAPACITOR DISCHARGE (CD) FLYWHEEL MAGNETO
65 HP — 1972 AND 1973
70 HP — 1974 THRU 1978
75 HP — 1974 THRU 1978
85 HP — 1973 THRU 1977
115 HP — 1973 THRU 1977
140 HP — 1977
150 HP — 1978
175 HP — 1977 AND 1978
200 HP — 1976 THRU 1978
235 HP — 1978

DESCRIPTION

READ AND BELIEVE. A battery installed to crank the engine **DOES NOT** mean the engine is equipped with a battery-type ignition system. A magneto system uses the battery only to crank the engine. Once the engine is running, the battery has absolutely no affect on engine operation. Therefore, if the battery is low and fails to crank the engine properly for starting, the engine may be cranked manually, started, and operated. Under these conditions, the key switch must be turned to the **ON** position or the engine will not start by hand cranking.

A magneto system is a self-contained unit. The unit does not require assistance from an outside source for starting or continued operation. Therefore, as previously mentioned, if the battery is dead, the engine may be cranked manually and the engine started.

The capacitor discharge (CD) magneto ignition system consists of the flywheel and ring gear assembly; timer base and sensor assembly installed under the flywheel; a Power Pack installed on the starboard side of the powerhead; and three ignition coils mounted at the rear of the powerhead on 3-cylinder engines. On V4 engines, the powerpack is installed at the rear of the engine with four coils, to on the port side and two on the starboard side. On the V6 engines, a powerpack is installed port and starboard, with three coils port and starboard. An alternator stator and charge coils assembly is installed directly under the flywheel. The spark plugs might be considered a part of the ignition system.

Repair of these components is not possible. Therefore, if troubleshooting indicates a part unfit for further service, the entire assembly must be removed and replaced in order to restore the outboard to satisfactory performance. As an example the coil and coil wire leading to the spark plug is one assembly. If the coil or wire is found to be faulty the coil and wire must be replaced as an assembly.

Before performing maintenance work on the system, it would be well to take time to read and understand the introduction information presented in Section 5-1 thru 5-4, at the beginning of this chapter, the Description at the start of this section, and the Theory of Operation in the following paragraphs.

THEORY OF OPERATION

This system generates approximately 30,000 volts which is fed to the spark plugs without the use of a point set or an outside voltage source.

To understand how high voltage current is generated and reaches a spark plug, imagine the flywheel turning very slowly. As the flywheel rotates, flywheel magnets induce current in the alternator stator and also

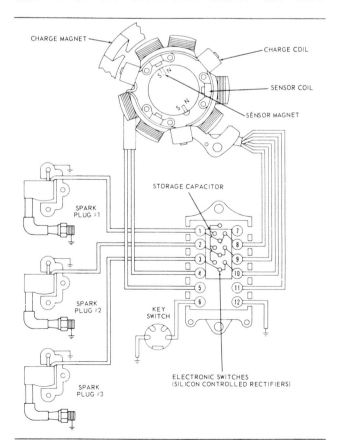

Functional diagram of a typical Type I capacitor discharge (CD) flywheel magneto ignition system.

5-8 IGNITION

generate about 300 volts AC in the charge coils. Therefore, no external voltage source is required. The 300 volts AC is converted to DC in the Power Pack, and is stored in the Power Pack capacitor. Sensor magnets are a part of the flywheel hub. Gaps exist between the sensor magnets. As one gap passes the sensor coil, voltage is generated. This small voltage generated in the sensor coil activates one of two electronic switches in the Power Pack. The switch discharges the 300 volts stored in the capacitor into one of the ignition coils. The ignition coil steps the voltage up to approximately 30,000 volts. This high voltage is fed to the spark plug igniting the fuel/air mixture in the cylinder.

Now, as the flywheel continues to rotate, the next sensor magnet gap on the flywheel hub, which is opposite in polarity from the first, generates a reverse polarity voltage in the sensor coil. This voltage activates the second electronic switch in the Power Pack, and discharges the capacitor into the other ignition coil. The voltage is stepped up to approximately 30,000 volts and fed to the next spark plug. The cycle is repeated as the flywheel continues to rotate.

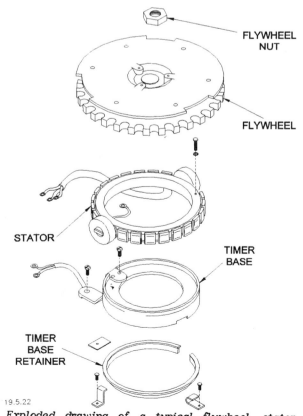

Exploded drawing of a typical flywheel, stator, timer base, and timer base retainer arrangement.

SPECIAL EQUIPMENT AND ADVICE

There is no way on this green earth to properly or accurately check the system without the following test equipment.

Continuity Meter
Ohmmeter
Timing Light
S-80 or M-80 neon test light
Neon spark tester

SAFETY WORDS

This CD ignition system generates approximately 30,000 volts which is fed to the spark plugs. Therefore, perform each step of the troubleshooting procedures exactly as presented as a precaution against personal injury.

The following safety precautions should always be observed:

DO NOT attempt to remove any of the potting in the back of the Power Pack. Repair of the Power Pack is impossible.

DO NOT attempt to remove the high tension leads from the ignition coil.

DO NOT open or close any plug-in connectors, or attempt to connect or disconnect any electrical leads while the engine is being cranked or is running.

DO NOT set the timing advanced any further than as specified.

DO NOT hold a high tension lead with your hand while the engine is being cranked or is running. Remember, the system can develop approximately 30,000 volts which will result in a severe shock if the high tension lead is held. **ALWAYS** use a pair of approved insulated pliers to hold the leads.

DO NOT attempt any tests except those listed in this troubleshooting section.

DO NOT connect an electric tachometer to the system unless it is a type which has been approved for such use.

DO NOT connect this system to any voltage source other than given in this troubleshooting section.

ONE MORE WORD: Each cylinder has its own ignition system in a flywheel-type ignition system. This means if a strong spark is observed on any one cylinder and not at another, only the weak system is at fault. However, it is always a good idea to check and service all systems while the flywheel is removed.

TESTING TYPE I CD 3-CYLINDER 5-9

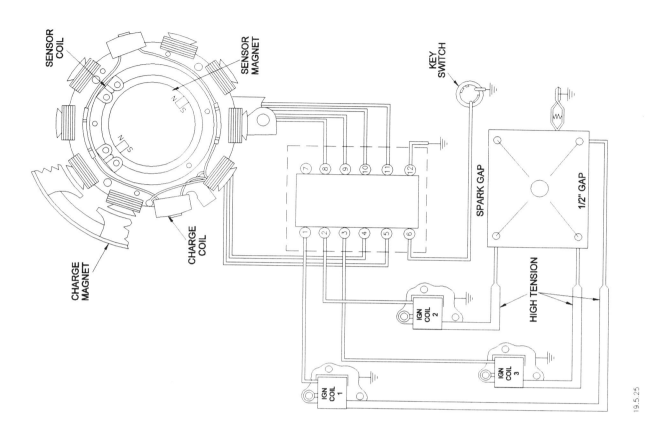

Test No. 1 Ignition Coil Output Check

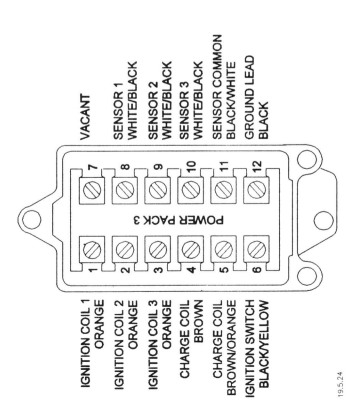

Power Pack Terminal Identification 3-Cylinder Engine

5-10 IGNITION

Preliminary Test

The first area to check on a CD flywheel magneto ignition system is the system ground.

Also check the battery charge and to verify the battery cables are connected properly for correct polarity. If the battery cables are connected backwards, testing will indicate **NO** spark.

Also check the battery charge. If the battery is below a full charge it will not be possible to obtain full cranking speed during the tests.

5-7 TROUBLESHOOTING
TYPE I CAPACITOR DISCHARGE (CD)
FLYWHEEL MAGNETO
65 HP – 1972 AND 1973
70 HP – 1974 THRU 1978
75 HP – 1974 THRU 1978

BEFORE TESTING

Before starting the troubleshooting work, it would be well worth the time and effort to read and understand the information presented in Sections 5-1 thru 5-4 and to conduct the general troubleshooting procedures listed in Section 5-5. Considerable time, money, and possible frustration can be saved by having a thorough knowledge of the system and appreciating why the tests are conducted in a particular sequence.

Test No. 1 (Diagram Page 5-9)
Ignition Coil Output Check

Use a standard spark tester and check for spark at each cylinder. If a spark tester is not available, use a pair of insulated pliers and hold the plug wire about 1/2-inch (12.7 mm) from the engine.

Turn the flywheel with a electrical starter and check for spark. A strong spark over a wide gap must be observed when testing in this manner, because under compression a strong spark is necessary in order to ignite the air/fuel mixture in the cylinder. This means it is possible to think you have a strong spark, when in reality the spark will be too weak when the plug is installed.

If there is no spark, or if the spark is weak, from one coil, proceed directly to Test No. 2, Trigger Coil Input Check. If there is no spark or the spark is weak from the coils, proceed directly to Test No. 3, Charge Coil Output Check. If the spark is strong and steady, across the 1/2-inch (12.7 mm) gap, or the spark tester indicates the spark is satisfactory, then the problem is in the spark plugs, the fuel system, or the compression is weak.

Test No. 2 (Diagram Page 5-11)
Trigger Coil Input Check

Remove the Power Pack cover.

QUICK WORD

Always identify leads prior to disconnecting to ensure they will be connected to the proper terminal after the test. Sometimes the leads are the same color. Therefore, without a tag or other identification, the lead may not be connected properly.

Disconnect the sensor leads from the No. 8, No. 9, No. 10, and No. 11 terminals at the powerpack Disconnect the ignition coil lead from the powerpack at No. 2 and No. 3 terminals. Obtain a neon tester, No. S-80 or M-80. Connect the black lead of the neon tester to the No. 8 terminal and the blue lead to the No. 11 terminal. Set the neon light selector to the No. 3 position.

Crank the engine with the electric starter motor, and at the same time momentarily depress (tap) the "B" load button rapidly.

a- If adequate spark was observed from the coil in the previous tests, the problem is in the sensor. To test the sensor assembly, see Test No. 6.

b- If no spark was observed from the coil during the previous tests, the charge coil must be checked as outlined in Test No. 3.

c- Check the second coil: Disconnect the ignition coil lead from the No. 1 terminal on the powerpack. Connect the No. 2 coil wire to the powerpack and repeat the above test.

d- Check the third coil: Disconnect the lead from the No. 2 terminal and connect the lead from the No. 3 coil to the powerpack to the No. 3 terminal. Repeat the above test.

e- If spark was not observed on all three coils, the Power Pack must be checked, see Test No. 4.

TESTING TYPE I CD 3-CYLINDER 5-11

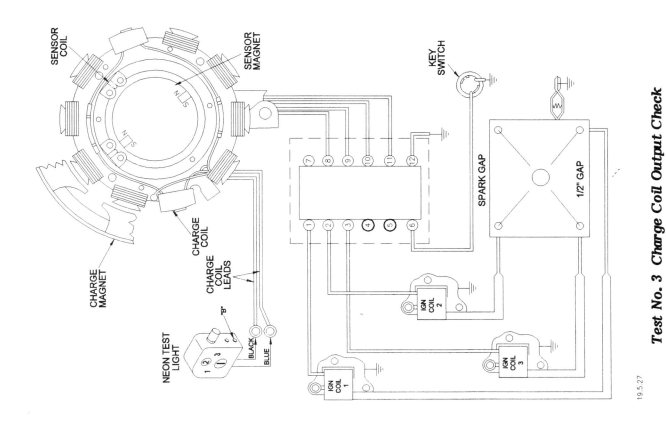

Test No. 3 Charge Coil Output Check

Test No. 2 Trigger Coil Input Check

Test No. 3 (Diagram Page 5-11)
Charge Coil Output Check

Identify the leads, then disconnect at the No. 4 and No. 5 terminals. Connect the tester lead across the charge coil leads to the timer base. Turn the neon tester to position No. 2.

Crank the engine and at the same time, depress load button **"B"** on the tester. Observe the light. If the light glows steadily, the charge coils are good.

Check the Power Pack output. If the light comes on intermittently, check for a broken or shorted wire. If the wires are in good shape, replace the charge coil. If the light fails to come on, replace the charge coil and stator assembly.

Test No. 4 (Diagram Page 5-13)
Power Pack Output Check

Check all connections at the Power Pack to be sure they are clean and secure. Disconnect the lead from the No. 1, No. 2, and No. 3 terminals. If the neon tester is used, connect the black lead to the No. 1 Power Pack terminal and the blue lead to a good ground.

Move the neon tester switch to the No. 1 position, and then depress load button **"A"** and crank the engine with the electric starter motor. Observe the light.

Now, move the black lead from the No. 1 terminal to the No. 2 terminal. Crank the engine and repeat the above test.

Move the black lead from the No. 2 terminal to the No. 3 terminal. Crank the engine and repeat the above test.

If the light is strong and steady on all three outputs, replace the faulty coil or coils. The No. 1 terminal connection is the top coil, the No. 2 terminal is the center coil, and the No. 3 terminal is the bottom coil.

If no light is visible on any of the three outputs, check the key switch as outlined in Test No. 5.

If a steady light was observed on one or two output, but no light on the other, replace the Power Pack.

If the light was very dim or intermittent on any of the three outputs during the tests, the Power Pack must be replaced.

Connect the wires disconnected at the start of the tests.

Test No. 5 (Diagram Page 5-13)
Key Switch Check

A CD magneto key switch operates in **REVERSE** of any other type key switch. When the key is moved to the **OFF** position, the circuit is **CLOSED** between the CD magneto and ground. In some cases, when the key is turned to the **OFF** position the points are grounded. For this reason, an automotive-type switch **MUST NEVER** be used, because the circuit would be opened and closed in reverse, and if 12-volts should reach the coil, the coil will be **DESTROYED**.

Connect the spark tester to the high tension leads. Disconnect the lead at the No. 6 Power Pack terminal. This is the wire from the key. Crank the engine with the key and observe the spark.

If there is no indication of spark on either coil or on only one, replace the Power Pack.

If spark is indicated on all three coils, the problem is most likely in the lead from the key.

If the key switch leads appear to be in good condition, replace the key switch.

Test No. 6 (Diagram Page 5-14)
Sensor Coil Low Ohm Check

Identify the leads with a tag or other means, and then disconnect them at the No. 8, No. 9, No. 10, and No. 11 terminals. Disconnect the sensor leads from the Power Pack terminals No. 1, No. 2, and No. 3. Connect the ohmmeter leads to the sensor coil leads alternately (the white lead with the black stripes and the black lead with white stripes).

Use the low ohm scale and observe the reading. The meter should indicate 8.5 ± 1.0 ohms at room temperature ($70°$ F). If the ohmmeter reading is not satisfactory, proceed as follows:

Sensor Coil High Ohm Check

Move the ohmmeter to the High Ohm scale. With the sensor leads still disconnected from the Power Pack, as in Low Ohm Test, connect the red ohmmeter test lead to either one of the sensor leads. Connect the black ohmmeter lead to a good ground. The meter should indicate infinity. A **ZERO** reading indicates a short to ground.

If the test fails, the sensor coil and timer base must be replaced as a unit.

If the test is successful disconnect the

TESTING TYPE I CD 3-CYLINDER 5-13

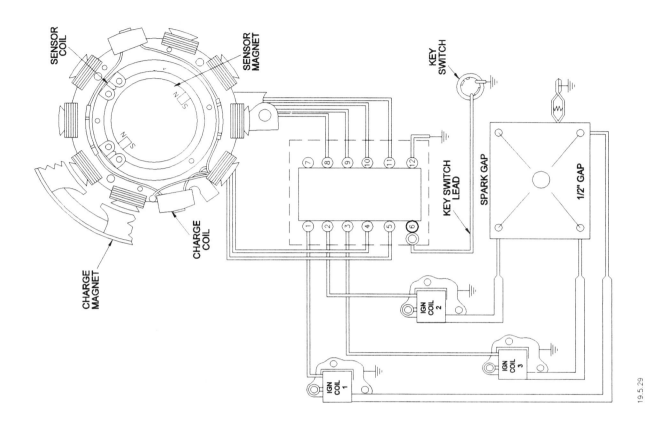

Test No. 5 Key Switch Check

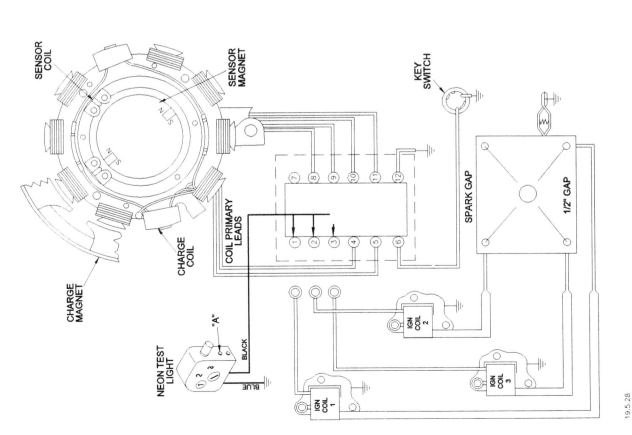

Test No. 4 Power Pack Output Check

5-14 IGNITION

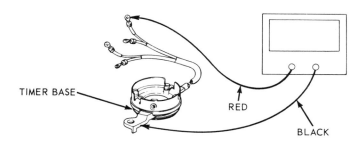

Test No. 6 Sensor Coil Low Ohm Check

ohmmeter and connect the leads at the No. 8, No. 9, No. 10, and No. 11 Power Pack terminals. Connect the leads at the No. 1, No. 2, and No. 3 terminals.

For component replacement, see Section 5-14.

For synchronizing procedures, see Section 5-15.

For detailed timing instructions, see Section 5-17.

5-8 TROUBLESHOOTING
TYPE I CAPACITOR DISCHARGE (CD)
FLYWHEEL MAGNETO
85 HP — 1973 THRU 1977
115 HP — 1973 THRU 1977
135 HP — 1973 THRU 1976
140 HP — 1977

BEFORE TESTING

Before starting the troubleshooting work, it would be well worth the time and effort to read and understand the information presented in Sections 5-1 thru 5-4 and to conduct the general troubleshooting procedures listed in Section 5-5. Considerable time, money, and possible frustration can be saved by having a thorough knowledge of the system and appreciating why the tests are conduted in a particular sequence.

Outboard motors have entered the electronic realm with modern sophisticated automobile ignition equipment. Therefore, there is no way on this green earth to properly or accurately check the system without the necessary test equipment.

Test No. 1 (Diagram Page 5-15)
Ignition Coil Output Check

Use a standard spark tester and check for spark at each cylinder. If a spark tester is not available, use a pair of insulated pliers and hold the plug wire about 1/2-inch (12.7 mm) from the engine. Turn the flywheel with a electrical starter and check for spark. A strong spark over a wide gap must be observed when testing in this manner, because under compression a strong spark is necessary in order to ignite the air/fuel mixture in the cylinder. This means it is possible to think you have a strong spark, when in reality the spark will be too weak when the plug is installed. If there is no spark, or if the spark is weak, from one coil, proceed directly to Test No. 2, Trigger Coil Input Check.

If there is no spark or the spark is weak from the coils, proceed directly to Test No. 3, Charge Coil Output Check. If the spark is strong and steady, across the 1/2-inch (12.7 mm) gap indicating it is satisfactory, then the problem is in the spark plugs, the fuel system, or the compression is weak.

Test No. 2 (Diagram Page 5-15)
Trigger Coil Input Check

Remove the Power Pack cover.

QUICK WORD

Always identify leads prior to disconnecting to ensure they will be connected to the proper terminal after the test. Sometimes the leads are the same color. Therefore, without a tag or other identification, the lead may not be connected properly.

Disconnect the sensor leads from the No. 2, No. 4, No. 9, and No. 12 terminals at the powerpack. Disconnect the ignition coil lead from the powerpack at No. 3 and No. 5 terminals. Obtain a neon tester, No. S-80 or M-80. Connect the black lead of the neon tester to the No. 9 terminal and the blue lead to the No. 12 terminal. Set the neon light selector to the No. 3 position.

Crank the engine with the electric starter motor, and at the same time momentarily depress (tap) the "B" load button rapidly. Now reverse the leads on No. 9 and No. 12, and then repeat the test. If both coils fire at the same time, the powerpack is defective.

a- If adequate spark was observed from the coil in the previous tests, the problem is in the sensor. To test the sensor assembly, see Test No. 6.

b- If no spark was observed from the coil during the previous tests, the charge coil must be checked as outlined in Test No. 3.

TESTING TYPE I CD V4 5-15

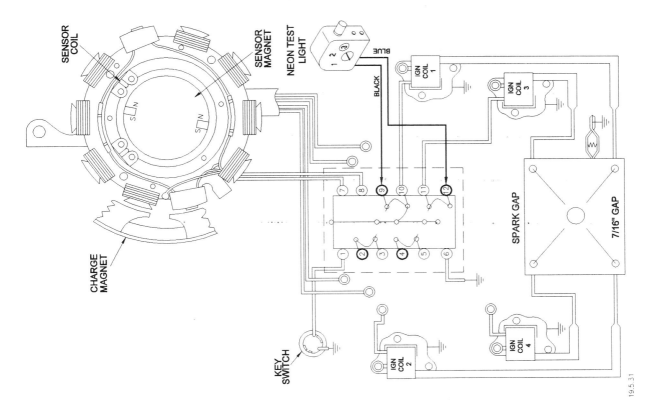

Test No. 2 Trigger Coil Input Check

Test No. 1 Ignition Coil Output Check

5-16 IGNITION

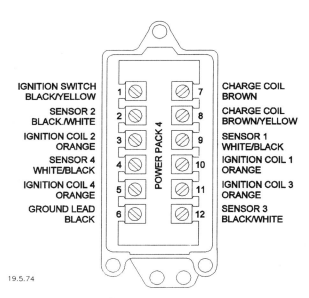

Terminal identification on a typical Power Pack for a V4 engine covered in this section. **ALWAYS** *identify and tag the electrical lead before making the disconnection to ensure the proper lead is connected back to the correct terminal following a test or service. The few moments spent in tagging the wire will remove any doubt as to where the lead should be connected.*

Repeat the complete step thus far by removing the ignition coil leads from terminals No. 10 and No. 11, and then connecting ignition coils No. 2 and No. 4 to terminals No. 3 and No. 5 respectively. Connect the black tester lead to terminal No. 2 and the blue tester lead to terminal No. 4. Crank the engine. Momentarily depress (tap) load button "B" rapidly and at the same time observe the spark. Reverse the leads of the tester on the power pack terminals No. 2 and No. 4 and repeat the test. If both coils fire at the same time, the Power Pack is defective.

a- If adequate spark was observed from the coil in the previous tests, the problem is in the sensor. To test the sensor assembly, see Test No. 6.

b- If no spark was observed from the coil during the previous tests, the charge coil must be checked as outlined in Test No. 3.

Test No. 3 (Diagram Page 5-17)
Charge Coil Output Check

Identify the leads, then disconnect at the No. 7 and No. 8 terminals. Connect the tester lead across the charge coil leads to the timer base. Turn the neon tester to position No. 2.

Crank the engine and at the same time, depress load button "B" on the tester. Observe the light. If the light glows steadily, the charge coils are good.

Check the Power Pack output. If the light comes on intermittently, check for a broken or shorted wire. If the wires are in good shape, replace the charge coil. If the light fails to come on, replace the charge coil and stator assembly.

Test No. 4 (Diagram Page 5-17)
Power Pack Output Check

Carefully check all connections at the Power Pack to be sure they are clean and secure.

Disconnect the ignition coil lead from the No. 3, and No. 4 terminals.

If the neon tester is used, connect the black lead to the No. 10 Power Pack terminal and the blue lead to a good ground.

Move the neon tester switch to the No. 1 position, and then depress load button "A" and crank the engine with the electric starter motor. Observe the light.

Now, move the black lead from the No. 10 terminal to the No. 3 terminal. Crank the engine and repeat the above test.

Move the black lead from the No. 3 terminal to the No. 5 terminal. Crank the engine and repeat the above test.

If the light is strong and steady on all four outputs, replace the faulty coil or coils. The No. 10 terminal connection is the No. 1 coil; the No. 3 terminal is the No. 2 coil; the No. 11 terminal is the No. 3 coil; and the No. 5 terminal is the No. 4 coil.

If no light is visible on any of the four outputs, check the key switch as outlined in Test No. 5.

If a steady light was observed on one or two or three output, but no light on the other, replace the Power Pack.

If the light was very dim or intermittent on any of the four outputs during the tests, the Power Pack must be replaced.

Connect the wires disconnected at the start of the tests.

TESTING TYPE I CD V4   5-17

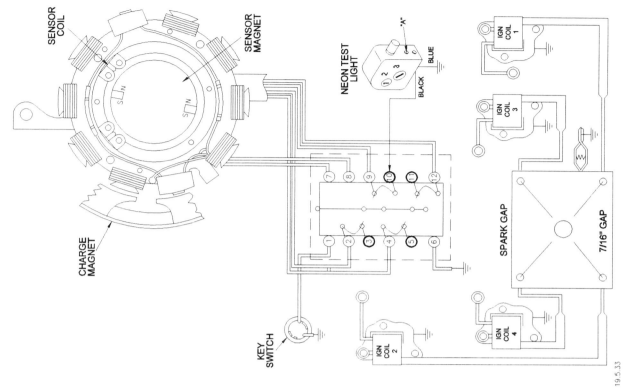

Test No. 4 Power Pack Output Check

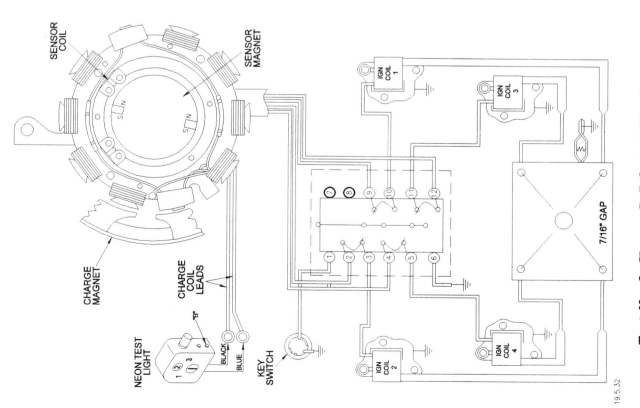

Test No. 3 Charge Coil Output Check

5-18 IGNITION

Test No. 5 (Diagram Page 5-19)
Key Switch Check

A CD magneto key switch operates in **REVERSE** of any other type key switch. When the key is moved to the **OFF** position, the circuit is **CLOSED** between the CD magneto and ground. In some cases, when the key is turned to the **OFF** position the points are grounded. For this reason, an automotive-type switch **MUST NEVER** be used, because the circuit would be opened and closed in reverse, and if 12-volts should reach the coil, the coil will be **DESTROYED**.

Connect the spark tester to the high tension leads. Disconnect the lead at the No. 1 Power Pack terminal. This is the wire from the key. Crank the engine with the key and observe the spark.

If there is no indication of spark on either coil or on only one, replace the Power Pack, or carefully repeat Test No. 2.

If spark is indicated on all four coils, the problem is most likely in the lead from the key.

If the key switch leads appear to be in good condition, replace the key switch.

Test No. 6 (Diagram This Page)
Sensor Coil Low Ohm Check

Identify the leads with a tag or other means, and then disconnect them at the sensor leads from the Power Pack terminals No. 2, No. 4, and No. 9, and No. 12. Connect the ohmmeter leads to the sensor leads No. 2 and No. 4. Observe and make a note of the meter reading. Disconnect the test leads and connect the ohmmeter leads to the sensor leads No. 9 and No. 12 (the white lead with the black stripes and the black lead with white stripes). Again observe and note the meter reading.

The meter should indicate 8.5 ± 2.0 ohms at room temperature ($70° F$). If the ohmmeter reading is not satisfactory, proceed as follows:

Sensor Coil High Ohm Check

Move the ohmmeter to the High Ohm scale. With the sensor leads still disconnected from the Power Pack, as in Low Ohm Test. Connect the black ohmmeter lead to a good ground. Connect the red ohmmeter test lead to either one of the sensor leads, alternately. The meter should indicate infinity. A **ZERO** reading indicates a short to ground.

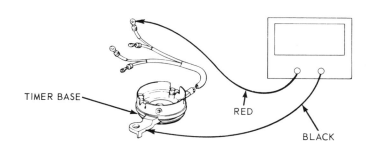

Test No. 6 Sensor Coil Low Ohm Check

If the test fails, the sensor coil and timer base must be replaced as a unit.

If the test is successful disconnect the ohmmeter and connect the leads at the Power Pack sensor at the No. 2, No. 4, No. 9, and No. 12 terminals.

For component replacement, see Section 5-14.

For synchronizing procedures, see Section 5-15.

For detailed timing instructions, see Section 5-17.

5-9 TROUBLESHOOTING
TYPE I CAPACITOR DISCHARGE (CD)
FLYWHEEL MAGNETO
150 HP — 1978
175 HP — 1977 AND 1978
200 HP — 1976 THRU 1978
235 HP — 1978

BEFORE TESTING

Before starting the troubleshooting work, it would be well worth the time and effort to read and understand the information presented in Sections 5-1 thru 5-4 and to conduct the general troubleshooting procedures listed in Section 5-5. Considerable time, money, and possible frustration can be saved by having a thorough knowledge of the system and appreciating why the tests are conducted in a particular sequence.

SPECIAL NOTE

All V6 engines contain two Power Packs. One pack services three cylinders and the second pack the other three cylinders. The following tests assume both packs and all cylinders are being checked. Therefore, two spark testers, etc, would be required, one for each bank, to test the complete system at one time.

TESTING TYPE I CD V6   5-19

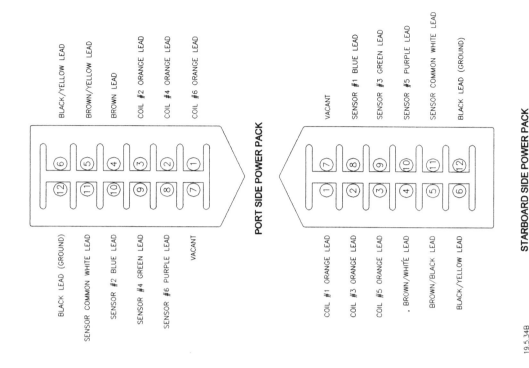

PORT SIDE POWER PACK

Terminal	Lead
12 (6)	BLACK/YELLOW LEAD
11 (5)	BROWN/YELLOW LEAD
10 (4)	BROWN LEAD
9 (3)	COIL #2 ORANGE LEAD
8 (2)	COIL #4 ORANGE LEAD
7 (1)	COIL #6 ORANGE LEAD
1	BLACK LEAD (GROUND)
2	SENSOR COMMON WHITE LEAD
3	SENSOR #2 BLUE LEAD
4	SENSOR #4 GREEN LEAD
5	SENSOR #6 PURPLE LEAD
6	VACANT

STARBOARD SIDE POWER PACK

Terminal	Lead
7 (1)	VACANT
8 (2)	SENSOR #1 BLUE LEAD
9 (3)	SENSOR #3 GREEN LEAD
10 (4)	SENSOR #5 PURPLE LEAD
11 (5)	SENSOR COMMON WHITE LEAD
12 (6)	BLACK LEAD (GROUND)
1	COIL #1 ORANGE LEAD
2	COIL #3 ORANGE LEAD
3	COIL #5 ORANGE LEAD
4	BROWN/WHITE LEAD
5	BROWN/BLACK LEAD
6	BLACK/YELLOW LEAD

*Terminal connections for port, left, and starboard, right, Power Packs on the V6 engines covered in this section. **ALWAYS** identify and tag the electrical lead before making the disconnection to ensure the proper lead is connected back to the correct terminal following a test or service. The few moments spent in tagging the wire will remove any doubt as to where the lead should be connected.*

Test No. 5 Key Switch Check

5-20 IGNITION

HOWEVER, it is possible to test one Power Pack and one bank of three cylinders independently, and then to test the other Power Pack and bank of cylinders.

Test No. 1 (Diagram Page 5-21)
Ignition Coil Output Check

Use a standard spark tester and check for spark at each cylinder. If a spark tester is not available, use a pair of insulated pliers and hold the plug wire about 7/16-inch (11.11 mm) from the engine. Turn the flywheel with a electrical starter and check for spark.

A strong spark over a wide gap must be observed when testing in this manner, because under compression a strong spark is necessary in order to ignite the air/fuel mixture in the cylinder. This means it is possible to think you have a strong spark, when in reality the spark will be too weak when the plug is installed.

If there is no spark, or if the spark is weak, from one coil, proceed directly to Test No. 2, Trigger Coil Input Check. If there is no spark or the spark is weak from the coils, proceed directly to Test No. 3, Charge Coil Output Check for that particular Power Pack. If the spark is strong and steady, across the 7/16-inch (11.11 mm) gap indicating it is satisfactory, then the problem is in the spark plugs, the fuel system, or the compression is weak.

If there is no spark at all six cylinders, proceed with Test No. 5, Key Switch Test.

Test No. 2 (Diagram Page 5-21)
Trigger Coil Input Check

Remove the Power Pack cover.

QUICK WORD

Always identify each individual lead prior to disconnecting to ensure all will be connected to the proper terminal after the test. Sometimes the leads are the same color. Therefore, without a tag or other identification, the lead may not be connected properly.

Disconnect the sensor leads from the No. 8, No. 9, No. 10, and No. 11 terminals at each Power Pack. Obtain a neon tester, No. S-80 or M-80. Connect the black lead of the neon tester to the No. 8 terminal and the blue lead to the No. 11 terminal, on each Power Pack. Set the neon light selector to the No. 3 position.

Crank the engine with the electric starter motor, and at the same time momentarily depress (tap) the **"B"** load button rapidly.

a- Spark present on the No. 1 coil, starboard side, or at the No. 6 coil, port side, indicates a problem in the sensor circuit. No spark on any other coil at this time, indicates the Power Pack is defective and **MUST** be replaced. Before replacing the Power Pack it would be good practice to check the sensor coil resistance for possible intermittent, short, or open circuit. Correct the problem as required.

b- If no spark was observed at the ignition coil, check the Power Pack output, Test No. 4.

c- Connect the black tester lead to terminal No. 9 and the blue tester lead to terminal No. 11. Crank the engine, tap the neon tester load button **"B"** rapidly. Observe for spark at the No. 3 ignition coil, starboard side, or the No. 4 coil, port side. Spark indicates a problem in the sensor cirucit. If spark was observed on any other coil at this time, the Power Pack is defective and **MUST** be replaced. Before replacing the Power Pack, check the sensor coil resistance for possible intermittent, short, or open circuit. Replace the sensor assembly if it is defective.

If there was absolutely no spark on any ignition coil, check the Power Pack output, Test No. 4.

Test No. 3 (Diagram Page 5-22)
Charge Coil Output Check

Identify the leads, then disconnect at the No. 4 and No. 5 terminals, at each Power Pack. Connect the tester lead across the charge coil leads to the timer base. Turn the neon tester to position No. 2.

Crank the engine and observe the light. If the light glows steadily, the charge coils are good. Check the trigger coil input. If the light comes on intermittently, check for a broken or shorted wire. If the wires are in good shape, replace the charge coil. If the light fails to come on, replace the charge coil and stator assembly.

Test No. 4 (Diagram Page 5-22)
Power Pack Output Check

Carefully check all connections at the Power Pack to be sure they are clean and secure. Identify and tag each individual lead at the Power Pack before making any disconnects. Sometimes the leads are the

TESTING TYPE I V6   5-21

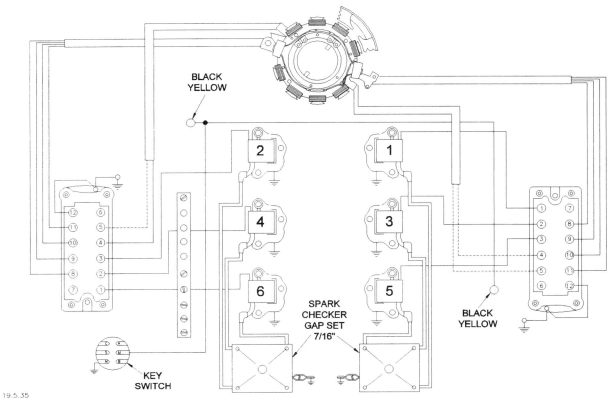

Test No. 1 Ignition Coil Output Check

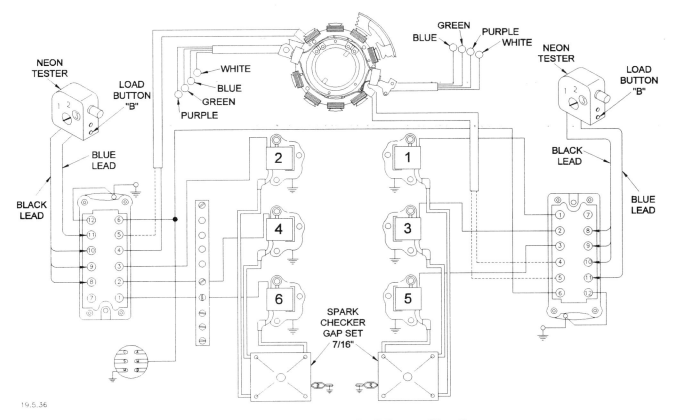

Test No. 2 Trigger Coil Input Check

5-22 IGNITION

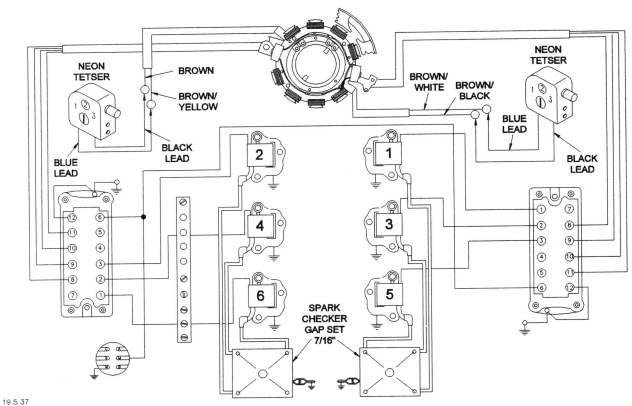

Test No. 3 Charge Coil Output Check

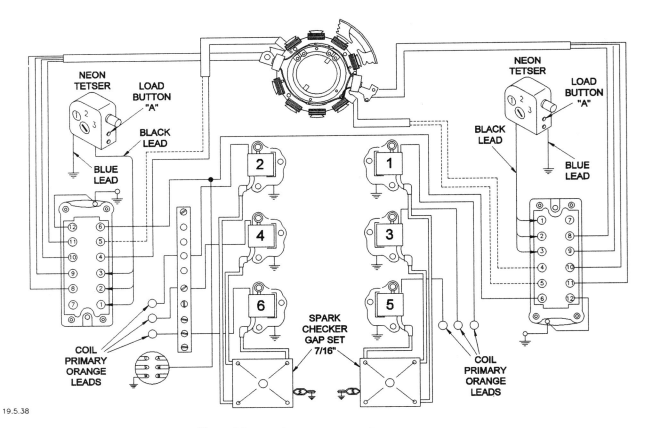

Test No. 4 Power Pack Output Check

TESTING TYPE I CD V6   5-23

same color. Therefore, taking time to tag will ensure all will be connected to the proper terminal after the test.

Disconnect the lead from the No. 1, No. 2, and No. 3 terminals. If the neon tester is used, connect the black lead to the No. 1 Power Pack terminal and the blue lead to a good ground.

Move the neon tester switch to the No. 1 position, and then depress load button "A" and crank the engine with the electric starter motor. Observe the light. Now, move the black lead from the No. 1 terminal to the No. 2 terminal. Crank the engine and repeat the above test.

Move the black lead from the No. 2 terminal to the No. 3 terminal. Crank the engine and repeat the above test.

If the light is strong and steady on all three outputs, replace the faulty coil or coils. On the starboard Power Pack, the No. 1 terminal is for the No. 1 coil; the No. 2 terminal is for the No. 3 coil; and the No. 3 terminal is for the No. 5 coil. On the port Power Pack, the No. 3 terminal is for the No. 2 coil; the No. 2 terminal is for the No. 4 coil; and the No. 1 terminal for the No. 6 coil.

If no light is visible on any of the three outputs of each Power Pack, check the key switch as outlined in Test No. 5.

If a steady light was observed on one or two output, but no light on the other, replace the Power Pack.

If the light was very dim or intermittent on any of the three outputs during the tests, the Power Pack must be replaced.

If no light was observed on all outputs of one Power Pack, check for a satisfactory ground to the Power Pack and Electrical Component Bracket. The check for a good ground can be accomplished by connecting the blue neon tester lead to terminal No. 12 of the Power Pack, and then repeating the Power Pack output test. Observe the results.

If there is a light on all outputs, repair or replace the defective ground wire on the Power Pack or the Electrical Component Bracket.

If there is no light indicated on any of the outputs, the ground wires are satisfactory. The Power Pack is defective and **MUST** be replaced.

Connect the wires disconnected at the start of the tests.

Test No. 5 (Diagram Page 5-24)
Key Switch Check

A CD magneto key switch operates in **REVERSE** of any other type key switch. When the key is moved to the **OFF** position, the circuit is **CLOSED** between the CD magneto and ground. In some cases, when the key is turned to the **OFF** position the points are grounded. For this reason, an automotive-type switch **MUST NEVER** be used, because the circuit would be opened and closed in reverse, and if 12-volts should reach the coil, the coil will be **DESTROYED**.

Two testers are required for this test. Connect the spark tester to the high tension leads. Disconnect the lead at the No. 6 terminal, at each Power Pack. This is the wire from the key. Crank the engine with the key and observe the spark.

If there is no indication of spark on either coil or on only one, replace the Power Pack.

If spark is indicated on all three coils, the problem is most likely in the lead from the key.

If the key switch leads appear to be in good condition, replace the key switch.

Test No. 6 (Diagram Page 5-24)
Sensor Coil Low Ohm Check

Six sensor coils are potted into the timer base. Therefore, the sensor coil and timer base are serviced as a unit. Four leads from these coils are routed to each Power Pack.

The **PORT** side leads consist of a common white lead connected to terminal No. 11; sensor No. 2 blue lead connected to No. 10 terminal; sensor No. 4 green lead connected to No. 9 terminal; and sensor No. 6 purple lead connected to terminal No. 8. See reference illustration "A".

The **STARBOARD** leads consist of a common white lead connected to terminal No. 11; sensor No. 5 purple lead connect to terminal No. 10; sensor No. 3 green lead connected to No. 9 terminal; and sensor No. 1 blue lead connected to terminal No. 8.

Set an ohmmeter to the Low Ohm scale. Connect the black test lead to the common ground white lead. Alternately make contact with the red meter lead to the blue, the green, and then the purple lead. The meter should indicate 18 +5 ohms at room temperature of 70° (21 C).

5-24 IGNITION

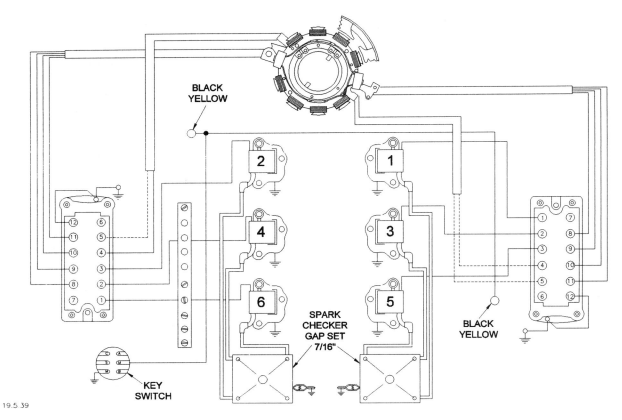

Test No. 5 Key Switch Check

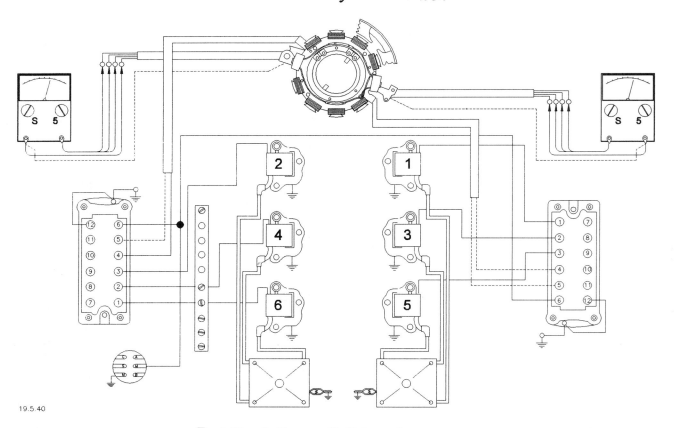

Test No. 6 Sensor Coil Low Ohm Check

Sensor Coil High Ohm Check

Set the ohmmeter to the high ohm scale. With the sensor leads still disconnected from the Power Pack, connect the meter leads alternately between each sensor lead and ground. An infinity reading on each sensor indicates the sensor is in satisfactory condition. A ZERO or any reading indicates the leads or the sensor coil is shorted to ground.

Under certain conditions, the sensors may be grounding out to the shield, but will not show up with an ohmmeter. This condition may be discovered using the condenser tester leakage test, where a voltage of approximately 500 volts is applied to the sensor.

Disconnect all sensor leads from the Power Pack. Connect the black tester lead to the timer base as a ground. Connect the red tester lead to one of the sensor leads, and apply the leakage test. Repeat the test on each sensor lead. Meter will indicate if the sensor is shorting. Check to be sure the sensor leads are not touching each other or making contact to a ground.

Test No. 7 (Diagram 5-26)
Engine Running Rough

If the engine is running rough, the cause may be interaction of the Power Packs due to faulty isolation of a diode in the Power Pack. The diode may be checked by using an ohmmeter set to the High Ohms scale.

The following procedures are to be repeated for the second Power Pack.

Disconnect the leads from terminals at No. 6 and No. 11 of the Power Pack. Connect the ohmmeter leads between the No. 6 and the No. 11 Power Pack terminals. Observe the reading.

Reverse the ohmmeter leads and again observe the meter reading. An infinite (very high) reading in both directions indicates the diode is open. A low ohms reading in both directions (approximately 100 ohms) indicates a shorted diode. A normal diode will read resistance in one direction and infinite reading in the other direction.

If the diode is defective, the Power Pack MUST be replaced.

For component replacement, see Section 5-14.

For synchronizing procedures, see Section 5-15.

For detailed timing instructions, see Section 5-17.

5-10 DESCRIPTION AND OPERATION TYPE II CAPACITOR DISCHARGE (CD) FLYWHEEL MAGNETO
70 HP AND 75 HP -- 1979 AND ON
ALL OTHERS COVERED IN THIS MANUAL -- 1978 AND ON

DESCRIPTION

READ AND BELIEVE. A battery installed to crank the engine **DOES NOT** mean the engine is equipped with a battery-type ignition system. A CD magneto system uses the battery only to crank the engine. Once the engine is running, the battery has absolutely no affect on engine operation.

A CD magneto system is a totally self-contained unit. The unit does not require assistance from an outside source for starting or continued operation. Therefore, if the battery is low and fails to crank the engine properly for starting, the engine may be cranked using jumper cables with good battery, started, and operated. Remember, connect the positive lead from the positive lead of the dead battery to the positive lead of the good battery **FIRST**. Then, connect the other lead from the negative good battery terminal to a good ground on the engine.

Components and Function

The Type II capacitor discharge (CD) ignition system consists of the following assemblies: flywheel, charge coil, sensor, power pack, and ignition coils.

The flywheel assembly contains two integral magnets installed 180° apart. These magnets are charged with opposite polarity.

The charge coil assembly is composed of a large coil of wire which generates alternating current (ac). The current is fed into the Power Pack.

The sensor assembly consists of a small coil of wire which generates the triggering voltage necessary for the Power Pack.

In simple terms, the Power Pack contains the electronic circuits required to produce ignition at the proper time.

The ignition coils generate approximately 30,000 volts which is fed to the spark plugs to ignite the fuel/air mixture in the cylinders.

Before performing maintenance work on the system, it would be well to take time to read and understand the introduction information presented in Section 5-1 thru 5-4 and

5-26 IGNITION

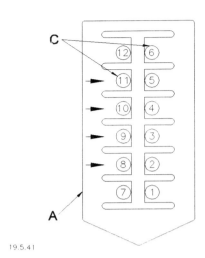

Test No. 7 Engine Running Rough
Port side Power Pack terminal identification. Connect a jumper wire between terminals No. 6 and No. 11 to perform the isolation diode test, as explained in the text.

the General Troubleshooting 5-5, at the beginning of this chapter, and the Description at the start of this section, and the Theory of Operation in the following paragraphs.

THEORY OF OPERATION

This system generates approximately 30,000 volts which is fed to the spark plugs without the use of a point set or an outside voltage source.

To understand how high voltage current is generated and reaches a spark plug, it is well to understand a couple of basic terms used throughout the following paragraphs and troubleshooting procedures.

Diode -- a small electrical part that allows current to flow in only one direction. On the accompanying diagrams the current flow is indicated by the direction of the arrow.

Silicon Controlled Rectifier -- commonly referred to as a SCR -- a small electrical part in which current can flow through in only one direction, and in which the current can flow only when a "Gate" in the part receives a positive (+) input to trigger or open the gate in the SCR. On the accompanying diagrams the direction of current flow is indicated by an arrow.

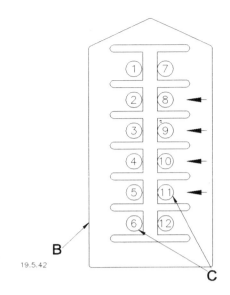

Test No. 7 Engine Running Rough
STARBOARD side Power Pack terminal identification. Connect a jumper wire between terminals No. 6 and No. 11 to perform the isolation diode test, as explained in the text.

Capacitor -- a part that stores voltage. One end of the capacitor is positive (+), and the other end is negative (-). The voltage is stored in the capacitor until the trigger circuit is activated.

BEFORE TESTING

Before starting the troubleshooting work, it would be well worth the time and effort to read and understand the information presented in Sections 5-1 thru 5-4 and to conduct the general troubleshooting procedures listed in Section 5-5. Considerable time, money, and possible frustration can be saved by having a thorough knowledge of the system and appreciating why the tests are conducted in a particular sequence.

CHECKING THE TYPE II SYSTEM

Outboard motors have entered the electronic realm with modern sophisticated automobile ignition equipment. The old days of only a magneto and spark plug to make the unit go are over. Today, there is no way on this green earth to properly or accurately check the system without the proper test equipment.

The test equipment listed at the top of the next page is required to troubleshoot the Type II CD Magneto Ignition System.

Spark Test -- with the air gap adjusted to 1/2 inch (12.7 mm).
Neon Test Light -- Model M-80 or S-80 with a 1-1/2 volt battery adapter. If a Model M-90 test light is available an adapter is not necessary. A Stevens or Electro Specialties C.D. Voltmeter Tester will also do the job.
Ohmmeter -- capable of indicating low ohms (RX1) and high ohms (RX1,000).
Jumper Wires -- four required. These jumper wires may be made using a piece of No. 16 solid wire about 8 inches long with the insulation stripped back about an inch from each end.

MORE GOOD WORDS

A magneto key switch operates in **REVERSE** of any other type key switch. When the key is moved to the **OFF** position, the circuit is **CLOSED** between the magneto and ground. In some cases, when the key is turned to the **OFF** position the system is grounded. For this reason, an automotive-type switch **MUST NEVER** be used, because the circuit would be opened and closed in reverse, and if 12-volts should reach the coil, the coil will be **DESTROYED**.

5-11 TROUBLESHOOTING 3-CYL. UNITS TYPE II CAPACITOR DISCHARGE CD FLYWHEEL MAGNETO
60 HP — 1985 AND ON
70 HP — 1979 AND ON
75 HP — 1979 THRU 1985

BEFORE TESTING

Before starting the troubleshooting work, it would be well worth the time and effort to read and understand the information presented in Sections 5-1 thru 5-4, and the Description in Section 5-10, and to conduct the general troubleshooting procedures listed in Section 5-5. Considerable time can be saved by having a thorough knowledge of the system and appreciating why the tests are conducted in a particular sequence.

Test No. 1 (Diagram This Page)
Key Switch Check

Separate the three- or four-prong connector between the Power Pack, the ignition coils, and the key switch.

Set the ohmmeter to the high ohm scale. Connect the ohmmeter red lead to the "D" terminal in the ignition coil end of the connector, and the ohmmeter black lead to a good ground on the engine.

The meter should indicate continuity with the switch in the **OFF** position. Turn the key switch to the **ON** position and the meter should indicate **NO** continuity.

If there is any needle movement, leave the ohmmeter connected and disconnect the black/yellow stripe wire from the key switch "M" terminal.

Now, if the meter indicates an open circuit with the wire disconnected from the key switch, replace the switch or perform Test No. 2 before replacing the switch. If a closed circuit (0 ohms) is indicated, repair or replace the black/yellow strip lead between the key switch and the ignition coil end of the four-prong connector.

Test No. 2 (Diagram This Page)
Three- or Four-Prong Connector Check

Test for a suspected faulty key switch. (The connector is in the lead from the Power Pack to the coils.)

Separate the connector. Four-prong connector: connect a jumper wire between the male and female part of the connector **"1"**, **"2"**, and **"3"**. Three-prong connector: connect the jumper wire between "1" and "2" connector..

Crank the engine with the starter motor. If spark is now present but was not with the connector connected, the key switch is faulty and must be replaced.

If there is no spark with this test, continue with the troubleshooting.

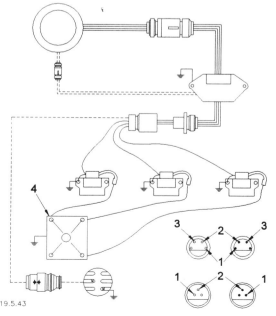

Test No. 1 Key Switch Check
Test No. 2 3- or 4-Prong Connector Test

IGNITION

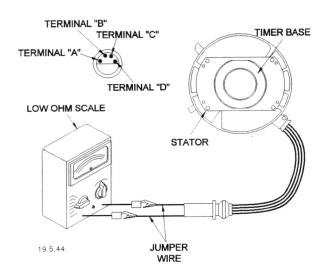

Test No. 3 Sensor Coil Resistance

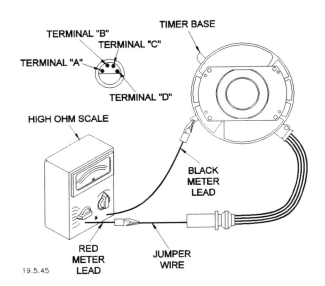

Test No. 4 Sensor Coil Test for Short

Test No. 3 (Diagram This Page)
Sensor Coil Resistance

Separate the four-prong connector. This is the connector for the leads from the Power Pack to the timer base.

Insert one end of a jumper wire into all four points of the portion of the connector to the timer base, **"A"**, **"B"**, **"C"**, and **"D"**.

Connect the test leads of an ohmmeter to each of the jumper wires to **"A"** and **"D"**. Set the ohmmeter to the low scale. Disconnect the ohmmeter lead to the **"A"** terminal and connect the test lead to the **"B"** jumper lead. Check the ohmmeter reading. Disconnect the meter lead to the **"B"** jumper lead and connect it to the **"C"** jumper lead. Again, check the meter reading. In each case the ohmmeter reading should read 17 ±5 ohms. If the resistance is not within this range, replace the sensor coil.

Test No. 4 (Diagram This Page)
**Sensor Coil
Testing for a Short**

Turn the ohmmeter to the Hi ohm scale. Connect the black ohmmeter test lead to the timer base. With the red meter lead, make momentary contact, one at-a-time, to the jumper wire connected to terminal **"A"**, **"B"**, **"C"**, and **"D"**. Any meter needle movement indicates the sensor coil or the leads are shorted to ground. A shorted sensor coil **MUST** be replaced. A shorted sensor coil lead can and **MUST** be repaired. Disconnect the jumper leads from the connector.

Test No. 5 (Diagram This Page)
**Charge Coil
Resistance Test**

Disconnect the two-wire connector leading to the timer base from the Power Pack. Insert a jumper wire into terminal **"A"** into the portion of the connector to the timer base.

Insert a jumper wire into terminal **"B"** of the connector.

Now, connect one meter lead to one jumper wire and the other test lead to the other jumper wire.

Set the ohmmeter to the Hi ohm scale. The meter should indicate 550 ohms ±75 ohms.

If the resistance is not within the limits given, the charge coil **MUST** be replaced.

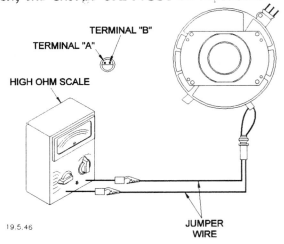

Test No. 5 Charge Coil Resistance Test

TESTING TYPE II CD 3-CYLINDER 5-29

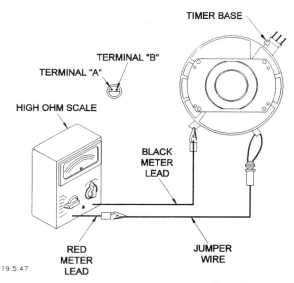

Test No. 6 Charge Coil Test for Short

Test No. 6 (Diagram This Page)
Charge Coil
Testing for a Short

Set the ohmmeter to the High Ohm scale. Connect the black ohmmeter lead to the timer base, or to a good ground. Connect a jumper wire to the **"A"** terminal and another to the **"B"** terminal of the connector. Connect the red meter lead one at-a-time to the **"A"** and **"B"** jumper wires. Any meter needle movement at either connections indicates the charge coil or the charge coil leads are shorted to ground.

A shorted charge coil **MUST** be replaced. A shorted charge coil lead can and **MUST** be repaired or replaced.

SPECIAL WORDS

A CD Voltmeter Tester is required for Test No. 7 This is a special tester available only through OMC. The test can only be performed using this piece of equipment.

BEFORE conducting the following test, the Spark Test, Key Switch Test and the Ohmmeter Tests should have been performed.

Test No. 7 (Diagram This Page)
Charge Coil Output

Set the meter switches to Negative and 500. Insert the red meter lead into the **"B"** cavity of the female part of the connector. Insert the black meter lead into the **"A"** cavity. Connect a test lead to **"A"** and **"B"**. Crank the engine and observe the meter reading.

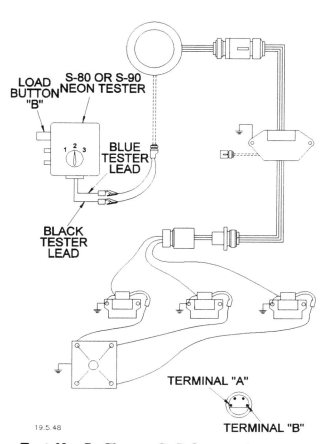

Test No. 7 Charge Coil Output Check

If the meter reading is less than 220, replace the charge coil. If the meter reading is 220 or higher, proceed with the Sensor Coil Output Test in the following paragraphs.

Test No. 8 (Diagram Page 5-30)
Sensor Coil Output Test

Set the meter switches to **"S"** and **"5"**. Insert the black tester lead into the **"D"** cavity of the part of the connector to the timer base. Insert the red meter lead into the **"A"** cavity.

Crank the engine with the starter motor and observe the meter reading. Move the red meter lead from **"A"** to **"B"**. Crank the engine and again observe the meter reading. Move the red meter lead from **"B"** to **"C"** and again crank the engine with the starter motor and observe the meter reading. If the meter reading is less than 0.4, replace the sensor coil. If the meter reading is 0.4 or higher, proceed to the Power Pack Output Test.

Assemble the four-prong connector.

5-30　IGNITION

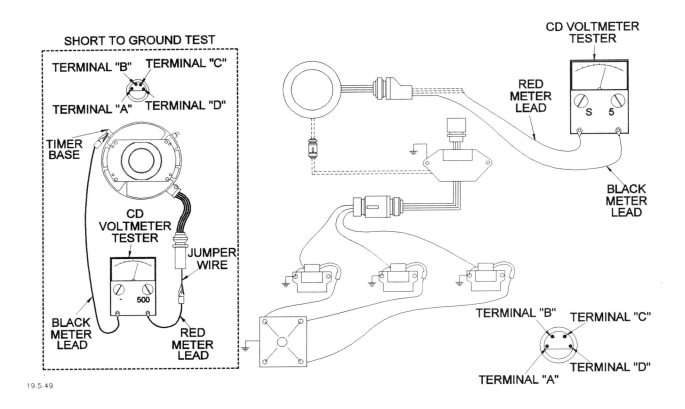

Test No. 8 Sensor Coil Output Check

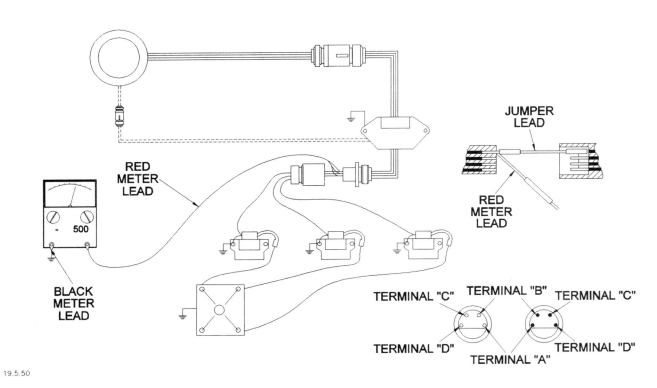

Test No. 9 Power Pack Output Test

Test No. 10 Another Power Pack Output Test

Test No. 9 (Diagram Page 5-30)
Power Pack Output Test

Set the meter switches to **NEGATIVE** and **500**. Disconnect the four-prong connector between the power pack and the ignition coils. Insert jumper leads between the "A", "B", and "C" terminals of both male and female halves of the connector -- "A" to "A", "B" to "B", and "C" to "C". Connect the black meter lead to a good ground on the engine. Connect the red meter lead to the metal part of the jumper lead coming from the power pack cavity "A" --the female connector. Crank the engine and observe the meter reading.

Now, connect the red meter lead to the metal part of the jumper lead in the "B" cavity. Crank the engine and observe the meter reading. Connect the red meter lead to the metal part of the jumper lead in the "C" cavity. Again crank the engine and observe the meter reading. The meter reading during each test should be 260 or higher. If the meter reading is satisfactory, check the ignition coil or coils. If the meter fails to indicate the required reading, the Power Pack is defective and must be replaced.

Test No. 10 (Diagram Page 5-30)
Another Power Pack Output Test

If the meter reading during the previous two checks was not 260 or higher, disconnect the jumper lead from the cavity on the Power Pack half of the connector which had the low reading (either "A", "B", or "C"). Insert the red meter lead into the cavity from which the jumper lead was just removed (either "A", "B", or "C"), on the Power Pack half of the connector.

Again, crank the engine and observe the meter reading. If the meter now reads 260 or higher, the ignition coil is defective and **MUST** be replaced. If the meter does **NOT** read 260 or higher, the Power Pack is defective.

For component replacement, see Section 5-14.
For synchronizing procedures, see Section 5-15.
For detailed timing instructions, see Section 5-17.

5-12 TROUBLESHOOTING V4 UNITS
TYPE II CAPACITOR DISCHARGE (CD)
FLYWHEEL MAGNETO
85 HP -- 1978 THRU 1980
90 HP -- 1981 AND ON
100 HP -- 1979 AND 1980
110 HP 1985 AND ON
115 HP -- 1978 THRU 1984
120 HP -- 1985 AND ON
140 HP -- 1978 AND ON

BEFORE TESTING
Before starting the troubleshooting work, it would be well worth the time and effort to read and understand the information presented in Sections 5-1 thru 5-4 and the Description in Section 5-10, and to conduct the general troubleshooting procedures listed in Section 5-5. Considerable time, money, and possible frustration can be saved by having a thorough knowledge of the system and appreciating why the tests are conducted in a particular sequence.

GOOD WORDS
The powerheads covered in this section all have two Power Packs. One serves the port bank of cylinders and the other the starboard bank. Therefore, two testers are required, one for each bank. However, if two testers are not available each Power Pack and bank of cylinders may be tested independently, **PROVIDED** each spark plug lead on each bank is grounded to the powerhead. If the cylinder banks are tested independently, repeat each test for the second Power Pack and other bank of cylinders.

MORE GOOD WORDS
A magneto key switch operates in **REVERSE** of any other type key switch. When the key is moved to the **OFF** position, the circuit is **CLOSED** between the magneto and ground. In some cases, then the key is turned to the **OFF** position, the system is grounded. For this reason, an automotive-type switch **MUST NEVER** be used, because the circuit would be opened and closed in reverse. If 12-volts should reach the coil, the coil will be **DESTROYED**.

Test No. 1 (Diagram Page 5-32)
Key Switch
Separate the four-prong connector between the Power Pack and the timer base on each side of the engine. One connector is located on the port side and the other on the starboard side.

5-32 IGNITION

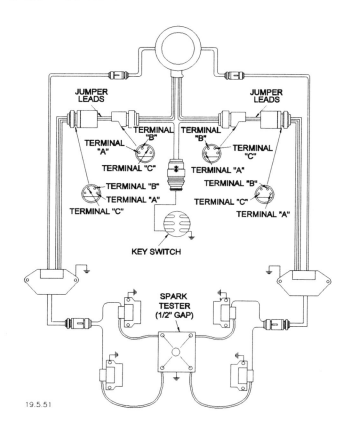

Test No. 1 Key Switch Check

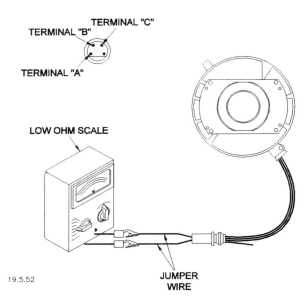

Test No. 2 Sensor Coil Resistance

Test No. 3 (Diagram This Page)
Sensor Coil
Testing for a Short

Turn the ohmmeter to the high ohm scale. Connect the black ohmmeter test lead to the timer base. With the red meter lead, make momentary contact, one at-a-time, to the jumper wire connected to terminal **"A"**, **"B"**, and **"C"**. Any meter needle movement indicates the sensor coil or the leads are shorted to ground. A shorted sensor coil **MUST** be replaced. A shorted sensor coil lead can and **MUST** be repaired. Disconnect the jumper leads from the connector.

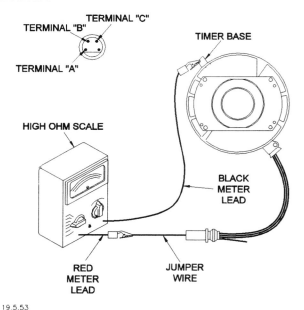

Test No. 3 Sensor Coil Test for Short

Test No. 1 Key Switch Check

Connect a jumper wire between the male and female part of the connector **"A"**, **"B"**, and **"C"**.

Crank the engine with the starter motor. If spark is now present but was not with the connector connected, the key switch is faulty and must be replaced.

If there is no spark with this test, continue with the troubleshooting.

Test No. 2 (Diagram This Page)
Sensor Coil Resistance

Separate the four-prong connector. This is the connector for the leads from the Power Pack to the timer base.

Insert one end of a jumper wire into all four points of the connector half with leads to the timer base, **"A"**, **"B"**, and **"C"**.

Set the ohmmeter to the low scale. Connect the test leads of the ohmmeter to each of the jumper wires to **"A"** and **"B"**. Observe the meter reading. Disconnect the ohmmeter lead to the **"B"** terminal and connect the test lead to the **"C"** jumper lead. Observe the ohmmeter reading. In each case the ohmmeter reading should read 40 ±10 ohms. If the resistance is not within this range, replace the sensor coil.

TESTING TYPE II CD V4

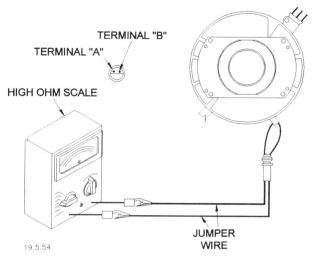

Test No. 4 Charge Coil Resistance

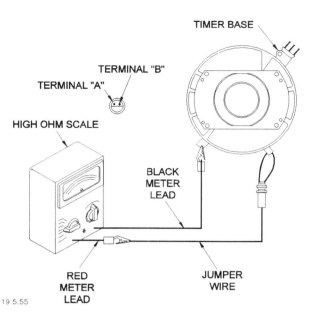

Test No. 5 Charge Coil Test for Short

Test No. 4 (Diagram This Page)
Charge Coil
Resistance Test

Disconnect the two-wire connector leading to the timer base from the Power Pack. Insert a jumper wire into terminal "A" into the connector half with leads to the timer base.

Insert a jumper wire into terminal "B" of the connector.

Now, connect one meter lead to one jumper wire and the other test lead to the other jumper wire.

Set the ohmmeter to the high ohm scale. The meter should indicate 560 ohms ± 75 ohms.

If the resistance is not within the limits given, the charge coil **MUST** be replaced.

Test No. 5 (Diagram This Page)
Charge Coil
Testing for a Short

Set the ohmmeter to the High Ohm scale. Connect the black ohmmeter lead to a good ground. Connect a jumper wire to the "A" terminal and another to the "B" terminal of the connector. Connect the red meter lead one at-a-time to the "A" and "B" jumper wires. Any meter needle movement at either connections indicates the charge coil or the charge coil leads are shorted to ground.

A shorted charge coil **MUST** be replaced. A shorted charge coil lead can and **MUST** be repaired or replaced.

SPECIAL WORDS

A CD Voltmeter Tester is required for Test No. 7 This is a special tester available only through OMC. The test can only be performed using this piece of equipment.

BEFORE conducting the following test, the Spark Test, Key Switch Test and the Ohmmeter Tests should have been performed.

Test No. 6 (Diagram Page 5-34)
Charge Coil Output

Disconnect both connectors, port and starboard, from the Power Pack to the timer base. Set the meter switches to Negative and 500. Insert the red meter lead into the "B" cavity of the connector half with leads to the timer base. Insert the black meter lead into the "A" cavity. Crank the engine and observe the meter reading.

If the meter reading is less than 150V, replace the charge coil. If the meter reading is 150V or higher, proceed with the Sensor Coil Output Test in the following paragraphs.

Test No. 7 (Diagram Page 5-35)
Sensor Coil Test for Short to Ground

Set the meter switches to "S" and "5". Disconnect the four-prong connector. Insert one end of a jumper wire into the "A" and "B" cavity of the connector half with leads to the timer base. Connect the black meter lead to the timer base or to a good ground on the engine. Now, connect the red meter lead to the two jumper wires, one at-a-time. Crank the engine and at the same time observe the meter reading during each connection.

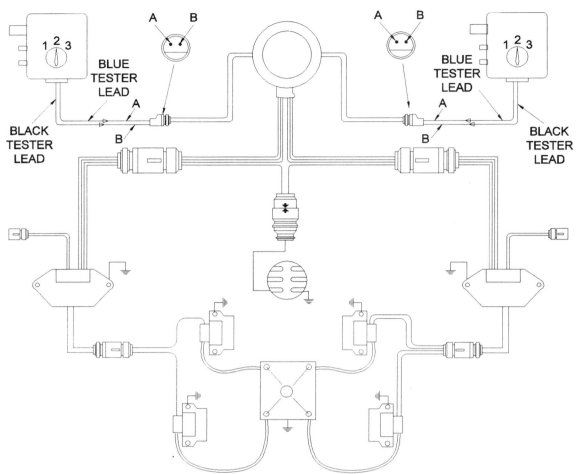

Test No. 6 Charge Coil Output Check

ANY reading during either test indicates the sensor coil is shorted to ground. Check it out. Find the short and repair it, or replace the sensor coil.

Test No. 8 (Diagram Page 5-35)
Sensor Coil Output Test

Set the meter switches to **"S"** and **"5"**. Insert the black tester lead into the **"B"** cavity of the connector half with leads to the timer base. Insert the red meter lead into the **"C"** cavity.

Crank the engine with the starter motor and observe the meter reading. Reverse the red and black meter leads. Crank the engine and again observe the meter reading. The meter should read 3 volts or 0.3. If the meter pegs, switch to 0-50 scale. If the reading is okay, check for short to ground. If the meter reading is zero or less than 0.3, replace the sensor coil. If spark appears at the spark gap tester during sensor test, replace the power pack.

Assemble the four-prong connector.

Test No. 9 (Diagram Page 5-36)
Power Pack Output Test

Set the meter switches to **NEGATIVE** and **500**. Disconnect the two-prong connector between the power pack and the ignition coils. Insert jumper leads between the **"A"** terminal of both male and female halves of the connector -- **"A"** to **"A"**. Insert a jumper wire into the **"B"** terminal of the connector half with leads to the Power Pack. Connect the black meter lead to a good ground on the engine. Connect the red meter lead to the metal part of the jumper lead coming from the power pack, terminal **"B"**. Crank the engine and observe the meter reading.

Now, switch the jumper leads from **"A"**

TESTING TYPE II CD V6   5-35

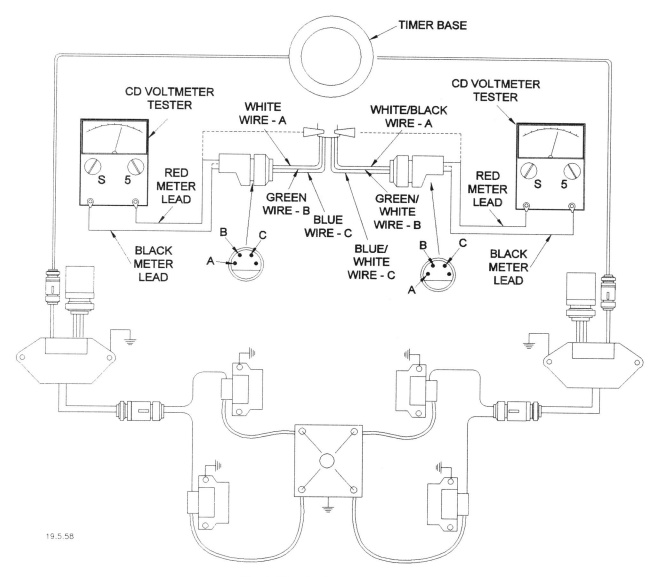

*Test No. 7 Sensor Coil Test
Short to Ground
Test No. 8 Sensor Coil Output Check*

to "B", and connect "B" to "B". Insert the red test lead into "A". Crank the engine and observe the meter reading. The meter reading during each test should be 170 or higher. If the meter reading is satisfactory, check the ignition coil or coils. If the meter fails to indicate the required reading, the Power Pack is defective and must be replaced.

For component replacement, see Section 5-14.

For synchronizing procedures, see Section 5-15.

For detailed timing instructions, see Section 5-17.

5-13 TROUBLESHOOTING V6 UNITS TYPE II CAPACITOR DISCHARGE (CD) FLYWHEEL MAGNETO
150 HP AND 175 HP — 1979 AND ON
200 HP — 1979 AND ON
225 HP — 1986 AND ON
235 HP — 1979 THRU 1985

BEFORE TESTING

Before starting the troubleshooting work, it would be well worth the time and effort to read and understand the information presented in Sections 5-1 thru 5-4 and the Description in Section 5-10, and to conduct the general troubleshooting procedures

5-36 IGNITION

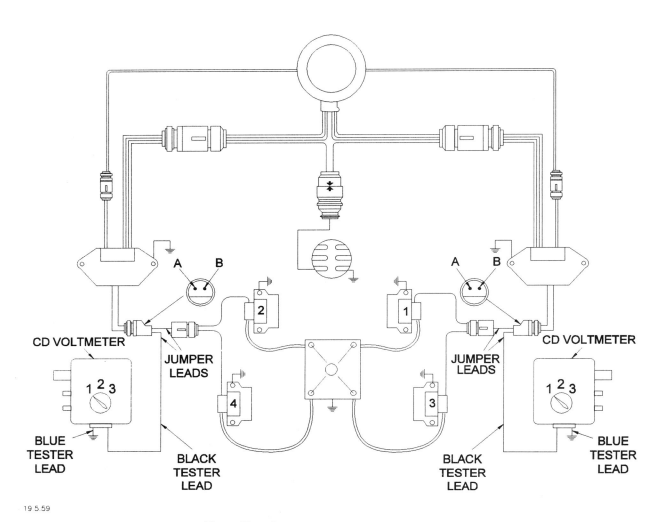

Test No. 9 Power Pack Output Test

listed in Section 5-5. Considerable time, money, and possible frustration can be saved by having a thorough knowledge of the system and appreciating why the tests are conducted in a particular sequence.

GOOD WORDS

The engines covered in this section all have two Power Packs. One serves the port bank of cylinders and the other the starboard bank. Therefore, two testers are required, one for each bank. However, if two testers are not available each Power Pack and bank of cylinders may be tested independently. If the cylinder banks are tested independently, repeat each test for second Power Pack and other bank of cylinders.

MORE GOOD WORDS

A magneto key switch operates in **RE-VERSE** of any other type key switch. When the key is moved to the **OFF** position, the circuit is **CLOSED** between the magneto and ground. In some cases, when the key is turned to the **OFF** position, the system is grounded. For this reason, an automotive-type switch **MUST NEVER** be used, because the circuit would be opened and closed in reverse. If 12-volts should reach the coil, the coil will be **DESTROYED**.

Test No. 1 (Diagram Page 5-37)
Key Switch

Separate the four-prong connector between the Power Pack and the timer base on each side of the engine. One connector is located on the port side and the other on the starboard side.

Connect a jumper wire between the male and female part of the connector "A", "B", and "C".

Crank the engine with the starter motor. If spark is now present but was not with the connector connected, the key switch is faulty and must be replaced.

If there is no spark with this test, continue with the troubleshooting.

TESTING TYPE II CD V6   5-37

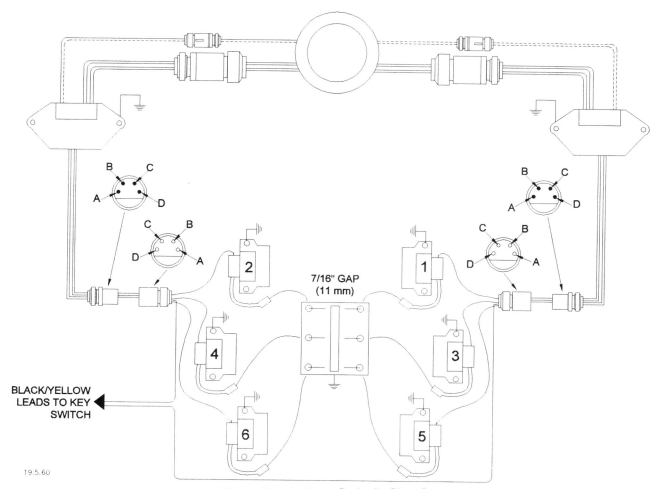

Test No. 1 Key Switch Check

Test No. 2 (Diagram Page 5-38)
Sensor Coil Resistance

Separate the four-prong connector. This is the connector for the leads from the Power Pack to the timer base.

Insert one end of a jumper wire into all four points of the connector half with leads to the timer base, **"A"**, **"B"**, **"C"**, and **"D"**.

Set the ohmmeter to the low scale. Connect the test leads of the ohmmeter to each of the jumper wires to **"A"** and **"D"**. Observe the meter reading. Disconnect the ohmmeter lead to the **""A"** terminal and connect the test lead to the **"B"** jumper lead. Observe the ohmmeter reading. Disconnect the meter lead to the **"B"** jumper lead and connect it to the **"C"** jumper lead. Again, observe the meter reading. In each case the ohmmeter reading should be: 150hp and 175hp -- 17 ±5 ohms; units larger than 175hp -- 40 ±10 ohms. If the resistance is not within the required limits, replace the sensor coil.

Test No. 3 (Diagram Page 5-38)
Sensor Coil
Testing for a Short

Turn the ohmmeter to the high ohm scale. Connect the black ohmmeter test lead to the timer base. With the red meter lead, make momentary contact, one at-a-time, to the jumper wire connected to terminal **"A"**, **"B"**, **"C"**, and **"D"**. Any meter needle movement indicates the sensor coil or the leads are shorted to ground. A shorted sensor coil **MUST** be replaced. A shorted sensor coil lead can and **MUST** be repaired. Disconnect the jumper leads from the connector.

Test No. 4 (Diagram 5-38)
Charge Coil
Resistance Test

Disconnect the two-wire connector leading to the timer base from the Power Pack. Insert a jumper wire into terminal **"A"** into

5-38 IGNITION

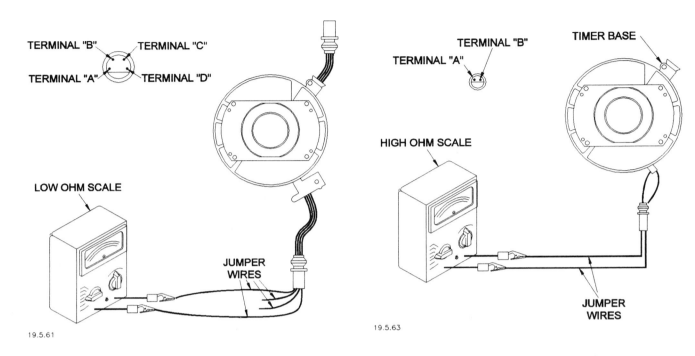

Test No. 2 Sensor Coil Resistance

Test No. 3 Sensor Coil Test for Short

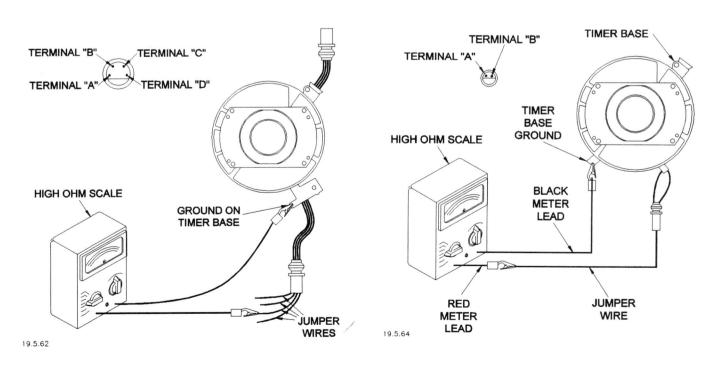

Test No. 4 Charge Coil Resistance

Test No. 5 Charge Coil Test for Short

TESTING TYPE II CD V6   5-39

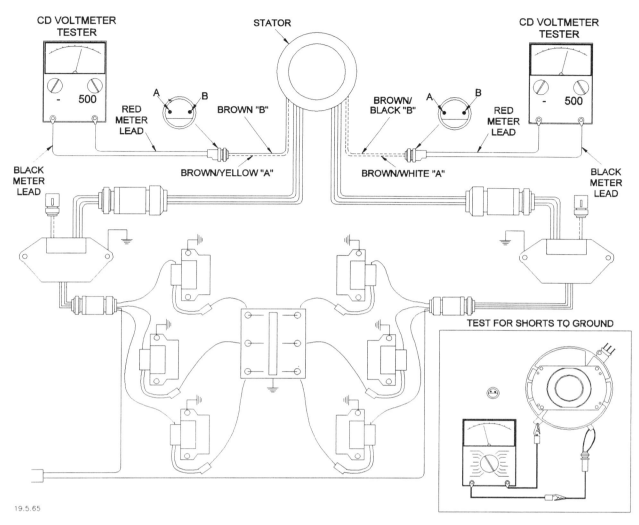

Test No. 6 Charge Coil Output Check
Test No. 7 Sensor Coil Test for Short to Ground

the connector half with leads to the timer base.

Insert a jumper wire into terminal **"B"** of the connector.

Now, connect one meter lead to one jumper wire and the other test lead to the other jumper wire.

Set the ohmmeter to the high ohm scale. The meter should indicate 970 ohms +75 ohms.

If the resistance is not within the limits given, the charge coil **MUST** be replaced.

Test No. 5 (Diagram Page 5-38)
Charge Coil
Testing for a Short

Set the ohmmeter to the Hi Ohm scale. Connect the black ohmmeter lead to a good ground. Connect a jumper wire to the **"A"** terminal and another to the **"B"** terminal of the connector. Connect the red meter lead one at-a-time to the **"A"** and **"B"** jumper wires. Any meter needle movement at either connection indicates the charge coil or the charge coil leads are shorted to ground.

A shorted charge coil **MUST** be replaced. A shorted charge coil lead can and **MUST** be repaired or replaced.

SPECIAL WORDS

A CD Voltmeter Tester is required for Test No. 7. This is a special tester available only through OMC. The test can only be performed using this piece of equipment.

BEFORE conducting the following test, the Spark Test, Key Switch Test, and the Ohmmeter Tests, should have been performed.

5-40 IGNITION

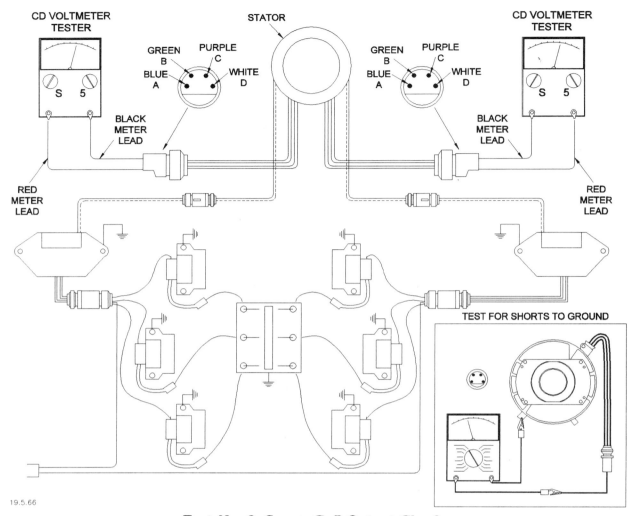

Test No. 8 Sensor Coil Output Check

Test No. 6 (Diagram Page 5-39)
Charge Coil Output

Disconnect both connectors, port and starboard, from the Power Pack to the timer base. Set the meter switches to Negative and 500. Insert the red meter lead into the "B" cavity of the connector half with leads to the timer base. Insert the black meter lead into the "A" cavity. Crank the engine and observe the meter reading.

If the meter reading is less than: 150hp and 175hp -- 200 volts; 200hp, 225hp, and 235hp --130 volts, replace the charge coil. If the meter reading is higher, proceed with the Sensor Coil Output Test in the following paragraphs.

Test No. 7 (Diagram This Page)
Sensor Coil Test for Short to Ground

Set the meter switches to "S" and "5". Disconnect the four-prong connector. Insert one end of a jumper wire into each cavity of the connector half with leads to the timer base. Connect the black meter lead to the timer base or to a good ground on the engine. Now, connect the red meter lead to the two jumper wires, one at-a-time, "A", "B", "C", and "D". Crank the engine and at the same time observe the meter reading during each connection.

ANY reading during either test indicates the sensor coil is shorted to ground. Check it out. Find the short and repair it, or replace the sensor coil.

Test No. 8
Sensor Coil Output Test

Set the meter switches to "S" and "5". Insert the black tester lead into the "D" cavity of the connector half with leads to

TESTING TYPE II CD V6   5-41

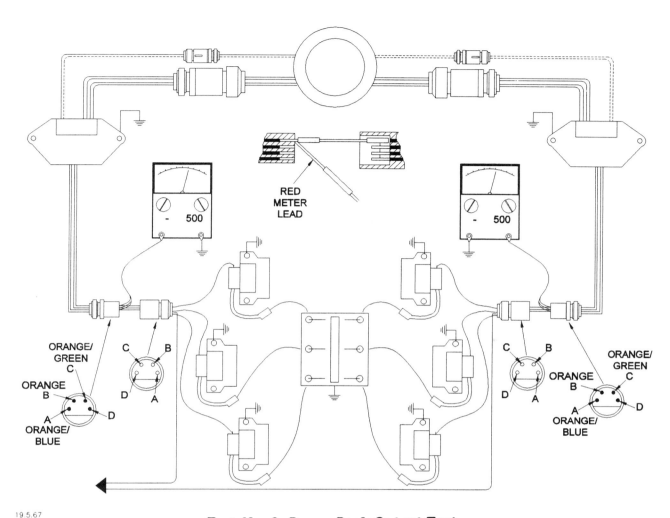

Test No. 9 Power Pack Output Test
Test No. 10 Another Power Pack Output Check

the timer base. Insert the red meter lead into the **"A"** cavity.

Crank the engine with the starter motor and observe the meter reading. Move the red meter lead from **"A"** to **"B"**. Crank the engine and again observe the meter reading. Move the red meter lead from **"B"** to **"C"** and again crank the engine with the starter motor and observe the meter reading. If the meter reading is less than 0.25, replace the sensor coil. If the meter reading is 0.25 or higher, proceed to the Power Pack Output Test.

Assemble the four-prong connector.

Test No. 9 (Diagram this page.)
Power Pack Output Test

Set the meter switches to **NEGATIVE** and **500**. Disconnect the two-prong connector between the power pack and the ignition coils. Insert jumper leads between the **"A"**, **"B"**, amd **"C"** terminals of both male and female halves of the connector -- **"A"** to **"A"**, **"B"** to **"B"**, and **"C"** to **"C"**. Connect the black meter lead to a good ground on the engine. Connect the red meter lead to the metal part of the jumper lead coming from the power pack, terminal **"A"**. Crank the engine and observe the meter reading.

Now, connect the red meter lead to the metal part of the jumper lead in the **"B"** cavity. Crank the engine and observe the meter reading. Connect the red meter lead to the metal part of the jumper lead in the **"C"** cavity. Again, crank the engine and observe the meter reading. The meter reading during each test should be at least: 150hp and 175hp -- 175 volts; 200hp, 225hp, and 235hp -- 100volts. If the meter reading

5-42 IGNITION

is satisfactory, check the ignition coil or coils. If the meter fails to indicate the required reading, the Power Pack is defective and must be replaced.

Test No. 10 (Diagram Page 5-41)
Another Power Pack Output Test

If the meter reading during the previous two checks was not 170 or higher, disconnect the jumper lead from the cavity on the Power Pack half of the connector which had the low reading (either **"A"**, **"B"**, or **"C"**). Insert the red meter lead into the cavity from which the jumper lead was just removed (either **"A"**, **"B"**, or **"C"**), on the Power Pack half of the connector.

Again, crank the engine and observe the meter reading. If the meter now reads 170 or higher, the ignition coil is defective and **MUST** be replaced. If the meter does **NOT** read 170 or higher, the Power Pack is defective and **MUST** be replaced.

For component replacement, see Section 5-14.

For synchronizing procedures, see Section 5-15.

For detailed timing instructions, see Section 5-17.

5-14 SERVICING THE TYPE I AND TYPE II CAPACITOR DISCHARGE (CD) FLYWHEEL MAGNETO

General Information

CD magneto systems installed on outboard engines will usually operate over extremely long periods of time without requiring adjustment or repair. However, if ignition system problems are encountered, and the usual corrective actions, such as replacement of spark plugs, does not correct the problem, the magneto output should be checked to determine if the unit is functioning properly.

CD magneto overhaul procedures may differ slightly on various outboard models but the following general basic instructions will apply to all Johnson/Evinrude high speed flywheel-type CD magnetos.

STATOR AND CHARGE COIL REPLACEMENT

Removal

1- Hold the flywheel with a proper tool and remove the flywheel nut. Obtain special tool OMC No. 378103 or an equivalent

SERVICING TYPE I AND TYPE II   5-43

puller. **NEVER** use a puller that exerts a force on the rim or ring gear of the flywheel. Remove the flywheel. Observe closely how the stator and trigger base is secured to the powerhead. You may elect to follow the practice of professional mechanics and take a picture with a polaroid-type camera as an aid during the assembling and installation work.

2- If servicing a Type I ignition system, disconnect the wires at the terminal block and Power Pack. From the charge coil, follow the leads down to the Power Pack. Identify and tag the wires. Disconnect them from the Power Pack.

If servicing a Type II ignition system, simply disconnect the quick disconnect.

Remove the four screws securing the stator to the powerhead. Lift the stator and charge coil assembly free of the powerhead. The stator **CANNOT** be repaired. Therefore, if troubleshooting and testing indicates the stator or the charge coil are unfit for service, the complete unit must be replaced as an assembly.

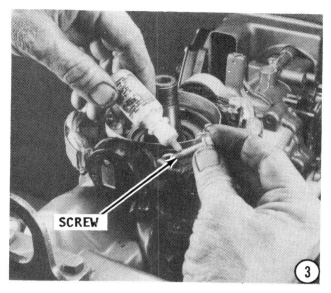

Installation

3- Slide the stator assembly down the crankshaft and into place on the powerhead. Apply just a drop of Locktite to the threads of the attaching screws, and then install and tighten them **ALTERNATELY** and **EVENLY** to the torque value given in the Appendix.

4- Check the crankshaft and flywheel tapers for any traces of oil, burrs, nicks, or other damage. Clean the tapered surfaces with solvent, and then blow them dry with compressed air. These two surfaces **MUST** be absolutely dry. Slide the flywheel onto the crankshaft with the slot in the flywheel indexed over the Woodruff key in the crankshaft. Thread the flywheel nut onto the

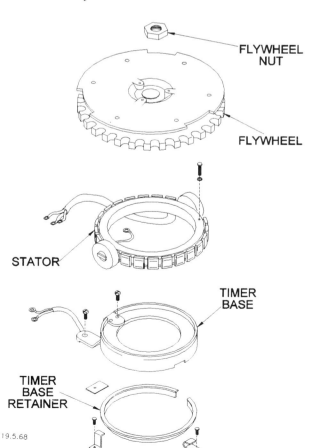

Exploded drawing of a typical flywheel, stator, timer base, and timer base retainer arrangement.

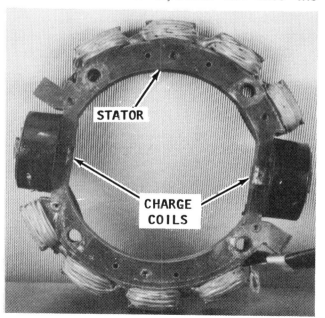

Stator assembly showing the two charge coils.

5-44 IGNITION

crankshaft. Hold the flywheel from rotating with a proper tool and tighten the nut to the torque value given in the Appendix.

If servicing a Type I system, connect the wires to the Power Pack.

If servicing a Type II system, connect the quick disconnect.

TIMER BASE AND SENSOR ASSEMBLY REPLACEMENT

Removal

1- If servicing a Type I ignition system, disconnect the necessary leads from the Power Pack.

If servicing a Type II system, separate the quick disconnect. Remove the stator and charge coil assembly as outlined in the previous section. Remove the four retaining clips and screws. The clips engage in a Delrin ring which fits around the timer base. Lift the timer base and Delrin free of the powerhead.

GOOD WORDS

A brass bushing is an integral part of the timer base. This bushing has a very close tolerance with the upper bearing and seal assembly. The bushing rotates as the spark is advanced or retarded. After the assembly has been removed, make a careful check for dirt, chips, or damage which might prevent the timer base from rotating freely, reference illustration "A" and "B".

Installation

2- Coat the upper bearing and seal assembly with OMC Sea-Lube (Trade Mark) or equivalent. Apply a coating of light-weight oil to the Delrin ring. Slip the ring into place on the timer base. Position the timer base assembly into position on the powerhead and secure it with the four retaining clips and screws. If the Woodruff key on the crankshaft was removed, insert it into the keyway with the outer edge of the key parallel to the centerline of the crankshaft.

Install the stator assembly and flywheel.

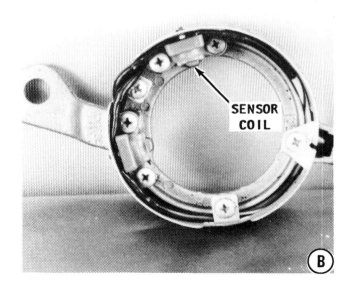

SYNCHRONIZING 5-45

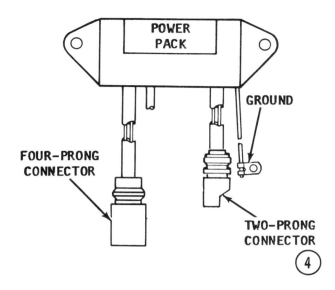

POWER PACK REPLACEMENT TYPE I

Removal

3- Disconnect the battery from the engine. Identify, tag, and then disconnect all leads at the Power Pack terminal board. Remove the attaching hardware and lift the Power Pack free of the powerhead.

Installation

Place the new Power Pack in place on the powerhead. Secure it with the attaching hardware. Connect the electrical leads to the terminal board following the color code designations given on the Power Pack cover. Install and secure the cover in place.

POWER PACK REPLACEMENT TYPE II

Removal

4- Separate the two-prong connector and the four-prong connector. Remove the two attaching bolts securing the Power Pack to the powerhead, and then lifting the Power Pack free.

Installation

Place the new Power Pack in position on the powerhead and securing it with the two attaching bolts. Mate the two-prong connector and the four-prong connector.

5-15 SYNCHRONIZATION FUEL AND IGNITION SYSTEMS

Timing is adjustable with a flywheel CD magneto system. The fuel and ignition systems **MUST** be carefully synchronized to achieve maximum performance from the engine. In simple terms, synchronization is timing the carburetion to the ignition. This means, as the throttle is advanced to increase engine rpm, the carburetor and ignition systems are both advanced equally and at the same rate.

Therefore, any time the fuel system or the ignition system is serviced to replace a faulty part, or any adjustments are made for any reason, the engine synchronization must be carefully checked and verified.

Before making any adjustments with the synchronization, the ignition system should be thoroughly checked according to the procedures outlined in this chapter and the fuel checked according to the procedures outlined in Chapter 4.

GOOD WORDS

When making the synchronization adjustment, it is well to know and understand exactly what to look for and why. The critical time when the throttle shaft in the carburetor begins to move is of the utmost importance. First, realize that the time the cam follower makes contact with the cam is not the time the throttle shaft starts to move. Instead, the critical time is when the follower hits the designated position (as described in the next five paragraphs) and the throttle shaft **AT THE CARBURETOR** begins to move.

TIMING POINTER ADJUSTMENT ALL UNITS COVERED IN THIS MANUAL

1- A special tool (OMC No. 384887 or a dial indicator) is required to accurately make this adjustment. Remove all spark plugs from the block. Install the special tool or the dial indicator into the No. 1 (top starboard bank), spark plug opening.

Rotate the flywheel **CLOCKWISE** until the piston crown makes contact with the dial indicator or the special tool at 12° TDC (top dead center). Lock the tool or the dial indicator. At that point, scribe a mark on the flywheel in line with the pointer. **SLOWLY** rotate the flywheel **COUNTERCLOCKWISE** (under normal conditions we say **NEVER** to rotate the flywheel counterclockwise but here it is permissable), until the piston again makes contact with the dial indicator or the special tool.

2- Scribe another mark on the flywheel in line with the pointer. Measure the distance between the two marks just scribed on the flywheel. The midway point between the two marks is top dead center. Loosen

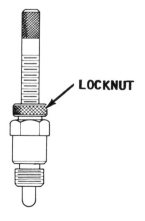

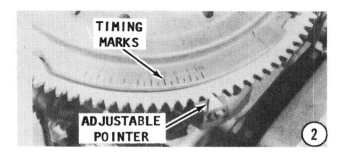

the screw and adjust the pointer to align with the midway point between the two marks.

SYNCHRONIZING ALL 3-CYLINDER UNITS COVERED IN THIS MANUAL

Pickup Timing

1- On the starboard side of the powerhead: loosen the screw on the roller arm of the throttle shaft and move the roller until the short mark on the throttle cam is aligned with the center of the cam follower roller **JUST** as the roller makes contact with the cam. Tighten the screw securely to hold the adjustment.

Mount the engine in a test tank, or move the boat into a body of water. **NEVER** use a flush device while making the primary pickup adjustments. connect a timing light to the No. 1 (top) spark plug lead.

CAUTION

Water must circulate through the lower unit to the powerhead anytime the powerhead is operating to prevent damage to the water pump in the lower unit. Just five seconds without water will damage the water pump impeller.

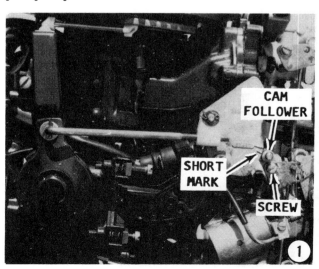

SYNCHRONIZING   5-47

Start the powerhead and allow it to warm to operating temperature.

With the outboard in **FORWARD** gear and running at idle speed, aim the timing light at the timing decal on the flywheel.

The primary pickup timing for the 70hp is 3 - 5° ATDC; for all other 3-cylinder units, the timing is 2° ATDC to TDC.

If an adjustment is required, lengthen the throttle control rod to increase the timing degrees or shorten the rod to decrease the timing degrees.

Maximum Spark Advance

2- Shut down the powerhead. Manually advance the throttle lever to the WOT position. Adjust the WOT stop screw until all three roll pins on the carburetor shafts are in the vertical position, as indicated in the accompanying illustration.

Start the powerhead. Shift the unit into **FORWARD** gear, and then increase engine speed to 5,000 rpm. Aim the timing light at the timing decal. The maximum spark advance for 3-cylinder units is 18-20° BTDC. If an adjustment is necessary, shut down the powerhead and loosen the locknut on the timing adjustment screw located portside under the flywheel just aft of the cranking motor. Rotate the screw one turn **CLOCKWISE** to retard the timing 1° and **COUNTERCLOCKWISE** one turn to advance the timing 1°.

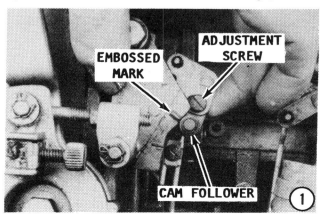

Idle Speed Adjustment

3- With the powerhead running, adjust the idle speed screw to obtain the correct idle speed listed in the Appendix for the unit being serviced. Rotating the screw **COUNTERCLOCKWISE** will decrease rpm and rotating the screw will increase rpm.

SYNCHRONIZING
V4 AND V6 POWERHEADS
PRIOR TO 1986

Pickup Timing

1- On the starboard side of the carburetor: check to be sure the throttle valves are closed. Work the adjustment screw until the center of the cam follower roller is aligned with the upper mark on the throttle cam **JUST** as the throttle begins to open.

Tighten the screw securely to hold the adjustment.

Mount the engine in a test tank, or move the boat into a body of water. **NEVER** use a flush device while making the primary pickup adjustments. Connect a timing light to the No. 1 (top) spark plug lead.

CAUTION

Water must circulate through the lower unit to the powerhead anytime the powerhead is operating to prevent damage to the water pump in the lower unit. Just five seconds without water will damage the water pump impeller.

Start the powerhead and allow it to warm to operating temperature.

With the outboard in **FORWARD** gear and running at idle speed, aim the timing light at the timing decal on the flywheel.

2- With the powerhead operating, **CAREFULLY** adjust the position of the spark advance lever until the timing pointer

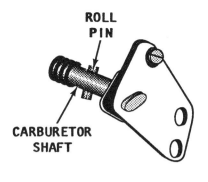

5-48 IGNITION

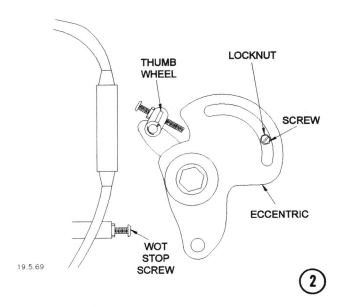

aligns as indicated in the table in the Appendix, column headed "Throttle Pickup".

Observe the location of the cam follower. The embossed mark on the throttle cam should align with the center of the cam follower. If an adjustment is necessary, back off the locknut on the thumb wheel screw, and then rotate the thumb wheel **CLOCKWISE** to increase the timing or **COUNTERCLOCKWISE** to decrease the timing.

SPECIAL WORDS
V6 POWERHEADS PRIOR TO 1986

Before performing the following step make a preliminary adjustment to the WOT stop screw, see Illustration 3A. The screw length protruding from the threaded boss must be 3/8" to 7/16" (9 to 11mm).

Maximum Spark Advance

3- With the powerhead still operating and the timing still connected, shift the unit

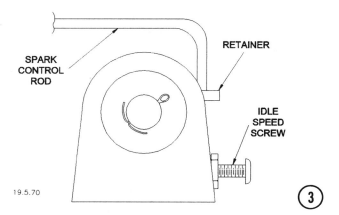

into **FORWARD** gear, and then increase powerhead speed to 5,000 rpm. Aim the timing light at the timing decal. The maximum spark advance should be as indicated in the table in the Appendix, column headed "Maximum Advance".

V4 Powerheads Prior to 1986

If an adjustment is necessary **AND** the correction is within 2° of the specified valve, loosen the locknut on the eccentric screw, see Illustration No. 2. Rotate the eccentric to bring the timing within specification, and then tighten the locknut to hold the adjustment.

If an adjustment is necessary **AND** the correction is greater than 2° of the specified value, remove the spark control rod from the retainer and position the rod into another hole in the retainer. Moving the rod forward to another hole will advance the timing. Conversely, moving the rod backward to another hole will retard the timing. For each hole the rod is moved, the timing will change 4°.

SPECIAL WORDS

The retainer block post is eccentric. If the block is rotated, three more rod locations are possible.

Wide Open Throttle Stop Adjustment

Shut down the powerhead. Manually advance the throttle lever to the WOT position. Adjust the WOT stop screw, see Illustration No. 2, until all four roll pins on the carburetor shafts are in the vertical position.

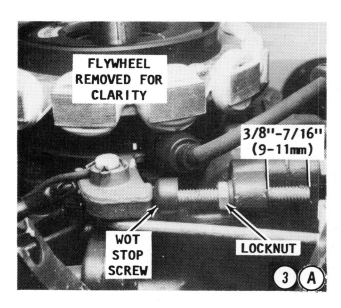

TIMING 5-49

V6 Powerheads Prior to 1986

Adjust the WOT stop screw, see Illustration No. 2, to obtain the specified timing. Rotate the screw one turn **CLOCKWISE** to retard the timing 1° and **COUNTERCLOCKWISE** one turn to advance the timing 1°.

V4 AND V6 POWERHEADS PRIOR TO 1986
Idle Speed Adjustment

With the powerhead running, adjust the idle speed screw, see Illustration No. 3, to obtain the correct idle speed listed in the Appendix for the unit being serviced. Rotating the screw **COUNTERCLOCKWISE** will decrease rpm and rotating the screw will increase rpm.

V4 AND V6 POWERHEADS — 1986 AND ON

Throttle Valve Synchronization

BEFORE making any adjustment to the throttle valve synchronization, check to be sure the horizontally mounted throttle cam and cam follower **DO NOT** make contact. If necessary, loosen the screw next to the cam follower and move the follower clear of the throttle cam.

Now, loosen the throttle shaft synchronizing screw, located on the upper portion of the throttle shaft, a couple full turns.

V6 Models

Loosen the **TOP** screw on the **PORT** throttle shaft connector and the **BOTTOM** screw on the **STARBOARD** throttle shaft connector.

V4 and V6 Models

Verify all throttle shutter valves are completely closed, and then tighten the following screws in the order listed:

V6 -- Bottom screw on Starboard throttle shaft connector.
V6 -- Top screw Port throttle shaft connector.
V4 & V6 - Throttle shaft synchronizing screw.
V4 & V6 - Cam follower adjusting screw.

Pickup Timing

1- Loosen the cam follower adjusting screw. Work the adjustment screw until the center of the cam follower roller is aligned with the embossed mark on the throttle cam **JUST** as the throttle shutter valves begin to open. Tighten the screw to hold the new adjustment.

WOT (Wide Open Throttle) Adjustment

2- Move the throttle arm to the WOT position. Loosen the locknut on the wide open throttle stop screw on the throttle arm. Rotate the screw until the carburetor throttle shutter valves are **JUST** approaching the wide open position. Tighten the locknut to hold the new adjustment.

Now, loosen the locknut on the throttle arm stop screw. Rotate the screw until the WOT mark on the cam follower bracket, see Illustration No. 4, is parallel to the edge of the air silencer box. Tighten the locknut to hold this new position.

Maximum Spark Advance

3- Mount the outboard in an adequate size test tank, or move the boat into a body of water. **NEVER** use a flush device while making the maximum spark advance adjustment. Connect a timing light to the No. 1 (top stbd. bank) spark plug lead.

CAUTION

Water must circulate through the lower unit to the powerhead anytime the power-

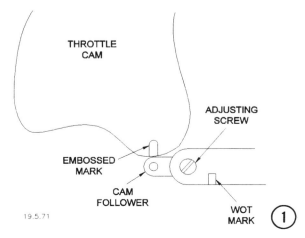
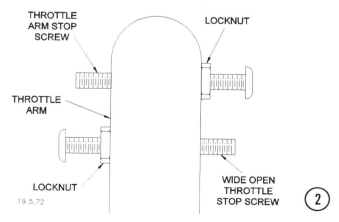

5-50 IGNITION

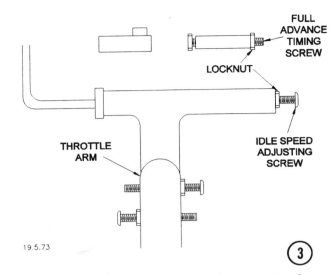

Operating a V4 engine in an adequate size test tank with the cowling removed in preparation to making adjustments.

head is operating to prevent damage to the water pump in the lower unit. Just five seconds without water will damage the water pump impeller.

Start the powerhead and allow it to warm to operating temperature.

Shift the unit into **FORWARD** gear, and then increase engine speed to 5,000 rpm. Aim the timing light at the timing decal. The maximum spark advance should be as indicated in the table in the Appendix, column headed "Maximum Advance".

If an adjustment is necessary, shut down the powerhead and loosen the locknut on the full advance timing screw located directly above the throttle arm. Rotate the screw one turn **CLOCKWISE** to retard the timing 1° and **COUNTERCLOCKWISE** one turn to advance the timing 1°.

Idle Speed Adjustment

With the powerhead running, adjust the idle speed screw to obtain the correct idle speed listed in the Appendix for the unit being serviced. Rotating the screw **COUNTERCLOCKWISE** will decrease rpm and rotating the screw **CLOCKWISE** will increase rpm.

6
ELECTRICAL

6-1 INTRODUCTION

The battery, gauges and horns, charging system, starter safety switch, and the cranking system, are all considered subsystems of the electrical system. Each of these units or subsystems will be covered in detail in this chapter beginning with the battery.

All engines covered in this manual have a magneto capacitor discharge (CD) ignition system. Engines equipped with a magneto ignition system may or may not have a generator installed for battery charging.

The starting circuit consists of a cranking motor and a starter-engaging mechanism. A solenoid is used as a heavy-duty switch to carry the heavy current from the battery to the starter motor. The solenoid is actuated by turning the ignition key to the **START** position. The choke system is activated by pushing the key inward while rotating it at the same time. The inward pressure activates the choke system and rotation of the key activates the cranking system.

It is possible to wrap a "pull rope" around the flywheel and if enough "muscle power" is available, the engine can be started by hand cranking.

6-2 BATTERIES

The battery is one of the most important parts of the electrical system. In addition to providing electrical power to start the engine, it also provides power for operation of the running lights, radio, electrical accessories, and possibly the pump for a bait tank.

Because of its job and the consequences, (failure to perform in an emergency) the best advice is to purchase a well-known brand, with an extended warranty period, from a reputable dealer.

The usual warranty covers a prorated replacement policy, which means you would be entitled to a consideration for the time left on the warranty period if the battery should prove defective before its time.

Do not consider a battery of less than 70-ampere hour capacity. If in doubt as to how large your boat requires, make a liberal estimate and then purchase the one with the next higher ampere rating.

MARINE BATTERIES

Because marine batteries are required to perform under much more rigorous conditions than automotive batteries, they are constructed much differently than those used in automobiles or trucks. Therefore, a marine battery, of at least 70-ampere hour

A fully charged battery, filled to the proper level with electrolyte, is the heart of the ignition system. Engine starting and efficient performance can never be obtained if the battery is below a fully charged rating.

capacity, should always be the No. 1 unit for the boat and other types of batteries used only in an emergency.

Marine batteries have a much heavier exterior case to withstand the violent pounding and shocks imposed on it as the boat moves through rough water and in extremely tight turns.

The plates in marine batteries are thicker than in automotive batteries and each plate is securely anchored within the battery case to ensure extended life.

The caps of marine batteries are "spill proof" to prevent acid from spilling into the bilges when the boat heels to one side in a tight turn, or is moving through rough water.

Because of these features, the marine battery will recover from a low charge condition and give satisfactory service over a much longer period of time than any type of automotive-type unit.

BATTERY CONSTRUCTION

A battery consists of a number of positive and negative plates immersed in a solution of diluted sulfuric acid. The plates contain dissimilar active materials and are kept apart by separators. The plates are grouped into what are termed elements. Plate straps on top of each element connect all of the positive plates and all of the negative plates into groups. The battery is divided into cells which hold a number of the elements apart from the others. The entire arrangement is contained within a hard-rubber case. The top is a one-piece cover and contains the filler caps for each cell. The terminal posts protrude through the top where the battery connections for the boat are made. Each of the cells is connected to its neighbor in a positive-to-negative manner with a heavy strap called the cell connector.

BATTERY RATINGS

Two ratings are used to classify batteries: one is a 20-hour rating at $80°F$ and the other is a cold rating at $0°F$. This second figure indicates the cranking load capacity and is referred to as the Peak Watt Rating of a battery. This Peak Watt Rating (PWR) has been developed to measure the cold-cranking ability of the battery. The numerical rating is embossed on each battery case at the base and is determined by multiplying the maximum current by the maximum voltage.

The ampere-hour rating of a battery is its capacity to furnish a given amount of amperes over a period of time at a cell voltage of 1.5. Therefore, a battery with a capacity of maintaining 3 amperes for 20 hours at 1.5 volts would be classified as a 60-ampere hour battery.

Do not confuse the ampere-hour rating with the PWR, because they are two unrelated figures used for different purposes.

A replacement battery should have a power rating equal or as close to the old unit as possible.

BATTERY LOCATION

Every battery installed in a boat must be secured in a well-protected ventilated area. If the battery area is not well ventilated, hydrogen gas which is given off during charging could become very explosive if the gas is concentrated and confined. Because of its size, weight, and acid content, the battery must be well-secured. If the battery should break loose during rough boat maneuvers, considerable damage could be done, including damage to the hull.

BATTERY SERVICE

The battery requires periodic servicing and a definite maintenance program to ensure extended life. If the battery should test satisfactorily, but still fails to perform properly, one of four problems could be the cause.

1- An accessory might have accidently been left on overnight or for a long period

The battery MUST be located near the engine in a well-ventilated area. It must be secured in such a manner that absolutely no movement is possible in any direction under the most violent actions of the boat.

during the day. Such an oversight would result in a discharged battery.

2- Slow speed engine operation for long periods of time resulting in an undercharged condition.

3- Using more electrical power than the generator or alternator can replace resulting in an undercharged condition.

4- A defect in the charging system. A faulty generator or alternator system, a defective regulator, or high resistance somewhere in the system, could cause the battery to become undercharged.

5- Failure to maintain the battery in good order. This might include a low level of electrolyte in the cells; loose or dirty cable connections at the battery terminals; or possibly an excessively dirty battery top.

Electrolyte Level

The most common practice of checking the electrolyte level in a battery is to remove the cell cap and visually observe the level in the vent well. The bottom of each vent well has a split vent which will cause the surface of the electrolyte to appear distorted when it makes contact. When the distortion first appears at the bottom of the split vent, the electrolyte level is correct.

Some late-model batteries have an electrolyte-level indicator installed which operates in the following manner:

One of the most effective means of cleaning the battery terminals is to use a wire brush designed for this specific purpose.

A transparent rod extends through the center of one of the cell caps. The lower tip of the rod is immersed in the electrolyte when the level is correct. If the level should drop below normal, the lower tip of the rod is exposed and the upper end glows as a warning to add water. Such a device is only necessary on one cell cap because if the electrolyte is low in one cell it is also low in the other cells. **BE SURE** to replace the cap with the indicator onto the second cell from the positive terminal.

During hot weather and periods of heavy use, the electrolyte level should be checked more often than during normal operation. Add colorless, odorless, drinking water to bring the level of electrolyte in each cell to the proper level. **TAKE CARE** not to overfill, because adding an excessive amount of water will cause loss of electrolyte and any loss will result in poor performance, short battery life, and will contribute quickly to corrosion. **NEVER** add electrolyte from another battery. Use only clean pure water.

Cleaning

Dirt and corrosion should be cleaned from the battery just as soon as it is discovered. Any accumulation of acid film or dirt will permit current to flow between the terminals. Such a current flow will drain the battery over a period of time.

Clean the exterior of the battery with a solution of diluted ammonia or a soda solution to neutralize any acid which may be present. Flush the cleaning solution off with clean water. **TAKE CARE** to prevent any of the neutralizing solution from entering the cells, by keeping the caps tight.

A poor contact at the terminals will add

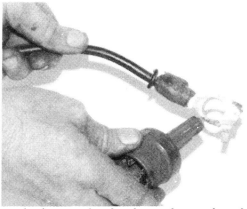

An inexpensive brush can be purchased and used to clean battery lead connectors to ensure a proper connection.

resistance to the charging circuit. This resistance will cause the voltage regulator to register a fully charged battery, and thus cut down on the alternator output adding to the low battery charge problem.

Scrape the battery posts clean with a suitable tool or with a stiff wire brush. Clean the inside of the cable clamps to be sure they do not cause any resistance in the circuit.

Battery Testing

A hydrometer is a device to measure the percentage of sulfuric acid in the battery electrolyte in terms of specific gravity. When the condition of the battery drops from fully charged to discharged, the acid leaves the solution and enters the plates, causing the specific gravity of the electrolyte to drop.

The following six points should be observed when using a hydrometer.

1- NEVER attempt to take a reading immediately after adding water to the battery. Allow at least 1/4 hour of charging at a high rate to thoroughly mix the electrolyte with the new water and to cause vigorous gassing.

2- ALWAYS be sure the hydrometer is clean inside and out as a precaution against contaminating the electrolyte.

3- If a thermometer is an integral part of the hydrometer, draw liquid into it several times to ensure the correct temperature before taking a reading.

4- BE SURE to hold the hydrometer vertically and suck up liquid only until the float is free and floating.

5- ALWAYS hold the hydrometer at eye level and take the reading at the surface of the liquid with the float free and floating.

Disregard the light curvature appearing where the liquid rises against the float stem. This phenomenon is due to surface tension.

6- DO NOT drop any of the battery fluid on the boat or on your clothing, it is extremely caustic. Use water and baking soda to neutralize any battery liquid that does accidently drop.

After withdrawing electrolyte from the battery cell until the float is barely free, note the level of the liquid inside the hydrometer. If the level is within the green band range, the condition of the battery is satisfactory. If the level is within the white band, the battery is in fair condition, and if the level is in the red band, it needs charging badly or is dead and should be replaced. If the level fails to rise above the red band after charging, the only answer is to replace the battery.

A check of the electrolyte in the battery should be on the maintenance schedule for any boat. A hydrometer reading of 1.300 or in the green band, indicates the battery is in satisfactory condition. If the reading is 1.150 or in the red band, the battery needs to be charged. Observe the six safety points given in the text when using a hydrometer.

A pair of pliers should be used to tighten the wingnuts, when they are used. Securing the wingnuts by hand is not adequate, the connections will vibrate loose.

JUMPER CABLES

If booster batteries are used for starting an engine the jumper cables must be connected correctly and in the proper sequence to prevent damage to either battery, or to the alternator diodes.

ALWAYS connect a cable from the positive terminal of the dead battery to the positive terminal of the good battery **FIRST**. **NEXT**, connect one end of the other cable to the negative terminal of the good battery and the other end to the **ENGINE** for a good ground. By making the ground connection on the engine, if there is an arc when you make the connection it will not be near the battery. An arc near the battery could cause an explosion, destroying the battery and causing serious personal **INJURY**.

DISCONNECT the battery ground cable before replacing an alternator or before connecting any type of meter to the alternator.

If it is necessary to use a fast-charger on a dead battery, **ALWAYS** disconnect one of the boat cables from the battery **FIRST**, to prevent burning out the diodes in the rectifier.

NEVER use a fast-charger as a booster to start the engine because the voltage regulator may be **DAMAGED**.

STORAGE

If the boat is to be laid up for the winter or for more than a few weeks, special attention must be given to the battery to prevent complete discharge or possible damage to the terminals and wiring. Before putting the boat in storage, disconnect and remove the batteries. Clean them thoroughly of any dirt or corrosion, and then charge them to full specific gravity reading. After they are fully charged, store them in a clean cool dry place where they will not be damaged or knocked over.

NEVER store the battery with anything on top of it or cover the battery in such a manner as to prevent air from circulating around the fillercaps. All batteries, both new and old, will discharge during periods of storage, more so if they are hot than if they remain cool. Therefore, the electrolyte level and the specific gravity should be checked at regular intervals. A drop in the specific gravity reading is cause to charge them back to a full reading.

In cold climates, care should be exercised in selecting the battery storage area. A fully-charged battery will freeze at about 60 degrees below zero. A discharged battery, almost dead, will have ice forming at about 19 degrees above zero.

DUAL BATTERY INSTALLATION

Three methods are available for utilizing a dual-battery hook-up.

Corroded battery terminals such as these result in high resistance at the connections. Such corrosion places a strain on all electrically operated devices on the boat and causes hard engine starting.

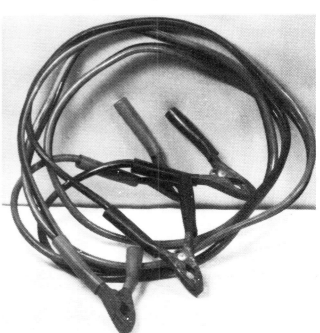

A common set of heavy-duty jumper cables. Observe the safety precautions given in the text when using jumper cables.

6-6 ELECTRICAL

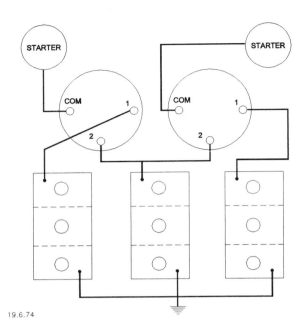

Schematic drawing of a three battery, two engine hookup.

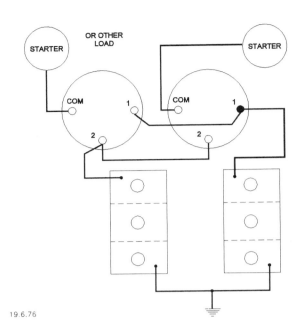

Schematic drawing for a two battery, two engine hookup.

1- A high-capacity switch can be used to connect the two batteries. The accompanying illustration details the connections for installation of such a switch. This type of switch installation has the advantage of being simple, inexpensive, and easy to mount and hookup. However, if the switch is accidently left in the closed position, it will cause the convenience loads to run down both batteries and the advantage of the dual installation is lost. The switch may be closed intentionally to take advantage of the extra capacity of the two batteries, or it may be temporarily closed to help start the engine under adverse conditions.

2- A relay, can be connected into the ignition circuit to enable both batteries to be automatically put in parallel for charging or to isolate them for ignition use during engine cranking and start. By connecting the relay coil to the ignition terminal of the ignition-starting switch, the relay will close during the start to aid the starting battery. If the second battery is allowed to run down,

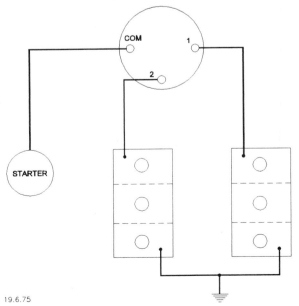

Schematic drawing for a two battery, one engine hookup.

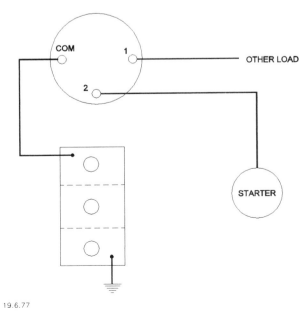

Schematic drawing for a single battery, one engine hookup.

this arrangement can be a disadvantage since it will draw a load from the starting battery while cranking the engine. One way to avoid such a condition is to connect the relay coil to the ignition switch accessory terminal. When connected in this manner, while the engine is being cranked, the relay is open. But when the engine is running with the ignition switch in the normal position, the relay is closed, and the second battery is being charged at the same time as the starting battery.

3- A heavy duty switch installed as close to the batteries as possible can be connected between them. If such an arrangement is used, it must meet the standards of the American Boat and Yacht Council, Inc. or the Fire Protection Standard for Motor Craft, N.F.P.A. No. 302.

Battery Selector Switch

Two differently designed selector switches are available. One design makes contact on the new side before breaking contact on the old; the other design breaks contact with the old before making contact with the new side.

DO NOT use the type which breaks contact with the old side before making contact with the new. This type switch will result in failure of the rectifier/regulator.

NEVER switch to the **OFF** position with the powerhead operating. Such action will result in failure of the rectifier/regulator.

When purchasing a battery selector switch, **INSIST** on the type which makes contact on the new side **BEFORE** breaking with the old side. Make the clerk prove it.

6-3 GAUGES AND HORNS

Gauges or lights are installed to warn the operator of a condition in the cooling and lubrication systems that may need attention. The fuel gauge gives an indication of the amount of fuel in the tank. If the engine overheats, a warning light will come on or a horn sound advising the operator to shut down the engine and check the cause of the warning before serious damage is done.

CONSTANT-VOLTAGE SYSTEM

In order for gauges to register properly, they must be supplied with a steady voltage. The voltage variations produced by the engine charging system would cause erratic gauge operation, too high when the generator or alternator voltage is high, and too low when the generator or alternator is not charging. To remedy this problem, a constant-voltage system is used to reduce the 12-14 volts of the electrical system to an average of 5 volts. This steady 5 volts ensures the gauges will read accurately under varying conditions from the electrical system.

SERVICE PROCEDURES

Systems utilizing warning lights do not require a constant-voltage system, therefore, this service is not needed.

Service procedures for checking the gauges and their sending units is detailed in the following sections.

TEMPERATURE GAUGES

The body of temperature gauges must be grounded and they must be supplied with 12 volts. Many gauges have a terminal on the mounting bracket for attaching a ground wire. A tang from the mounting bracket makes contact with the gauge. **CHECK** to be sure the tang does make good contact with the gauge.

Ground the wire to the sending unit and the needle of the gauge should move to the full right position indicating the gauge is in serviceable condition.

See Chapter 3, to test the sender unit.

WARNING LIGHTS

If a problem arises on a boat equipped with water and temperature lights or warning horn, the first area to check is the light assembly for loose wires or burned-out bulbs. Check the horn in the shift box for loose connections and proper grounding. When the ignition key is turned on, the light assembly is supplied with 12 volts and grounded through the sending unit mounted on the engine. When the sending unit makes contact because the water temperature is too hot, the circuit to ground is completed and the lamp should light or the horn sound.

Check The Bulb: Turn the ignition switch on. Disconnect the wire at the engine sending unit, and then ground the wire. The lamp on the dash should light or the horn sound. If it does not light, check for a burned-out bulb or a break in the wiring to the light. If the horn does not sound, check inside the shift box. Disconnect the horn wires, and then connect a good direct wire from the battery to the horn and another wire from the horn to a good ground. The horn should sound. If the horn fails to sound, the horn is defective. If the horn does sound, the wires or connections from the engine to the horn need attention.

THERMOMELT STICKS

Thermomelt sticks are an easy, inexpensive, and fairly accurate method of determining if the engine is running at approximately the temperature recommended by the manufacturer. Thermomelt sticks are available at your local marine dealer or at an automotive parts house.

Start the engine with the propeller in the water and run it for about 5 minutes at roughly 3000 rpm.

CAUTION: Water must circulate through the lower unit to the engine any time the engine is run to prevent damage to the water pump in the lower unit. Just five seconds without water will damage the water pump.

The 140 degree stick should melt when you touch it to the lower thermostat housing or on the top cylinder. If it does not melt, the thermostat is stuck in the open position and the engine temperature is too low.

Touch the 170 degree stick to the same spot on the lower thermostat housing or on the top cylinder. The stick should not melt. If it does, the thermostat is stuck in the closed position or the water pump is not operating properly because the engine is running too hot. For service procedures on the cooling system, see Chapter 8.

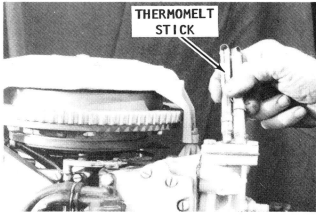

A thermomelt stick is a quick, simple, inexpensive, and fairly accurate method to determine engine running temperature.

6-4 FUEL SYSTEM

FUEL GAUGE

The fuel gauge is intended to indicate the quantity of fuel in the tank. As the experienced boatman has learned, the gauge reading is seldom an accurate report of the fuel available in the tank. The main reason for this false reading is because the boat is rarely on an even keel. A considerable difference in fuel quantity will be indicated by the gauge if the bow or stern is heavy, or if the boat has a list to port or starboard.

Therefore, the reading is usually low. The amount of fuel drawn from the tank is dependent on the location of the fuel pickup tube in the tank. The engine may cutout while cruising because the pickup tube is above the fuel level. Instead of assuming the tank is empty, shift weight in the boat to change the trim and the problem may be solved until you are able to take on more fuel.

FUEL GAUGE HOOKUP

The Boating Industry Association recommends the following color coding be used on all fuel gauge installations:

Black — for all grounded current-carrying conductors.

Pink — insulated wire for the fuel gauge sending unit to the gauge.

Red — insulated wire for a connection from the positive side of the battery to any electrical equipment.

Connect one end of a pink insulated wire to the terminal on the gauge marked **TANK** and the other end to the terminal on top of the tank unit.

Connect one end of a black wire to the terminal on the fuel gauge marked **IGN** and the other end to the ignition switch.

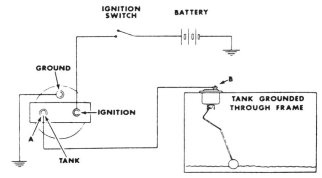

Schematic for a safe fuel tank gauge hookup.

FUEL SYSTEM 6-9

Connect one end of a second black wire to the fuel gauge terminal marked **GRD** and the other end to a good ground. It is important for the fuel gauge case to have a good common ground with the tank unit. Aboard an all-metal boat, this ground wire is not necessary. However, if the dashboard is insulated, or made of wood or plastic, a wire **MUST** be run from the gauge ground terminal to one of the bolts securing the sending unit in the fuel tank, and then from there to the **NEGATIVE** side of the battery.

FUEL GAUGE TROUBLESHOOTING

In order for the fuel gauge to operate properly the sending unit and the receiving unit must be of the same type and preferably of the same make.

The following symptoms and possible corrective actions will be helpful in restoring a faulty fuel gauge circuit to proper operation.

If you suspect the gauge is not operating properly, the first area to check is all electrical connections from one end to the other. Be sure they are clean and tight.

Next, check the common ground wire between the negative side of the battery, the fuel tank, and the gauge on the dash.

If all wires and connections in the circuit are in good condition, remove the sending unit from the tank. Run a wire from the gauge mounting flange on the tank to the flange of the sending unit. Now, move the float up-and-down to determine if the receiving unit operates. If the sending unit does not appear to operate, move the float to the midway point of its travel and see if the receiving unit indicates half full.

If the pointer does not move from the **EMPTY** position one of four faults could be to blame:

1- The dash receiving unit is not properly grounded.

2- No voltage at the dash receiving unit.

3- Negative meter connections are on a positive grounded system.

4- Positive meter connections are on a negative grounded system.

If the pointer fails to move from the **FULL** position, the problem could be one of three faults.

1- The tank sending unit is not properly grounded.

2- Improper connection between the tank sending unit and the receiving unit on the dash.

3- The wire from the gauge to the ignition switch is connected at the wrong terminal.

If the pointer remains at the 3/4 full mark, it indicates a six-volt gauge is installed in a 12-volt system.

If the pointer remains at about 3/8 full, it indicates a 12-volt gauge is installed in a six-volt system.

Preliminary Inspection

Inspect all of the wiring in the circuit for possible damage to the insulation or conductor. Carefully check:

1- Ground connections at the receiving unit on the dash.

2- Harness connector to the dash unit.

3- Body harness connector to the chassis harness.

4- Ground connection from the fuel tank to the tank floor pan.

5- Feed wire connection at the tank sending unit.

Gauge Always Reads Full when the ignition switch is **ON:**

1- Check the electrical connections at the receiving unit on the dash; the body harness connector to chassis harness connector; and the tank unit connector in the tank.

2- Make a continuity check of the ground wire from the tank to the tank floor pan.

3- Connect a known good tank unit to the tank feed wire and the ground lead. Raise and lower the float and observe the receiving unit on the dash. If the dash unit follows the arm movement, replace the tank sending unit.

Gauge Always Reads Empty when the ignition switch is **ON:**

Disconnect the tank unit feed wire and do not allow the wire terminal to ground. The gauge on the dash should read **FULL**.

If Gauge Reads Empty:

1- Connect a spare dash unit into the dash unit harness connector and ground the unit. If the spare unit reads **FULL**, the original unit is shorted and must be replaced.

2- A reading of **EMPTY** indicates a short in the harness between the tank sending unit and the gauge on the dash.

6-10 ELECTRICAL

If Gauge Reads Full:

1- Connect a known good tank sending unit to the tank feed wire and the ground lead.

2- Raise and lower the float while observing the dash gauge. If dash gauge follows movement of the float, replace the tank sending unit.

Gauge Never Indicates Full

This test requires shop test equipment.

1- Disconnect the feed wire to the tank unit and connect the wire to a good ground through a variable resistor or through a spare tank unit.

2- Observe the dash gauge reading. The reading should be **FULL** when resistance is increased to about 90 ohms. This resistance would simulate a full tank.

3- If the check indicates the dash gauge is operating properly, the trouble is either in the tank sending unit rheostat being shorted, or the float is binding. The arm could be bent, or the tank may be deformed. Inspect and correct the problem.

6-5 TACHOMETER

An accurate tachometer can be installed on any engine. Such an instrument provides an indication of engine speed in revolutions per minute (rpm). This is accomplished by measuring the number of electrical pulses per minute generated in the primary circuit of the ignition system. The proper tachometer **MUST** be installed. Be sure to check with your local marine dealer to ensure the proper unit is being installed. The wrong tachometer will cause serious damage to the ignition system.

Maximum engine performance can only be obtained through proper tuning using a tachometer.

The meter readings range from 0 to 6,000 rpm, in increments of 100. Tachometers have solid-state electronic circuits which eliminates the need for relays or batteries and contributes to their accuracy. The electronic parts of the tachometer susceptible to moisture are coated to prolong their life. Take time at the marine dealer to specify the unit being serviced when purchasing a new tachometer to ensure the correct unit is installed.

6-6 HORNS

The only reason for servicing a horn is because it fails to operate properly or because it is out of tune. In most cases, the problem can be traced to an open circuit in the wiring or to a defective relay.

Cleaning:

Crocus cloth and carbon tetrachloride should be used to clean the contact points. **NEVER** force the contacts apart or you will bend the contact spring and change the operating tension.

Check Relay and Wiring:

Connect a wire from the battery to the horn terminal. If the horn operates, the problem is in the relay or in the horn wiring.

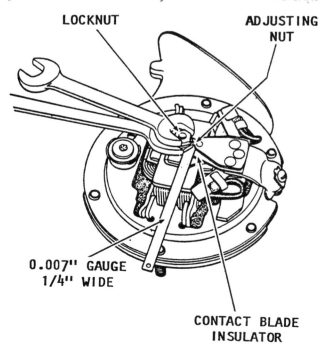

*The tone of a horn can be adjusted with a 0.007" feeler gauge, as described in the text. **TAKE CARE** to prevent the feeler gauge from making contact with the case, or the circuit will be shorted out.*

If both of these appear satisfactory, the horn is defective and needs to be replaced. Before replacing the horn however, connect a second jumper wire from the horn frame to ground to check the ground connection.

Test the winding for an open circuit, faulty insulation, or poor ground. Check the resistor with an ohmmeter, or test the condenser for capacity, ground, and leakage. Inspect the diaphragm for cracks.

Adjust Horn Tone

Loosen the locknut, and then rotate the adjusting screw until the desired tone is reached. On a dual horn installation, disconnect one horn and adjust each, one-at-a-time. The contact point adjustment is made by inserting a 0.007" (1.78 mm) feeler gauge blade between the adjusting nut and the contact blade insulator. **TAKE CARE** not to allow the feeler gauge to touch the metallic parts of the contact points because it would short them out. Now, loosen the locknut and turn the adjusting nut down until the horn fails to sound. Loosen the adjusting nut slowly until the horn barely sounds. The locknut **MUST** be tightened after each test. When the feeler gauge is withdrawn the horn will operate properly and the current draw will be satisfactory.

Typical starter motor mounted on the port side of the engine with three bolts.

6-7 ELECTRICAL SYSTEM GENERAL INFORMATION

Starter Motor Circuit

All outboard engines covered in this manual have an electric starter motor coupled with a mechanical gear mesh between the cranking motor and the engine flywheel. This arrangement is similar to the method used to crank an automotive engine. If the electric cranking system is inoperative for any reason, including a dead or weak battery, some of the large horsepower engines may still be cranked and started by hand. This is not an easy task, but it can be done.

The starter motor may be an American Bosch or Prestolite. Both operate on the same basic principle. The pinion gear of the starter motor moves upward and meshes directly with the flywheel ring gear.

The starter motor circuit consists of a cranking motor and a starter-engaging mechanism. A solenoid is used as a heavy-duty switch to carry the heavy current from the battery to the starter motor and safety switch. The solenoid is actuated by turning the ignition key to the **START** position. On some models, a pushbutton is used to actuate the solenoid. See Section 6-11 for detailed service procedures on the starter motor circuit.

Charging Circuit

Since the starting motor requires a large amount of electrical current, it is necessary to have a fully charged battery available for the starting system. If the boat is equipped with several electrical accessories, such as bait tank with circulating pump, radio, a number of running and accessory lights etc., the charging system must be performing properly to keep the battery charged.

All engines covered in this manual have an alternator installed to generate current while the engine is operating for charging purposes. The alternator may supply from 4-amps to 8-amps depending on the model installed. The alternator stator is mounted underneath the flywheel. The rotor is attached to the inside diameter of the flywheel. As the flywheel rotates, a current is generated and feed into the charging circuit.

Choke Circuit

The choke is activated by a solenoid.

This solenoid attracts a plunger to close the choke valves. The solenoid is energized when the ignition key is turned to the **START** position and the choke button is depressed. When using the electric choke, the manual choke **MUST** be in the **NEUTRAL** position.

Late model engines are equipped with a **"FULL ON"** choke. Electrical current passes through switch in the water jacket. As water temperature in the water jacket rises, the switch is opened and the choke opens, increasing the volume of air to the carburetor.

Late model V6 engines and some V4 engines have what is commonly termed "push to prime" choke system. The ignition switch is pushed inward at the same time it is rotated. Pushing in on the key switch activates a solenoid at each bypass cover. The solenoid operates a small plunger which acts as a pump, injecting fuel through the nozzles directly into the cylinder to assist engine start. If the engine is still at operating temperature when the key switch is activated, it is not necessary to push in on the switch for the choking action.

Complete service procedures for all types of chokes are given in Chapter 4.

6-8 ALTERNATOR CHARGING CIRCUIT

An alternator system, reference functional diagram illustration **"A"**, is installed on engines equipped with a battery or CD ignition system. The system replaces the current drained from the battery to start and operate the engine.

On engines with the magneto CD ignition system, the alternator is rated at approximately 2 to 3-amps above the requirements for operating the engine. Therefore, any accessories added to the engine or boat would require a higher output alternator. This higher output is usually accomplished through the installation of an alternator kit available from the local OMC dealer.

The alternator consists of a stator mounted underneath the flywheel, a rotor attached to the inside diameter of the flywheel, and a regulator on the larger output alternators. A set of positive and negative diodes is installed in the system to change the alternating current to direct (dc) current. The necessary wiring ties the system together.

Under normal operating conditions, very few problems are encountered with the alternator system. The most prevalent problem is connecting the battery backwards. Such action will damage the diodes in the system. The red cable from the starter motor solenoid terminal must be connected to the positive battery terminal. The black cable from engine ground must be connected to the negative battery terminal. Another problem area is the use of a charging system with the battery disconnected. This practice will damage the diodes or the voltage regulator.

OPERATION

Immediately after engine start, the ammeter, if installed, on the dash may indicate a full charging rate equal to the capacity of the alternator. As engine operation continues, the rate of charge will fall off, depending on the number of electrical accessories in use. Any high demand on the battery will result in an increase in the charging rate until the battery approaches a full or nearly full state of charge. The alternator system contains two distinct circuits, the field circuit and the charging circuit.

Field Circuit

When the ignition switch is turned passed the **ON** position to the **START** position, the field circuit is closed to cause a current flow from the positive pole of the battery directly to and through the field coils; on through the transistor; then through a

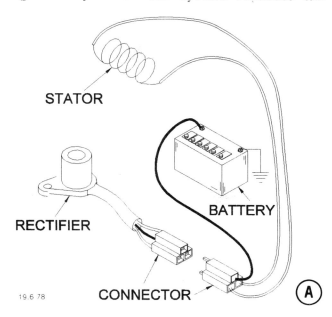

ground return to the negative pole of the battery to complete the circuit. The ignition switch is spring loaded for starting.

Once the engine starts, the key will return to the ON position and remain there while the engine is operating. An indicator light bridges the field circuit and glows GREEN when the key is set to the ON position. This light will continue to glow until the switch is turned to the OFF position when the engine is shut down.

Bear-in-mind, battery current continues flowing through the field circuit as long as the ignition switch is in the ON position. This is true regardless of whether the engine is operating or not. Therefore, if the switch should be accidently turned to or left in the ON position, the battery will discharge itself through the field and ignition circuits and continue doing so until the battery is completely discharged or, until the ignition switch is turned to the OFF position.

Charging Circuit

When the field coil is energized, as just described, the stator core becomes magnetized. The upper row of pole segments assumes a NORTH polarity and the lower row of segments a SOUTH polarity.

A magnetic field is built-up around the poles, shifting through the surrounding atmosphere, from north to south poles. Now, the irregularly shaped rotor attached to and revolving with the flywheel around the stator poles, passes through intermittent areas of variable magnetic field density. It does so in such a manner as to cause a current surge traveling in one direction to be induced in every other stator coil. At the next instant, a surge traveling in the opposite direction, is induced in the oppositely wound intervening coils as it enters and passes through the adjacent field, creating alternating (ac) current.

The stator includes 36 stator coils. Therefore, the current surges alternately 36 times per EACH revolution of the flywheel, which amounts to 36,000 such reversals per each 1000 motor rpm. At 4500 engine rpm, the rate is 162,000 alternating surges. This amount of current cannot be employed to charge the battery but must be first rectified or changed to direct (dc) current. This function is accomplished by the diodes installed in series with the charging circuit. Two diodes are of positive polarity and two

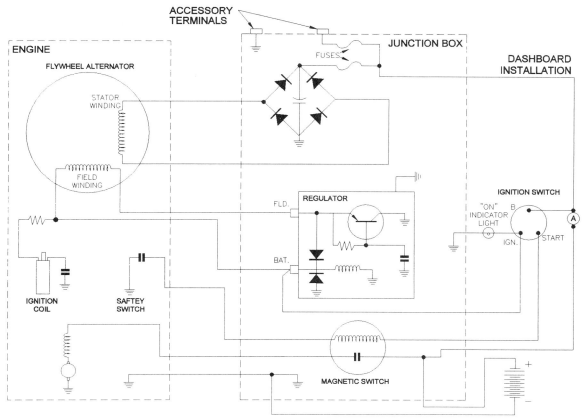

Schematic diagram for an engine with flywheel alternator.

are negative. The diodes permit voltage to pass in one direction and prevent passage in the opposite direction.

On engines equipped with a higher output alternator, a transistorized voltage regulator is installed in series with the battery field circuit. This regulator confines alternator voltage rise to within predetermined limits by automatically regulating intensity of the stator field.

6-9 ALTERNATOR WITH CD IGNITION ALL POWERHEADS COVERED IN THIS MANUAL

All powerheads covered in this manual were manufactured and distributed with a 6-amp to 10-amp alternator as standard factory equipment. However, due to the addition of accessories, these units may have as high as a 35-amp alternator installed. If a higher rated alternator is used, a voltage regulator must be installed.

TROUBLESHOOTING

The following troubleshooting procedures give detailed steps for checking the voltage regulator, alternator output, the rectifier, and the stator.

BEFORE assuming the alternator may be at fault, check all wiring and connections associated with the alternator circuit. Frayed wires, loose connections, corroded terminals, or a defective battery, will cause problems in the alternator circuit. The battery must be **FULLY** charged, the terminals clean, the cable connections tight, and the battery polarity properly connected, before any troubleshooting work is commenced on the alternator circuit. Proper polarity means the correct cables are connected to the positive and negative battery terminals. One end of the positive (red) cable should be connected to the positive battery terminal and the other end to the starter motor solenoid terminal. One end of the negative (black) cable should be connected to the negative battery terminal and the other end to a good ground on the engine.

Voltage Regulator Check

A voltage regulator is only installed with a 15-amp or higher alternator following engine purchase as an accessory. It is not a factory item nor is it original equipment.

Some 90hp, 100hp, and 110hp and all powerheads larger than 110hp are equipped with a combination rgulator/rectifier. The regulator portion of the unit may be tested as described in Step 1. The rectifier portion may be tested as described in Step 2. The unit installed may not ressemble illustration No. 3, but the test procedure is valid.

The voltage regulator installed with a CD ignition system is a solid state type. This means, if the regulator is proven to be defective, it must be discarded, it cannot be repaired. Therefore, if the following checks indicate the voltage regulator is defective, the only remedy to restore the circuit to satisfactory performance is to replace the regulator.

If the regulator is defective, it may allow too much voltage to pass through, or it may prevent sufficient voltage from passing to meet the engine and accessory demands to keep the battery fully charged.

Continually adding water to the battery at an unreasonable frequency, could be an indication the voltage regulator is allowing excessive voltage to pass through to the battery. If the battery cannot be maintained at a satisfactory charge, a defective regulator could be one of several items to blame.

1- Mount the engine in an adequate size test tank or in a body of water. If an ammeter is not installed on the dash, disconnect the red wire on the positive side of the starter solenoid. Connect one ammeter lead to this red wire and the other meter lead to the positive terminal of the starter solenoid. The meter is now in series with the alternator circuit. Start the engine and

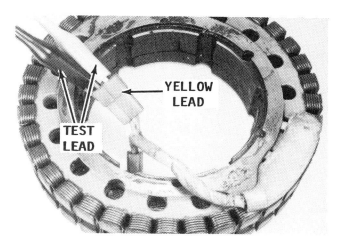

Proper test lead hookup in preparation to testing a typical stator.

ALTERNATOR CIRCUIT 6-15

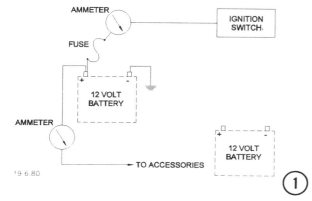

allow it to warm to operating temperature at 500 rpm.

CAUTION: Water must circulate through the lower unit to the engine any time the engine is run to prevent damage to the water pump in the lower unit. Just five seconds without water will damage the water pump.

Increase engine speed to approximately 1500 rpm. Observe the ammeter on the dash or the meter connected into the circuit. The ammeter should indicate maximum output for the alternator rating, then gradually fall back. If the ammeter fails to indicate the required current, shut down the engine and disconnect the voltage regulator from the circuit, either at the quick-disconnect at the rear of the engine, or from the terminal board. Again, start and operate the engine at approximately 1500 rpm and observe the ammeter. If the meter indicates adequate current passing through, the voltage regulator is defective and must be replaced. If the ammeter still fails to indicate adequate current, proceed with testing other components in the system.

Alternator Output Check

2- Connect one lead of an ac voltmeter to a good ground on the engine. Start the engine.

CAUTION: Water must circulate through the lower unit to the engine any time the engine is run to prevent damage to the water pump in the lower unit. Just five seconds without water will damage the water pump.

MOMENTARILY make contact with the other meter lead to first one of the yellow leads from the stator and then to the other yellow lead. This momentary contact will be made to the yellow leads inside the quick-disconnect fitting, or at the terminal board, depending on the model engine being serviced. The meter should indicate 12-volts when the meter contact is made to either yellow lead. If the meter fails to indicate 12-volts, the rectifier or the stator may be defective. Continue with the testing.

Rectifier and Rectifier/Regulator Checks

Four checks must be performed on the rectifier to determine if it is acceptable for further service. Two tests are performed for the negative diodes and two for the positive diodes.

Test 6-Amp Air Cooled Unit

3- Disconnect both the positive and negative cables from the battery terminals. Obtain an ohmmeter and set the selector to the high ohm scale. Disconnect the rectifier leads at the terminal board. (If a terminal board is not used, disconnect the rectifier leads at the connector.) Connect one ohmmeter lead to a good ground on the rectifier and the other lead to one of the Yellow or Yellow/Grey leads. Observe the meter reading. Reverse the connections and again observe the meter. The meter should indicate continuity in one direction but not in the other. Therefore, if a reading is obtained in both directions, or if a reading is not obtained in one direction, the diode in the rectifier is defective and the rectifier must be replaced. Repeat the test for the other Yellow

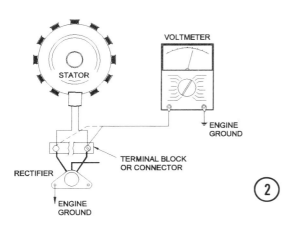

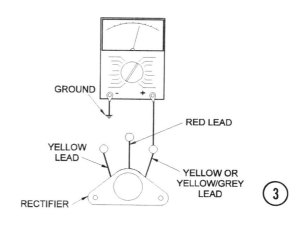

6-16 ELECTRICAL

lead. If the test fails, the other diode is defective and the rectifier must be replaced.

4- Connect one ohmmeter lead to one of the Yellow leads from the rectifier and the other lead to the Red or Purple wire. Observe the meter reading. Reverse the connections and again observe the meter. The meter should indicate continuity in only one direction. If the test fails, the diode in the rectifier is defective and the rectifier must be replaced. Leave one meter lead connected to the Red wire and connect the other meter lead to the other Yellow wire. Observe the meter reading. If continuity is observed in both directions or not in either direction, the diode in the rectifier is defective and the rectifier must be replaced.

Test 35-Amp Water Cooled Unit

The following tests must be performed with a variable load high rate discharge tester (carbon pile), capable of drawing at least 35-amps.

Begin by disconnecting the battery cables at the battery.

Next, connect one lead of a 0-40 ammeter to the rectifier/regulator Red lead and the other ammeter lead to the starter solenoid. Reconnect the leads to the battery terminals.

Connect the leads of a variable load tester (carbon pile), to the battery terminals.

Now, draw about 30-amps from the battery with the variable load tester. Start the powerhead and allow it to warm to operating temperature, and then bring the speed up to approximately 3000 rpm. The ammeter should indicate approximately 3- to 35-amps output.

Decrease the variable load tester from 30-amps to 0-amps. The ammeter will begin to show a decrease in current output. As the current decreases, the voltage across the battery should stabilize at 14 to 15-volts.

If the ammeter does not indicate as outlined in the above test, the stator should be checked. If the stator checks satisfactorily, then the rectifier/regulator must be replaced.

Stator Check

5- Check to be sure both cables are disconnected from the battery. Separate the quick-disconnect from the stator leads. Obtain an ohmmeter and set the selector to the low ohm scale. Connect the meter leads to the two yellow or yellow/grey leads going to the stator. The meter should indicate as follows: 6-amp alternator -- 1.3 ± 0.1 ohm; 9-amp alternator -- 0.75 ± 0.2 ohm; 15-amp alternator -- 0.4 ± 0.1 ohm; 35-amp alternator -- 0.17 ± 0.05 ohm.

SPECIAL WORDS

The 15-amp alternator installation has a voltage regulator, the 9-amp unit does not. Therefore, the alternator output rating can quickly be determined.

If the reading is not as indicated, the stator is defective and must be replaced.

6- Set the ohmmeter selector to the high ohm scale. Connect one meter lead to a good ground on the engine and the other lead to one of the yellow wires going to the stator. The meter should indicate no continuity. If continuity is indicated, the stator is defective and must be replaced.

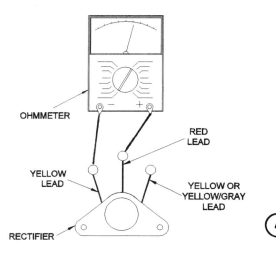

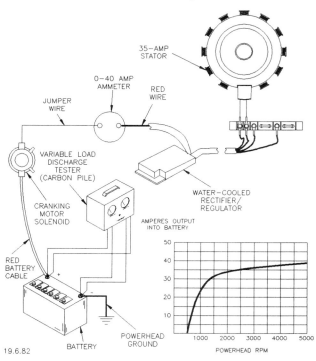

Electrical hook-up to test a 35-amp water-cooled rectifier/regulator, as described in the text.

6-10 CHOKE CIRCUIT SERVICE

This short section provides instructions to test the choke circuit. Complete service procedures for the "push to prime" choke system is presented in Chapter 4.
If the system fails the test, the attaching hardware can be removed and the choke assembly replaced.

Choke Circuit Testing

The choke circuit may be quickly tested to determine if it is functioning properly as follows:

a- Obtain an ohmmeter.

b- Connect the black meter lead to an unpainted portion of the engine block for a good ground.

c- Connect the red meter lead to the choke terminal.

d- Test the circuit using the Rx1 scale of the ohmmeter. A satisfactory reading is approximately 3 ohms.

e- After the test is completed, check to be sure the choke plunger is pulled into the choke solenoid.

6-11 STARTER MOTOR CIRCUIT SERVICE

CIRCUIT DESCRIPTION

As the name implies, the sole purpose of the starter motor circuit is to control operation of the starter motor to crank the engine until the engine is operating. The circuit includes a solenoid or magnetic switch to connect or disconnect the starter from the battery. The operator controls the switch with a pushbutton or key switch.

A cutout switch is installed in the system to prevent starting the engine if the throttle is advanced too far, beyond idle speed. When the throttle is advanced, the starter solenoid is not grounded and the starter motor will not rotate. The cutout switch is installed inside the shift box.

STARTER MOTOR DESCRIPTION

American Bosch and Prestolite, starter motors are used on the engines covered in this manual. Either one may be installed on the engine.

Marine starter motors are very similar in construction and operation to the units used in the automotive industry. Some marine starter motors use the inertia-type drive assembly. This type assembly is mounted on an armature shaft with external spiral splines which mate with the internal splines of the drive assembly.

The starter motor is a series wound electric motor which draws a heavy current from the battery. It is designed to be used only for short periods of time to crank the engine for starting. To prevent overheating

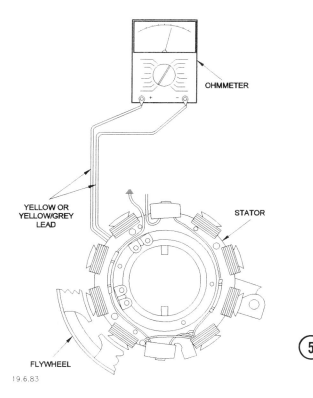

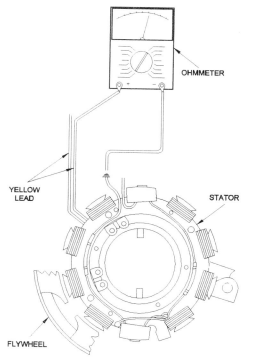

6-18 ELECTRICAL

Typical starter motor installation on the port side of a three-cylinder engine.

Two handy instruments for use in checking the generating circuit (left) and the starter motor circuit (right). These meters do not require any wire connections. A reading will be obtained by simply placing the meter on the line.

Theory of Operation

With this type of Bendix drive, power is transmitted from the starter motor to the engine flywheel directly through the Bendix drive. This drive has a pinion gear mounted on screw threads. When the motor is operated, the pinion gear moves up to mesh with the teeth on the flywheel ring gear.

When the engine starts, the pinion gear is driven faster than the shaft, and as a result, it screws out of mesh with the flywheel. A rubber cushion is built into the Bendix drive to absorb the shock when the pinion meshes with the flywheel ring gear. The parts of the drive **MUST** be properly assembled for efficient operation. If the drive is removed for cleaning, **TAKE CARE** to assemble the parts as shown in the accompanying illustration. If the screw shaft assembly is reversed, it will strike the splines and the rubber cushion will not absorb the shock.

the motor, cranking should not be continued for more than 30-seconds without allowing the motor to cool for at least three minutes. Actually, this time can be spent in making preliminary checks to determine why the engine fails to start.

Most starter motors operate in much the same manner and the service work involved in restoring a defective unit to service is almost identical. Therefore, the information in this chapter is grouped together for the major components of the starter under separate headings. Differences, where they occur, between the various manufacturers, are clearly indicated.

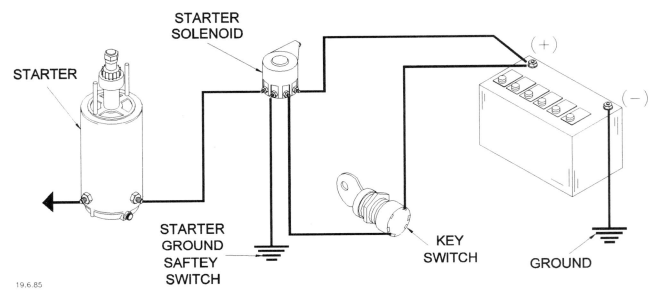

*Functional diagram to show current flow when the key switch is turned to the **START** position.*

The sound of the motor during cranking is a good indication of whether the starter motor is operating properly or not. Naturally, temperature conditions will affect the speed at which the starter motor is able to crank the engine. The speed of cranking a cold engine will be much slower than when cranking a warm engine. An experienced operator will learn to recognize the favorable sounds of the cranking engine under various conditions.

Faulty Symptoms

If the starter spins, but fails to crank the engine, the cause is usually a corroded or gummy Bendix drive. The drive should be removed, cleaned, and given an inspection.

If the starter motor cranks the engine too slowly, the following are possible causes and the corrective actions that may be taken:

a- Battery charge is low. Charge the battery to full capacity.

b- High resistance connections at the battery, solenoid, or motor. Clean and tighten all connections.

c- Undersize battery cables. Replace cables with sufficient size.

d- Battery cables too long. Relocate the battery to shorten the run to the starter solenoid.

Maintenance

The starter motor does not require periodic maintenance or lubrication EXCEPT

Typical Bendix spring arrangement on a starter motor. A small amount of oil on the shaft in the spring area will prolong satisfactory operation.

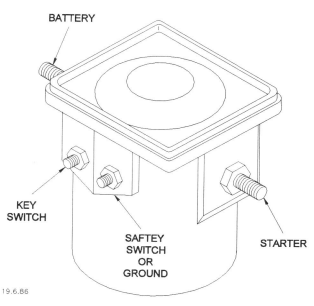

*Functional diagram of a starter motor solenoid. Notice the separate terminal for a ground wire. This solenoid is **NOT** grounded through the mounting bracket.*

just a drop of light-weight oil on the starter shaft to ease movement of the Bendix drive. If the motor fails to perform properly, the checks outlined in the previous paragraph should be performed.

The frequency of starts governs how often the motor should be removed and reconditioned. The manufacturer recommends removal and reconditioning every 1000 hours.

Naturally, the motor will have to be removed if the corrective actions outlined under **Faulty Symptoms** above, does not restore the motor to satisfactory operation.

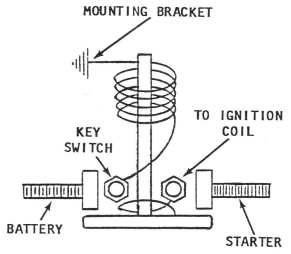

*Functional diagram of a "slave-type" starter motor solenoid used in four-cycle engine installations. This solenoid **CANNOT** be used on a two-cycle engine.*

STARTER MOTOR TROUBLESHOOTING

Before wasting too much time troubleshooting the starter circuit, the following checks should be made. Many times, the problem will be corrected.

   a- Battery fully charged.
   b- Throttle advanced too far (beyond fast idle speed).
   c- All electrical connections clean and tight.
   d- Wiring in good condition, insulation not worn or frayed.
   e- One of the cutout switches may be defective.

Two more areas may cause the engine to turn over slowly even though the starter motor circuit is in excellent condition: A tight or "frozen" engine; and, water in the lower unit causing the bearings to tighten up. The following troubleshooting procedures are presented in a logical sequence, with the most common and easily corrected areas listed first in each problem area. The connection number refers to the numbered positions in the accompanying illustrations.

Perform the following quick checks and corrective actions for following problems:

TESTING

FIRST THESE WORDS

The starter solenoid is actually nothing more than a switch between the battery and the starter motor. Several types of solenoids are used and many appear similar. **NEVER** attempt to use an automotive-type solenoid in a marine installation. Such practice will lead to more problems than can be imagined. An automotive-type solenoid has a completely different internal wiring circuit. If such a solenoid is connected into the starter system, and the system is activated, current will be directed to ground. The wires will be burned and the cutout switch will be burned and rendered useless. Therefore, when installing replacement parts in the starter or other circuits on a marine installation, always take time to obtain parts from a **MARINE** outlet to ensure proper service and to prevent damage to other expensive components.

SAFETY WORD

Before making any test of the cranking system, disconnect the spark plug leads at the spark plugs to prevent the engine from possibly starting during the test and causing personal injury.

The following tests are to be performed according to the faulty condition described. The numbers referenced in the steps are correlated with numbers on the accompanying circuit diagram on Page 6-21, to identify exactly where the connection or test is to be made.

Starter Motor Turns Slowly

   a- Battery charge is low. Charge the battery to full capacity.
   b- Electrical connections corroded or loose. Clean and tighten.
   c- Defective starter motor. Perform an amp draw test. Lay an amp draw-gauge on the cable leading to the starter motor No. 5. Turn the key to the **START** position and attempt to crank the engine. If the gauge indicates an excessive amperage draw, the starter motor **MUST** be replaced or rebuilt.

Starter Motor Fails To Rotate
Voltage Check

   a- Check the voltage at No. 2, the battery and ground.
   b- If satisfactory voltage is indicated at the battery, check the voltage at No. 3, the positive side of the starter solenoid. Weak, or no voltage at this point indicates corroded battery terminals, poor connection at the solenoid, or defective wiring between the battery and the solenoid.
   c- Test the voltage at No. 4, the key. A full 12-volt reading should be registered at the key. Weak or no voltage at the key indicates a poor connection at the solenoid, or a broken wire between the starter solenoid and the key.
   d- If satisfactory voltage is indicated during Steps a, b, and c, connect a voltmeter at No. 5 and ground, and then turn the key switch to the **START** position. If 12-volts is registered at No. 5 and the starter still fails to operate, the starter is defective and requires service. If voltage is **NOT** present at No. 5, proceed to the next section, Testing Starter Solenoid.

Testing Starter Solenoid

   a- Remove the heavy starter cable at No. 5, at the starter. This cable **MUST** be disconnected prior to performing this test to prevent the starter motor from turning and cranking the engine. Connect a voltmeter to No. 6 (the starter solenoid), and ground.

Turn the key to the **START** position. The meter should indicate 12-volts. If voltage is not present at No. 6, the key switch is defective, or the wire is broken between the key switch and the starter solenoid.

b- If voltage is present at No. 6, connect a voltmeter at No. 3 and to No. 7. Connect one end of a jumper wire to No. 2, the positive terminal of the battery and **MOMENTARILY** make contact with the other end at No. 6, the starter solenoid. If voltage is indicated through the starter solenoid, the solenoid is satisfactory and the problem has been corrected while making the tests. Sometimes, when working with electrical circuits, corrective action has been taken almost accidently, a bad connection has been made good, etc. If the solenoid test failed, it does not necessarily mean the solenoid is defective. The solenoid may not be properly grounded through the cutout switch. Therefore, the cutout switch may be defective and should be checked as outlined later in this section.

c- With the voltmeter still connected at No. 3 and No. 7, connect one end of a jumper wire at No. 8, the starter solenoid, and the other lead to a good ground. Connect a second jumper wire at No. 2, the positive terminal of the battery, to No. 6, the starter solenoid. The voltmeter should indicate voltage is present. If voltage is not present, the starter solenoid is defective and **MUST** be replaced.

Testing Throttle Advance Cutout Switch

a- Remove the existing wire from the No. 1 switch terminal. Connect one probe lead of an ohmmeter to the terminal. Connect the other test probe lead to a good ground. Depress the switch button and the ohmmeter should indicate continuity. If continuity is not indicated, the switch is defective and **MUST** be replaced. Connect the heavy cable at No. 5, the starter motor.

6-12 STARTER DRIVE GEAR SERVICE ALL ENGINES COVERED IN THIS MANUAL

STARTER REMOVAL

Before beginning any work on the starter motor, disconnect the positive (+) lead from the battery terminal. Remove the hood. Disconnect the red cable at the starter motor terminal.

1- Remove the three 1/2" bolts securing the starter motor to the powerhead. One of

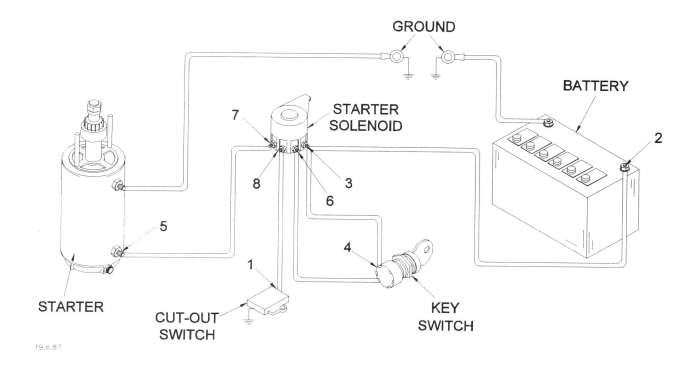

Use this diagram and the numbers shown for tests outlined in the text.

6-22 ELECTRICAL

these bolts is located on the port side just above the carburetor. The brackets are welded to the starter motor housing. Once the bolts have been removed, the starter motor can be lifted free of the powerhead.

DISASSEMBLING

Starter Drive Gear

Prevent the armature from turning by holding it with the proper size wrench on

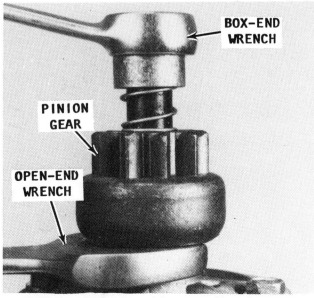

Removing the starter drive assembly using an open-end wrench and a box-end wrench to remove the shaft nut.

Removing the drive gear using a pair of pliers to hold the gear and a box-end wrench to remove the shaft nut.

the hex nut provided for this purpose on the opposite end from the shaft nut. If the hex nut is not provided, hold the drive assembly with a pair of water pump pliers. Remove the shaft nut, spring retainer, spring, and then the drive assembly. The shaft nut should be replaced and **NOT** used a second time. The manufacturer **STRONGLY** recommends against using any type of self-locking nut on the shaft.

The exploded drawing accompanying this section will be helpful in assembling the starter motor in the proper sequence.

CLEANING AND INSPECTING

Inspect the drive gear teeth for chips, cracks, or a broken tooth. Check the spline

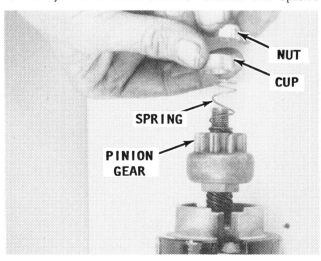

Removal sequence of parts when disassembling the drive gear.

The teeth of the pinion gear must be in good condition for satisfactory service. The pinion gear should be carefully inspected as outlined in the text. Clean and undamaged internal threads will ensure the pinion will ride properly on the armature shaft. Damaged internal pinion threads will not only give poor service but will eventually cause damage to the armature shaft.

inside the drive gear for burrs and to be sure the drive gear moves freely on the armature shaft. Check to be sure the return spring is flexible and has not become distorted. Clean the armature shaft with crocus cloth.

ASSEMBLING

Starter Drive Gear

Begin by assembling the following parts in the order given. The accompanying illustration will be most helpful in assembling the parts in the proper sequence. First, slide the drive gear onto the shaft, then the spring, spring retainer, and then a **NEW** locking nut. Prevent the armature shaft from turning by holding it with the proper size wrench on the hex nut provided for this purpose on the opposite end from the shaft nut. If the armature hex nut is not provided, hold the drive assembly with a pair of water pump pliers. Tighten the shaft nut securely.

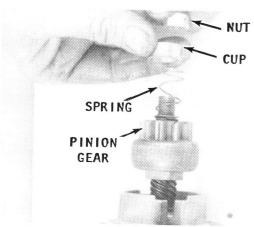

Installation sequence of parts when assembling the drive gear.

To test the complete starter motor, proceed directly to Section 6-15.

To install the starter motor onto the engine, if no further work is to be performed, proceed directly to Section 6-16.

6-13 PRESTOLITE SERVICE

The unit is installed on the port side of the engine. A bracket is welded to the starter and the bracket is then secured to the powerhead.

REMOVAL
PORT SIDE INSTALLATION
DRIVE GEAR ON ARMATURE SHAFT

1- Remove the three attaching bolts from the powerhead. One is partially hidden just behind the carburetor and the other two are visible. Lift the starter motor free of the powerhead.

GOOD NEWS

If the only motor repair necessary is replacement of the brushes, the drive gear does not have to be removed. All starter motors have thru-bolts securing the upper and lower cap to the field frame assembly. In all cases both caps have some type of mark or boss. These marks are used to properly align the caps with the field frame assembly.

DISASSEMBLING

2- Observe the caps and find the identifying mark or boss on each. If the marks are not visible, make an identifying mark prior to removing the thru-bolts as an essential aid during assembling. Remove the thru-bolts from the starter motor.

3- Use a small hammer and **CAREFULLY** tap the lower cap free of the starter motor.

4- Pull on the armature shaft from the drive gear end and remove it from the field frame assembly. Remove the brushes from their holders, and then remove the brush springs. Lift the white plastic retainer free from the frame. Observe the location of the notch on the retainer in relation to the frame. The retainer must be installed in the same position.

ARMATURE TESTING

Testing for a Short

1- Position the armature on a growler, then hold a hacksaw blade over the armature core. Turn the growler switch to the

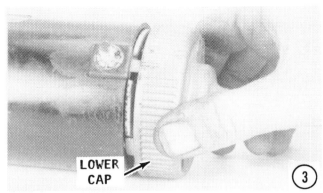

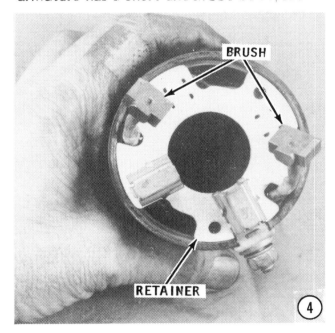

ON position. Slowly rotate the armature. If the hacksaw blade vibrates, the armature or commutator has a short. Clean the grooves between the commutator bars on the armature. Perform the test again. If the hacksaw blade still vibrates during the test, the armature has a short and **MUST** be replaced.

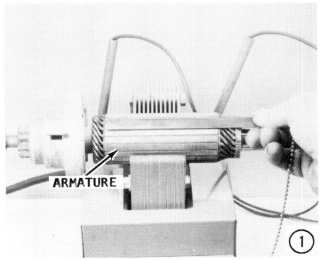

PRESTOLITE SERVICE 6-25

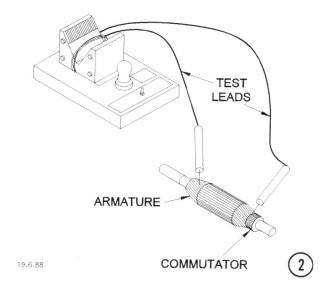

Testing for a Ground
2- Obtain a test lamp or continuity meter. Make contact with one probe lead on the armature core and the other probe lead on the commutator bar. If the lamp lights, or the meter indicates continuity, the armature is grounded and **MUST** be replaced.

Checking the Commutator Bar
3- Check between or check bar-to-bar as shown in the accompanying illustration. The test light should light, or the meter should indicate continuity. If the commutator fails the test, the armature **MUST** be replaced.

Turning the Commutator
4- True the commutator, if necessary, in a lathe. **NEVER** undercut the mica because the brushes are harder than the insulation. Undercut the insulation between the commutator bars 1/32" (0.79 mm) to the full width of the insulation and flat at the bottom. A triangular groove is not satisfactory. After the undercutting work is completed, clean out the slots carefully to remove dirt and copper dust. Sand the commutator lightly with No. 00 sandpaper to remove any burrs left from the undercutting. Check the armature a second time on the growler for possible short circuits.

Positive Brushes
5- Notice how the positive brush lead is attached to the terminal on the end of the frame. This is the same terminal to which the heavy battery cable is attached. The terminal may be removed from the frame. Pull the terminal free of the frame.

Obtain an ohmmeter. Connect one test lead of an ohmmeter to the brush and the other test lead to the terminal. Continuity should be indicated on the ohmmeter. If continuity is not indicated, the brush must be replaced. The brush and terminal are sold as an assembly, eliminating the necessity for soldering.

Negative Brushes
6- The complete terminology for Prestolite negative brushes is: Field Coil -- Negative Brush.

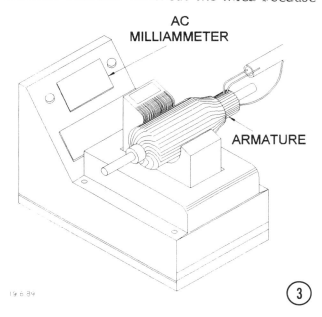

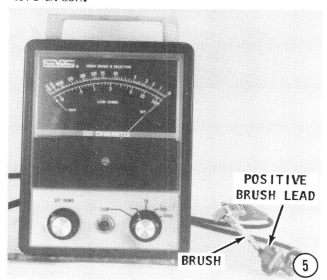

6-26 ELECTRICAL

Obtain an ohmmeter. Make contact with one test lead to the negative brush and make contact with the other lead to the starter frame. If the meter does not indicate continuity, the field coils are open and **MUST** be replaced.

Check to be sure the soldered connections are **NOT** touching the frame. The fields must not be grounded. If the connections make contact with the frame, the fields would be grounded.

CLEANING AND INSPECTING

Clean the field coils, armature, commutator, armature shaft, brush-end plate, and drive-end housing with a brush or compressed air. Wash all other parts in solvent and blow them dry with compressed air.

Inspect the insulation and the unsoldered connections of the armature windings for breaks or burns.

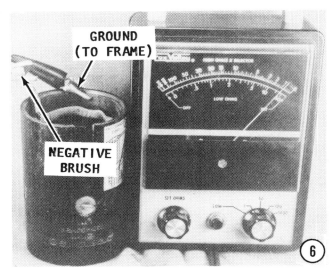

Perform electrical tests on any suspected defective part, according to the procedures outlined earlier in this section.

Check the commutator for run-out. Inspect the armature shaft and both bearings for scoring.

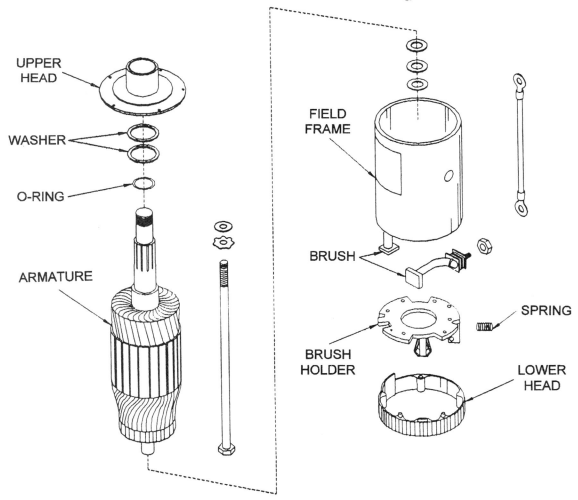

Exploded drawing of a Prestolite starter motor.

Turn the commutator in a lathe if it is out-of-round by more than 0.005" (1.27 mm).

Check the springs in the brush holder to be sure none are broken. Check the spring tension and replace if the tension is not 32-40 ounces. Check the insulated brush holders for shorts to ground. If the brushes are worn down to 1/4" (6.35 mm) or less, they must be replaced.

Check the field brush connections and lead insulation. A brush kit and a contact kit are available at your local marine dealer, but all other assemblies must be replaced rather than repaired.

The armature, fields, and brush holders, must be checked before assembling the starter motor. See the testing section in this chapter for detailed procedures to test the starter motor.

ASSEMBLING THE PRESTOLITE

1- Slide the plastic terminal and brush lead retainer into the groove in the frame with the small protrusion on one side facing **DOWNWARD**. Continue pushing the retainer into the groove until it is fully seated.

Work the brush retainer down on top of the frame with the positive lead through the cutaway in the retainer plate. Check to be sure the field coil negative brush passes through the cutaway in the plate.

2- Install the spring into the retainer. Push the negative brush into its retainer and then, wrap a fine piece of wire around the front side of the brush and the back side of the retainer. Tighten the wire snugly. This wire will hold the brush in the retainer. Repeat the procedure for the positive brush.

Check to be sure the plate is secured onto the frame and the cutaway is over the

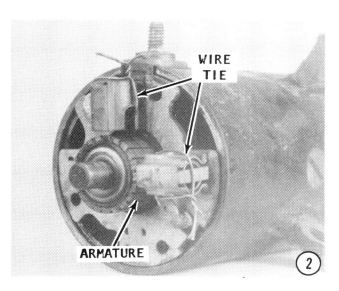

protrusion of the positive plastic terminal.

Clamp the armature in a vise equipped with soft jaws with the drive gear facing **DOWNWARD**. Install the thrust washers onto the end of the armature shaft. Lower the frame assembly down over the armature until the brushes are over the commutator.

3- After the armature is in place, cut and remove the wire wrapped around the brushes to hold them in place. The brushes should then make firm contact with the commutator.

4- Install the end cap onto the end of the starter motor. Observe three small nipples on the inside of the end cap. These nipples **MUST** index with matching dimples in the retaining plate. Align the mark on the side of the end cap with the terminal. Lower the cap onto the frame, and seat it **GENTLY**. **NEVER** tap with a hammer or other tool, because the nipples may not be indexed with the dimples and the tapping may cause damage.

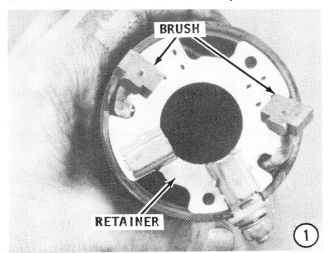

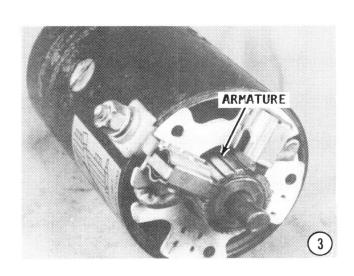

6-28 ELECTRICAL

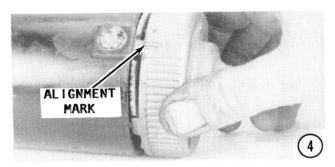

Align the end cap notch or mark, with the mark on the frame, and the upper cap mark with its mark.

Starter Motor Installed on Port Side

Install the thru-bolts through the end cap and frame.

To test the assembled starter motor, see the Section 6-15.

To install the starter motor, proceed directly to Section 6-16.

6-14 AMERICAN BOSCH STARTER MOTOR SERVICE

REMOVAL

1- Before beginning any work on the starter motor, disconnect the positive (+) lead from the battery terminal. Remove the engine hood. Disconnect the red cable at the starter motor terminal. Remove the three attaching bolts securing the starter motor to the engine. One bolt is very near the carburetor and the other two bolts are on the side. Remove the starter motor from the engine.

GOOD NEWS

If the only motor repair necessary is replacement of the brushes, the drive gear does not have to be removed. All starter motors have thru-bolts securing the upper

and lower cap to the field frame assembly. In all cases both caps have some type of mark or boss. These marks are used to properly align the caps with the field frame assembly.

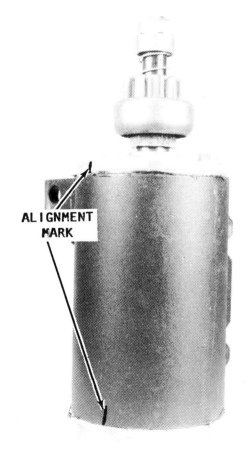

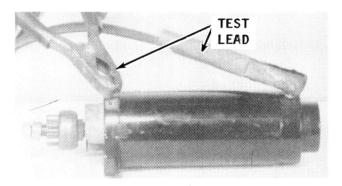

Hookup to test an assembled starter motor.

BOSCH SERVICE

commutator has a short. Clean the grooves between the commutator bars on the armature. Perform the test again. If the hacksaw blade still vibrates during the test, the armature has a short and **MUST** be replaced.

Testing for a Ground

2- Obtain a test lamp or continuity meter. Make contact with one probe lead on the armature core and the other probe lead on the commutator bar. If the lamp lights, or the meter indicates continuity, the armature is grounded and **MUST** be replaced.

DISASSEMBLING

2- Observe the caps and find the identifying mark or boss on each. If the marks are not visible, make an identifying mark prior to removing the thru-bolts as an essential aid during assembling. Remove the thru-bolts from the bracket and the starter motor.

3- Use a small hammer and **CARE-FULLY** tap the lower cap free of the starter motor. On the Bosch starter motor, the brushes are mounted in the end cap. Take care not to lose the four springs and four brushes when the end cap is removed and the brushes pop out.

4- Pull on the armature shaft from the drive gear end and remove it from the field frame assembly.

ARMATURE TESTING

Testing for a Short

1- Position the armature on a growler, then hold a hacksaw blade over the armature core. Turn the growler switch to the **ON** position. Slowly rotate the armature. If the hacksaw blade vibrates, the armature or

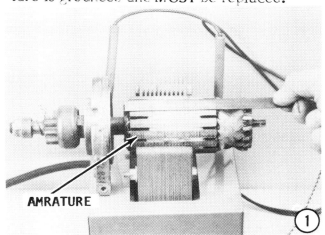

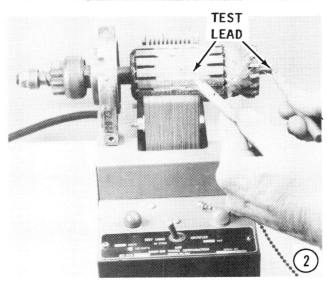

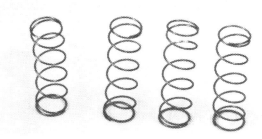

Typical brush springs. If the springs have turned blue in color, they must be replaced.

6-30 ELECTRICAL

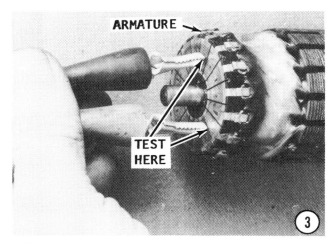

Checking the Commutator Bar

3- Check between or check bar-to-bar as shown in the accompanying illustration. The test light should light, or the meter should indicate continuity. If the commutator fails the test, the armature **MUST** be replaced.

Turning the Commutator

4- True the commutator, if necessary, in a lathe. **NEVER** undercut the mica because the brushes are harder than the insulation. Undercut the insulation between the commutator bars 1/32" to the full width of the insulation and flat at the bottom. A triangular groove is not satisfactory. After the under-cutting work is completed, clean out the slots carefully to remove dirt and copper dust. Sand the commutator lightly with No. "00" sandpaper to remove any burrs left from the undercutting. Check the armature a second time on the growler for possible short circuits.

Positive Brushes

5- The positive brushes can always be identified as the brush with the lead connected to the starter terminal.

Obtain an ohmmeter. Connect one lead of the meter to the positive terminal of the cap and the other lead alternately to the positive brushes. The ohmmeter **MUST** indicate continuity between the brush and the terminal. If the meter indicates any resistance, check the lead to the brush and

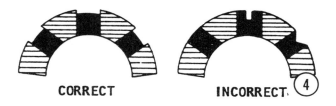

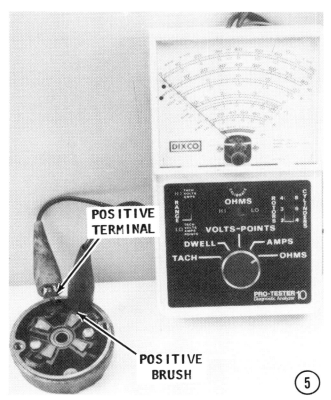

the lead to the positive terminal solder connection. If the connection cannot be repaired, the brush **MUST** be replaced.

Negative Brush

6- The negative brush can always be identified because the lead is connected to the starter cap.

Obtain an ohmmeter. Make contact with one lead to the starter motor frame and the other lead alternately to the negative brushes. If the meter does not indicate continuity, the field coils open and **MUST** be replaced.

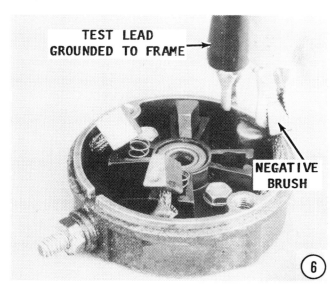

BOSCH SERVICE 6-31

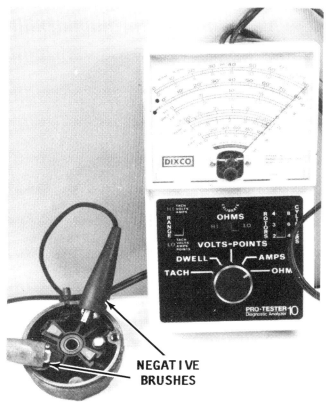

Hookup to check the continuity between the negative brushes in a Bosch starter motor, as explained in the text.

CLEANING AND INSPECTING

Clean the field coils, armature, commutator, armature shaft, brush-end plate and

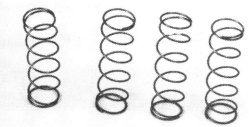

The length and condition of all brush springs must be equal for proper operation. If the springs have been stretched, or appear bluish in color, the complete set should be replaced. Good shop practice calls for replacing a complete set, even if only one spring is damaged.

drive-end housing with a brush or compressed air. Wash all other parts in solvent and blow them dry with compressed air.

Inspect the insulation and the unsoldered connections of the armature windings for breaks or burns.

Perform electrical tests on any suspect-

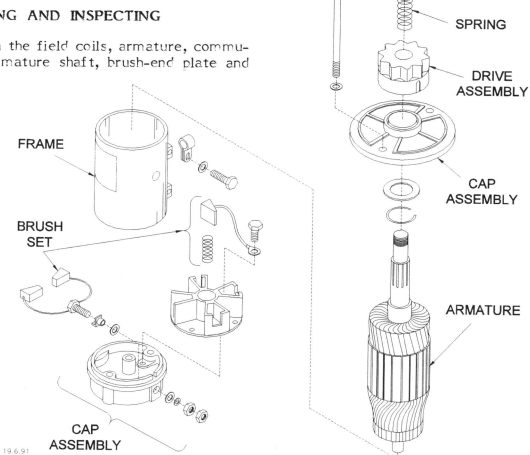

Exploded drawing showing arrangement of principle parts for the Bosch starter motor.

6-32 ELECTRICAL

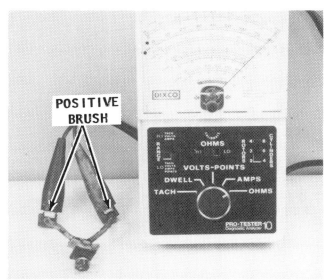

Checking the positive brushes on a Bosch starter motor. Each test lead is connected to a positive brush. Continuity must be indicated between the brushes.

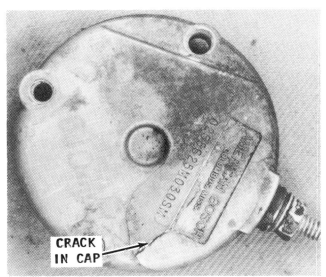

Cracked lower end cap of a Bosch starter motor. This damage was caused by salt water corrosion.

ed defective part, according to the procedures outlined in Section 6-15.

Check the commutator for run-out. Inspect the armature shaft and both bearings for scoring.

Turn the commutator in a lathe if it is out-of-round by more than 0.005".

Check the springs in the brush holder to be sure none are broken. Check the spring tension and replace if the tension is not 32-40 ounces. Check the insulated brush holders for shorts to ground. If the brushes are worn down to 1/4" or less, they must be replaced.

Check the field brush connections and lead insulation. A brush kit and a contact kit are available at your local marine dealer, but all other assemblies must be replaced rather than repaired.

The armature, fields, and brush holders must be checked before assembling the starter motor. See the testing section in this chapter for detailed procedures to test the starter motor.

ASSEMBLING THE AMERICAN BOSCH

Brush Installation

1- Both the positive and negative brushes on a Bosch starter motor are mounted in the lower cap. The positive brushes are attached to the positive terminal and are sold as an assembled set. The negative brushes are attached to the lower cap with a bolt.

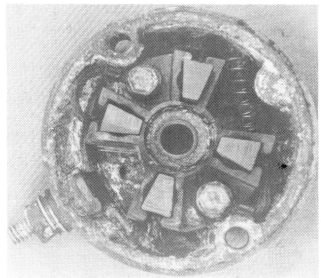

Badly corroded starter motor end cap and brushes. Such damage can only be corrected by purchasing and installing new parts.

To remove the positive brushes, slip the terminal out of the slot in the cap. The negative brushes are removed by simply removing the two bolts attaching the brush lead to the lower cap.

Installation of the new positive brushes is accomplished by sliding the new positive terminal into the slot of the end cap. Install the negative brushes by positioning them in place in the lower cap, and then securing the leads with the attaching bolts.

Assembling a Bosch Using Special Tool

Make a tool as shown in the accompanying illustration to prevent the brushes from being damaged during installation of the commutator end cap. If a special tool is not possible, see the next section, Assembling a Bosch Without a Special Tool.

1- Slide the brush springs into the brush holders, and then install the positive and negative leads. Position the special tool over the cap and brushes, to hold the brushes in place.

2- Clamp the drive gear in a vise equipped with soft jaws and with the drive gear down. Lower the frame assembly over the armature. Align the marks on the frame assembly with the marks on the upper end cap.

3- Position the lower end cap onto the frame assembly. Lower the cap as far as it will go, and then remove the special tool. Now, align the mark on the cap with the mark on the frame, and then install the thru-bolts and tighten them securely.

4- Clamp the starter motor in a vise equipped with soft jaws, as shown, and test its operation.

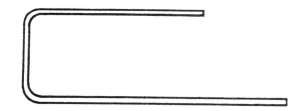

Special tool required to install the end cap on a Bosch starter motor. If this tool is not available, special instruction and illustrations are included later in this section to accomplish the end cap installation work.

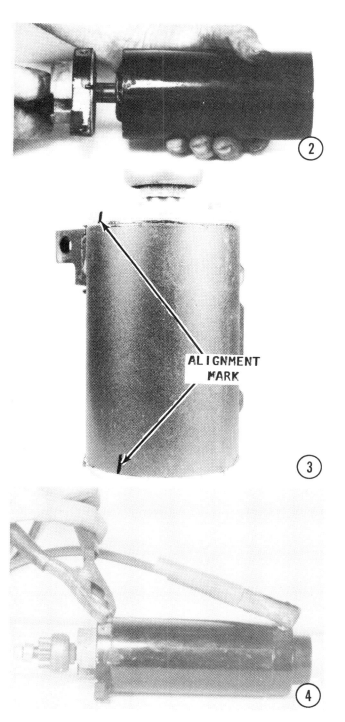

6-34 ELECTRICAL

5- Install the starter motor onto the engine and secure it in place with the clamps and mounting bolts. If the starter motor has the mounting flanges permanently attached, then position the motor in place and start the three bolts to attach the motor to the engine. Now, tighten the three bolts **ALTERNATELY** and **EVENLY** until all bolts are tight. The bolts **MUST** be tightened alternately to prevent binding and possibly bending the bolts.

Assembling a Bosch Starter Without Special Tool

GOOD WORDS

The drive gear assembly must have been removed in order to assemble the starter using this method.

1- Install the brush springs into the brush holder, and then place each brush on top of the springs. Lay the lower cap on the bench with the brush facing up. Pickup the armature and place the commutator on top of the brushes. Lower the armature and at the same time, work each brush into its holder. Continue to lower the armature until the full weight of the armature is on the brushes.

2- Now, very **CAREFULLY** lower the frame assembly down over the armature. **TAKE CARE** because the magnets in the frame assembly will tend to pull against the armature.

3- When the frame makes contact with the lower cap, align the marks on the cap and the frame.

4- Slide the upper cap washer onto the shaft.

5- Install the upper cap with the mark on the cap aligned with the mark on the frame.

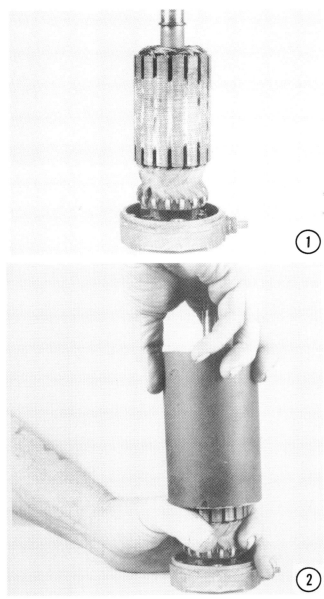

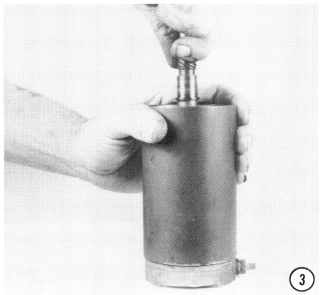

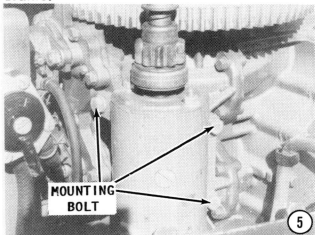

TESTING 6-35

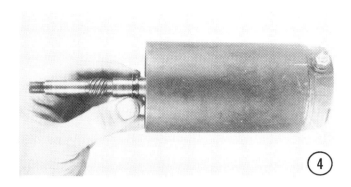

6- Install the thru-bolts and tighten them securely. Clamp the starter motor in a vise equipped with soft jaws. To test the complete starter motor, proceed directly to the next Section.

7- Install the starter motor onto the engine and secure it in place with the mounting bolts, as described in Section 6-16.

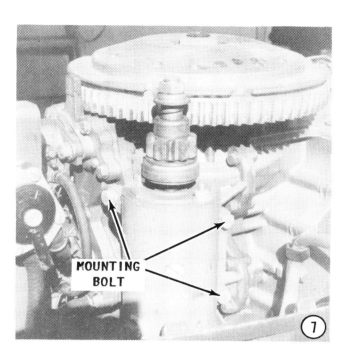

6-15 STARTER MOTOR TESTING

All Engines
Flanges Welded to Motor Housing
This unit is held together by the thru-bolts threaded into the upper cap.

Testing
Clamp the starter motor in a vise equipped with soft jaws, as shown.

SAFETY WORDS
The armature will turn rapidly during this test. Therefore, the starter motor **MUST** be well **SECURED** before making the

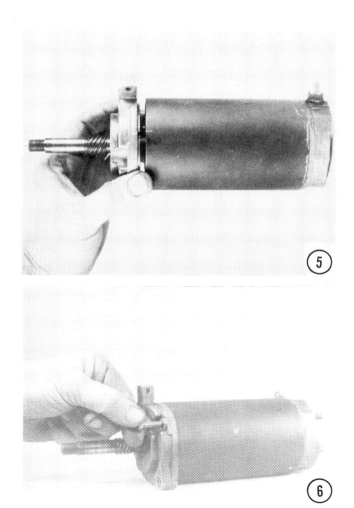

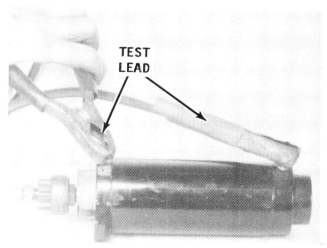

Testing an assembled starter motor, with the positive lead from the battery connected to the positive terminal of the motor and the negative lead connected to the motor ground. The motor must be well secured before making this test.

test to prevent personal **INJURY** or damage to the starter motor.

Firmly connect one end of a heavy-duty jumper wire to the **POSITIVE** terminal of a battery. Firmly connect the other end of the jumper lead to the starter motor terminal.

Connect a second heavy-duty jumper wire to the negative terminal of the battery. Now, **MOMENTARILY** make contact with the other end of the second jumper lead anywhere to the frame of the starter motor. **NEVER** make the momentary contact with the positive lead to the terminal, because any arcing at the terminal may damage the terminal threads and the nut may not take to the damaged threads. The motor should turn rapidly. If the starter motor fails to rotate, the starter motor must be disassembled again and the service work carefully checked. Sorry about that, but some phase of the rebuild task was not performed properly.

6-16 STARTER MOTOR INSTALLATION

All Engines

1- The starter motor has the mounting flanges permanently attached. Therefore, position the motor in place and start the three bolts to attach the motor to the engine. The hole for one bolt is hidden behind the carburetor. Tighten the three bolts **ALTERNATLY** and **EVENLY** until all bolts are tight. The bolts **MUST** be tightened alternately to prevent binding and possibly bending the flanges.

Connect the positive red lead to the starter motor. Connect the electrical lead to the battery. Test the completed work by cranking the engine with the starter motor.

DO NOT, under any circumstances, start the engine unless it is mounted in an adequate size test tank or body of water.

CAUTION: Water must circulate through the lower unit to the engine any time the engine is run to prevent damage to the water pump in the lower unit. Just five seconds without water will damage the water pump.

Starter motor mounted on the port side of a V6 engine.

A Bosch starter motor mounted on the port side of a three-cylinder engine.

7
REMOTE CONTROLS

7-1 INTRODUCTION

Boat accessories are seldom obtained from the original equipment manufacturer.

Shift boxes, steering, bilge pumps, blowers, and other similar equipment may be added after the boat leaves the plant. Because of the wide assortment, styles, and price ranges of such accessories, the distributor, dealer, or customer, has a wide selection from which to draw when outfitting the boat.

This chapter covers two types of shift units, a hydro-electric shift installed on the 65 hp engines, 1972 and a mechanical shift unit installed as standard equipment with the engine on all other units covered in this manual.

Complete procedures for removal, installation, and adjustment, of both shift mechanisms are covered in this chapter. The hydro-electric unit in Section 7-3 and the mechanical shift in Section 7-4.

7-2 SHIFT BOXES
DESCRIPTION

Undoubtedly, the most used accessory on any boat is the shift control box. This unit is a remote-control device for shifting the outboard and at the same time controlling the throttle. Therefore, on the engines covered in this manual, only rarely will the installation have other than an OMC installed unit.

Because the cable length requirements cannot be known for each installation, the shift and throttle cables must be purchased separately. When the cables are purchased, the cable ends will have the end fittings attached and ready for installation at the shift box and the engine. A kit is also available enabling the owner of old-style cables to adapt to the new shift boxes.

OMC equipped boats may be equipped with one of two different type shift boxes, either the hydro-electric type or the manual mechanical shift unit. The new improved hydro-electric unit incorporates an electrical harness, key switch, and a "hot horn".

The shift box installed with Johnson engines has one handle for shifting and throttle control. Another lever, considered a "warmup" lever, is installed at the rear, or at the side of the box. This warmup lever may be adjusted for low and fast idle speeds.

Outboard models are equipped with a cutout switch in the cranking system to open the circuit to the starter solenoid. This arrangement prevents the cranking system from operating unless the throttle is in the proper idle range. Stating it another

A Johnson Hydro-Electric Drive shift box ready for installation.

way, the throttle MUST be in the idle position or the starter system will not operate. The position of the shift lever does not affect the starting motor circuit. In most cases this cutout switch is located in the shift box. All shift box models have a means of advancing the throttle without moving the shift lever into gear. This device is commonly known as the "warm-up" lever and may be adjusted for low and fast idle speeds.

7-3 HYDRO-ELECTRIC SHIFT BOX
65 HP — 1972 ONLY

The hydro-electric shift box is mounted on the starboard side of the boat. The unit contains a shift select lever, an ignition and choke switch, temperature warning horn, and a warmup speed control lever. An electric switch activates the **NEUTRAL, FORWARD,** and **REVERSE** solenoids for shift movement. With no current to the switch, the lower unit will automatically move into the **FORWARD** position. Therefore, current must be present and the unit shifted into **NEUTRAL** and **REVERSE**. Both the neutral and reverse solenoids are activated to shift into **REVERSE**. The box is equipped with a friction adjustment to hold the shift lever and throttle in desired engine speed position. A warmup lever is installed to advance the throttle without the need to move the shift lever. This lever is provided with an adjustment to obtain maximum efficiency during engine startup. The lever has a cutout feature to prevent current from passing to the starter motor circuit once engine speed has reached a predetermined rpm. The front of the shift box contains a blocking diode. This diode is used to block current to the lower unit when the key switch is in the **OFF** position. Detailed testing procedures of the diode are presented in Chapter 8.

TROUBLESHOOTING

The following paragraphs provide a logical sequence of tests, checks, and adjustments, designed to isolate and correct a problem in the Johnson single lever shift box with the warmup lever to the rear and the Evinrude single lever pushbutton shift box operation.

The procedures and suggestions are keyed by number to matching numbered illustrations as an aid in performing the work.

1- Difficult Shift Operation
If difficult shifting is experienced, the problem is usually in the shift box. The friction knob may be adjusted too tightly, the pad that works the friction knob could be excessively worn, or the advance arm on the starboard side of the engine could be inoperative.

2- Amp Draw Test
Turn the ignition switch to the **ON** position and note the ammeter reading. There will be no amp draw when the shift lever is in the **FORWARD** position. Now, operate the shift control lever to the **NEUTRAL** and

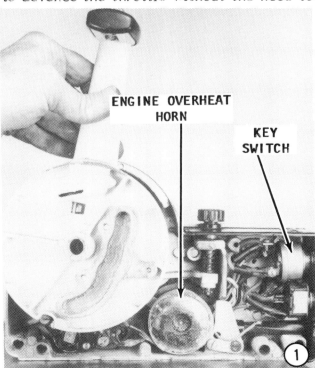

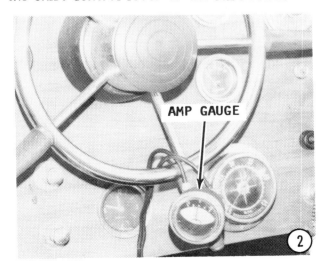

then to the **REVERSE** position. Note how much the ammeter reading increased each time the shift lever was moved. If the reading was more than 2.5 amperes for either shift positions, continue with the following checks. If the boat is not equipped with an ampere gauge, then temporarily disconnect the **GREEN** and **BROWN** (or **RED**) wires from the back side of the key switch and temporarily install an amp gauge for the test. Replace the wires after the test is completed.

Disconnect the shift leads at the rear of the engine. Temporarily lay a piece of cloth or other insulating material under the wires to prevent them from shorting out during the following tests.

Again operate the shift lever and note the current loss. If the current draw is still more than 2.5 amperes, then check for a short in the control box switch or wiring. If the current draw is normal with the leads disconnected from the engine, then check for a short in the gear case coil(s) or wiring. If the coil leads are shorted to each other, both shift coils would be energized, stalling the engine or causing serious damage to the driveshaft.

3- Shift Solenoid Test

Testing the shift solenoids is accomplished by first disconnecting the blue and green wires to the lower unit at the rear of the engine. Next, connect an ohmmeter first to one solenoid lead and ground, and then to the other in the same manner. A reading of more than 5.0 to 6.0 ohms indicates a short in the solenoid or lead. No reading at all indicates an open circuit. If the results of this test indicate a short in the circuit, the lower unit must be disassembled and inspected, see Chapter 8.

4- Testing Shift Switch — Forward

The shift switch is tested by making meter connections on the blue and green wires running to the key switch. These are the wires that were disconnected in Test No. 3 at the engine terminal. (To test at the shift box would involve cutting wires in order to make the meter connections.)

To test the switch for the **FORWARD** position, make contact with one probe of a voltmeter to a good ground on the engine. Move the shift lever to the **FORWARD** position and the key switch to **ON**. Make contact with the other meter probe to first the green and then to the blue wires. The voltmeter should indicate **ZERO** volts. If voltage is indicated, the shift switch is defective and must be replaced.

5- Testing Shift Switch — Reverse

To test the shift switch for **REVERSE**, make contact with one probe of a voltmeter to a good ground on the engine. Move the shift lever to the **REVERSE** position and the key switch to **ON**. Make contact with the other meter probe to the green and blue

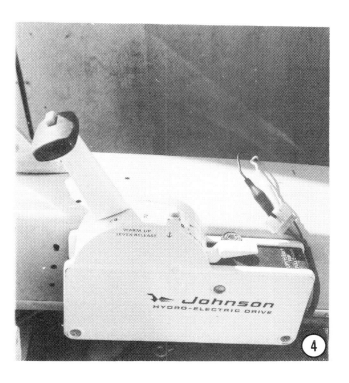

7-4 REMOTE CONTROLS

wires. The meter should indicate 12-volts. If voltage is not present, the shift switch is defective and must be replaced.

6- Testing Shift Switch — Neutral

After the **FORWARD** and **REVERSE** tests have been completed, check for continuity with the shift lever in the **NEUTRAL** position. With one voltmeter probe still connected to a good ground on the engine and the key switch still at the **ON** position, make contact with the other meter probe to the green and blue wires. When the meter probe makes contact with the green wire, the meter should indicate 12-volts. When the probe makes contact with the blue wire, the meter should indicate **ZERO** volts. If these meter readings are not satisfactory, the shift switch is defective and must be replaced.

7- Cranking System Inoperative

If the starter fails to crank the engine, check to be sure the throttle lever is in the idle position. If the throttle is advanced more than 1/4 forward, the cutout switch, attached to the armature plate, will open the circuit to the starter solenoid. If the cranking system fails to operate the starter properly when the throttle lever is in the **IDLE** position, check the 20-ampere fuse between the ignition switch **BAT** terminal and the ammeter **GEN** terminal.

DISASSEMBLING

GOOD WORDS

Before starting work on the shift system,

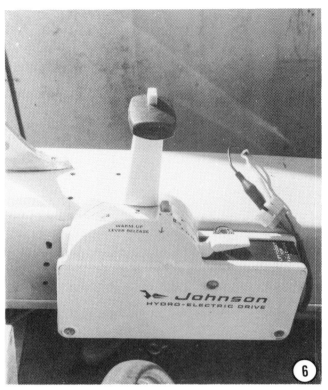

disconnect both battery cables at the battery terminals.

1- Remove the three attaching bolts and then move the box away from the side of the boat. Remove the three screws from the back side of the box. The throttle cable can be removed and serviced without removing it from the boat.

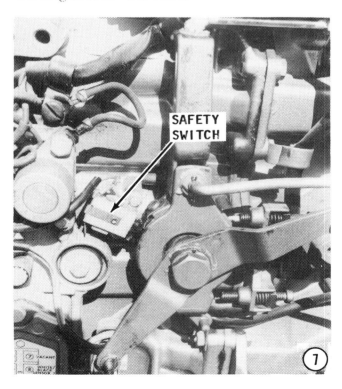

TROUBLESHOOTING 7-5

2— Lift off the front cover, but keep the warmup lever with the back cover. Observe the warmup lever and the throttle cable in the front cover. Remove the screw from the top of the box securing the arm to the warmup lever.

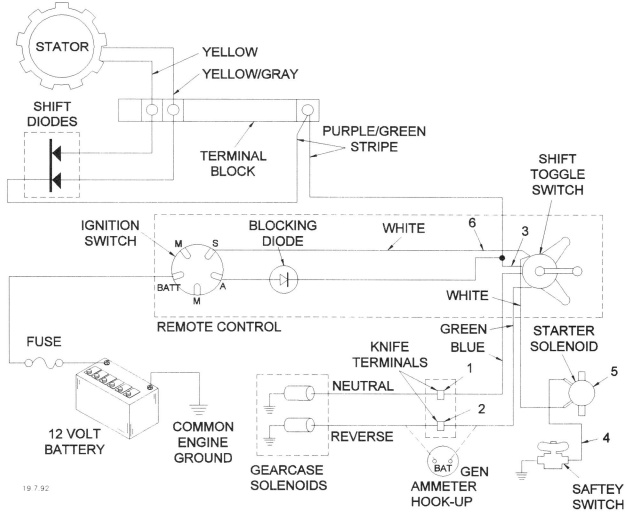

Troubleshooting Schematic Diagram of the Shift Circuit

7-6 REMOTE CONTROLS

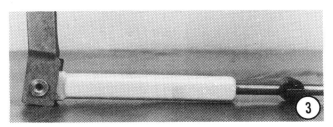

3- Lift the arm and the throttle cable from the shift box housing. **TAKE CARE** not to lose the two caps from the back side of the cable. One cap is located underneath the cable and the other on top of the cable.

4- Remove the Allen screws from the slider on the cable end. Pull the cable free of the slider.

5- Remove the two warmup lever retaining clips.

6- Depress the release for the warmup lever and rotate the release up about halfway. Work the warmup lever upward and at the same time, be careful not to lose the spring and detent located in the warmup lever. Once the warmup lever begins to move upward place a towel or cloth over the lever to prevent the spring and detent from becoming lost. The warmup lever, detent, and spring are all sold as separate items.

SPECIAL WORDS

The shift box housing contains the shift lever, shift switch, warning horn, diode, choke, key switch, and the wiring harness. To service any of these items, simply remove the attaching hardware and cut or disconnect the attaching wires, reference illustration "A", this page. The friction knob on top of the shift box may be removed by pulling it free. After the knob is free the "L" shaped bracket may be removed if the wiring needs service.

Any time the wires from the key switch are removed, the new connection **MUST** be

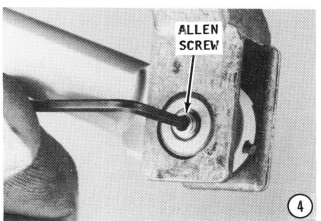

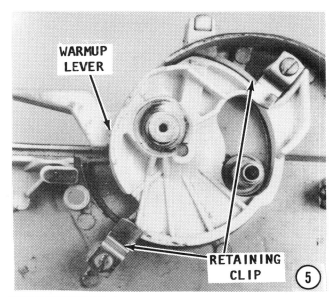

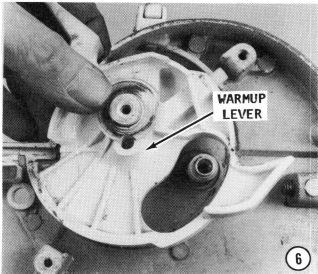

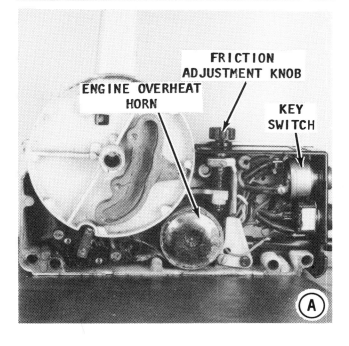

coated with Neoprene sealer as a protection against moisture and corrosion.

CLEANING AND INSPECTING

Clean the box halves thoroughly inside and out with solvent and blow them dry with compressed air. Apply a thin coat of engine oil on all metal parts.

If the throttle cable is not to be replaced, now is an excellent time to lubricate the inner wire.

Throttle Cable Lubrication
Long-Life Cables Only

Not Applicable For Snap-in Type Cables

7- To lubricate the inner wire, remove the casing guide from the cable at both ends. Attach an electric drill to one end of the wire. Momentarily turn the drill on and off to rotate the wire and at the same time allow lubricant to flow into the cable, as shown.

Checking the Shift Diode

8- Disconnect both wires from the shift diode. These wires have a rubber seal where they are routed underneath the horn. Peel back the seal and attach one lead of an ohmmeter to one of the wires, and the other meter lead to the other wire. Observe the ohmmeter for a reading. Reverse the meter leads to the wires. Again observe the meter reading. The meter should indicate continuity when the leads are connected one way to the wires, and no continuity when the leads are connected the other way. If there was a

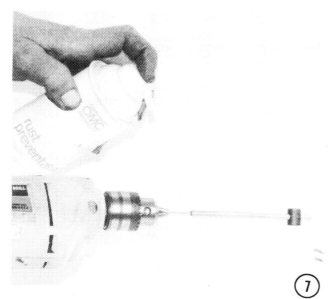

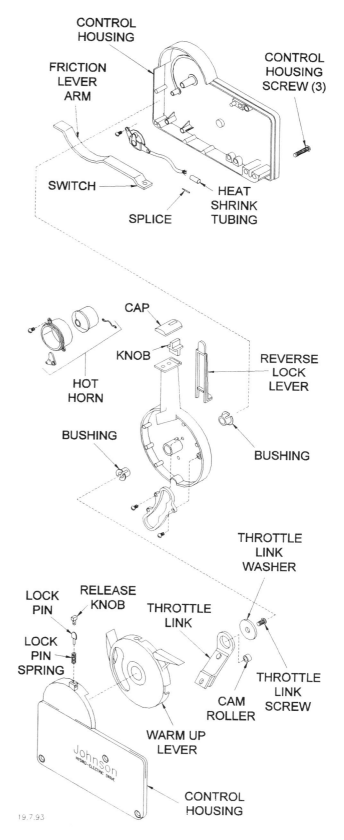

Exploded view of a switch box with principle parts identified.

meter reading when the leads were connected both ways, the diode is defective. If the meter did not indicate continuity when the meter leads were connected either way, the diode is defective. Stating it another way, the meter should indicate continuity when the meter leads are connected to the wires **ONLY** one way.

ASSEMBLING

1- Lubricate the warmup lever with light-weight oil and then just start it into place in the housing. Install the spring and detent, and then push the warmup lever fully into place in the housing. Secure the warmup lever in place with the two retaining clips.

CRITICAL WORDS

Check the end of the throttle cable to determine if the temper has been removed. If the end has a bluish appearance, it has been heated at an earlier date and the temper removed. The temper **MUST** be removed to permit the holding screw to make a crimp in the wire to hold an adjustment. If the wire has not been tempered, heat the end, but not enough to melt the wire.

2- Work the cable end into the slider and through the anchor hole. Adjust the cable end to be flush with the surface of the slider. Tighten the **TOP** holding screw

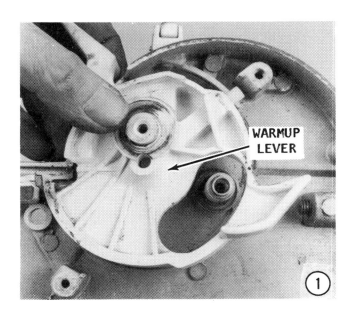

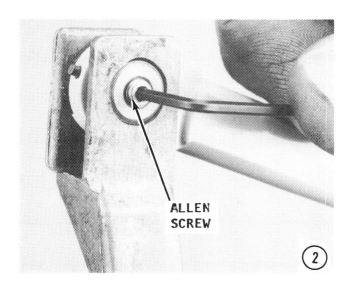

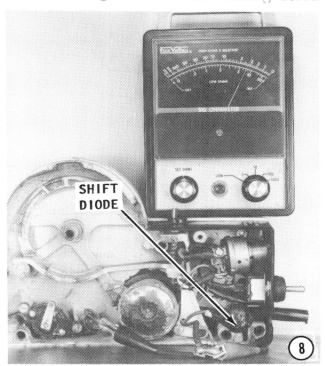

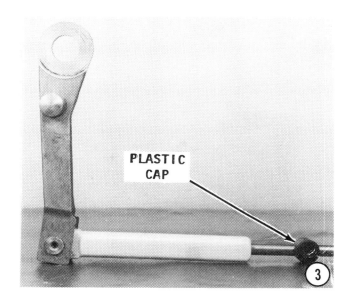

ASSEMBLING 7-9

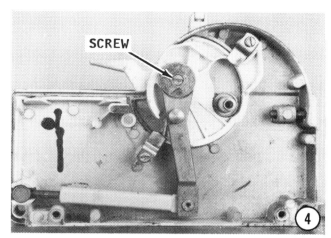

enough to make a definite crimp in the wire, as shown. If this screw is not tightened to make the crimp, the wire will slip during operation and the adjustment will be lost. After the top screw has been fully tightened, bring the other screw up tight against the wire. It is not necessary for this second screw to make a crimp in the wire.

3- Place one of the plastic caps in place in the shift box housing. Lower the slider and cable into position in the housing. Place the other plastic cap on top of the cable.

4- Secure the arm to the warmup lever with the washer (having a concave surface on the inside diameter) and the screw.

5- Release the friction adjustment all the way. Set the shift handle over the protrusion in the inner box.

6- Coat all wiring connections and screws with Neoprene Dip, as a protection against corrosion and a possible short.

7- Bring the two halves of the shift box together. Notice the cut-a-way on the shift handle **MUST** index with a matching protrusion on the warmup lever. Work the two halves completely together and secure them with the three screws in the back. Install the shift box into place in the boat.

7-4 SINGLE-LEVER REMOTE CONTROL SHIFT BOX
ALL ENGINES COVERED IN THIS MANUAL EXCEPT 65 HP — 1972

Description

This unit is a single throttle and gear shift lever shift box with a warmup lever on the side of the box. The unit has the ignition key switch, and choke built-in. Some models may have additional built-in features such as a motor overheat horn, a start-in neutral only switch, and an ignition ON light. A "Kill Safety" switch is now available as an accessory. A cap with a line fits over the switch. One end of the line is attached to the cap and the other end to the helmsman's clothing. With this arrangement, if the helmsman should be accidently thrown overboard, the cord will remove the key from the ignition and the engine will immediately shut down.

The control cables connected between the motor and the remote control lever at the shift box open the throttle after the desired gear is engaged. A throttle friction adjustment is provided to permit the operator to release his grip on the control lever without a change in engine speed.

The warmup lever mounted on the side of the shift box opens the throttle enough to start the engine and to control the fast idle speed for warmup after the engine has started.

On models equipped with the start-in-neutral only switch the starting circuit is completed only when the control lever is in the **NEUTRAL** position. The switch opens the circuit when the control lever is in either **FORWARD** or **REVERSE** position making the ignition key switch inactive.

To start the engine, the control lever moved to the **NEUTRAL** position and the warmup lever to the **START** position. After engine has started and allowed to warm to normal operating temperature, the warmup lever should be moved to the **RUN** position.

A lockout knob is installed under the control lever handle. This knob **MUST** be depressed to permit the control lever to move to the **FORWARD** or to the **REVERSE** position. The control lever handle must be moved approximately 45° of its total travel for complete shift movement in the lower unit. If the control handle is moved past the 45° point, the throttle is advanced and engine speed increases.

A throttle friction adjustment knob installed on the front of the control box can be adjusted to permit the operator to release his grasp on the handle without the throttle "creeping" and thus changing engine speed. The friction knob should be adjusted only to the point to prevent the throttle from "creeping".

TROUBLESHOOTING

The following paragraphs provide a logical sequence of tests, checks, and adjustments, designed to isolate and correct a problem in the shift box operation.

The procedures and suggestions are keyed by number to matching numbered illustrations as an aid in performing the work.

The single-lever remote control shift boxes are fairly simple in construction and operation. Seldom do they fail creating problems requiring service in addition to normal lubrication.

Hard Shifting or Difficult Throttle Advance

Checking Throttle Side

Remove the throttle and shift control at

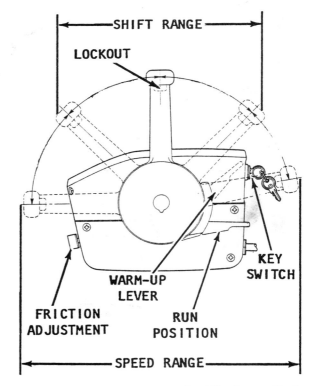

Single lever remote control shift box with key, choke, and "hot horn" incorporated.

the engine. Now, at the shift box, attempt to move the throttle or shift lever. If the lever moves smoothly, without difficulty, the problem is immediately isolated to the engine. The problem may be in the tower shaft between the connector of the throttle and the armature plate. The armature plate may be "frozen", unable to move properly. On the late model units, the "lever advance arm" located on the starboard side of the engine may be "frozen" and require disassembly and lubrication.

If the problem with shifting is at the engine, the first place to check is the area where the shift lever extends through the exhaust housing. The bushing may be worn or corroded. If the bushing requires replacement, the engine powerhead must be removed. Another cause of hard shifting is water entering the lower unit. In this case the lower unit must be disassembled, see Chapter 8.

If hard shifting is still encountered at the shift box when the controls are disconnected from the engine, one of two areas may be causing the problem: the cables may be corroded and require replacment; or, the teeth on the plastic shift lever assembly in the shift box may be worn or broken. This is a common area for problems.

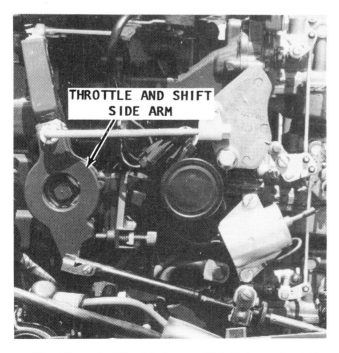

Throttle and shift side lever. This linkage should be checked for corrosion and freedom of movement.

Unable to Obtain Full Shift Movement or Full Throttle "Long-Life" Cables Only

Normally, this type of problem is the result of improper shift box installation. This area includes connection of the shift and throttle cables in the shift box. If the stainless steel inner wire was not heated and the clamp did not hold the inner cable (wire), the wire could slip inside the sleeve and the cable would be shortened. Therefore, if it is not possible to obtain full shift or full throttle, the shift box must be removed, opened, and checked for proper installation work. The inner wire could also slip at the engine end of the control, but problems at that end are very rare. Usually if improper installation work has been done at the engine end, the ability to shift at all is lost, or the throttle cannot be actuated. If the lower unit has previously been removed, the shift rod may not have been adjusted properly.

DISASSEMBLING

Throttle Cable and Shift Cable

Preparation Tasks: Disconnect the leads at the battery terminals and disconnect the spark plug wires at the plugs, as a safety precaution to prevent possible personal injury during the work. A professional mechanic can usually service the cables, including replacement, without removing the box from the side of the boat. However, the job is made much easier if the box is removed and laid on its side on the boat seat.

1- Use the proper size Allen wrench and loosen the screw by backing it out about three complete turns. Set a small center

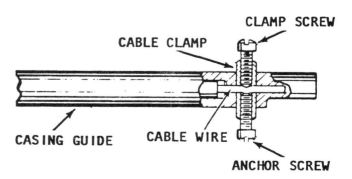

Cut-a-way view of the cable passing through the casing guide. Note the crimp made by the clamp screw. The anchor screw is brought up just tight as described in the text.

REMOTE CONTROLS

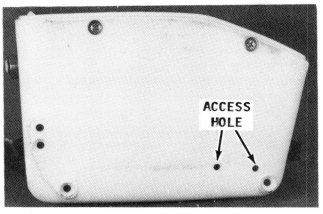

Backside of the shift box showing the two access holes through which an Allen wrench may be passed to tighten the inner cable retaining screws.

punch in the center of the screw, and then strike the punch a quick hard blow. The shock will break the handle loose from the splined shaft. Remove the Allen screw, and then the handle.

It is not necessary to remove the shift handle, however the work will progress easier if the handle is removed and out of the way.

2- Remove the three Phillips screws on the bottom, the side, or both ends of the panel on the lower section of the shift box. Observe inside the box and notice the attaching screws securing the box to the side of the boat. Remove the screws and lay the shift box on the boat seat.

3- Lift the electrical cable and grommet up out of the slot at the rear of the shift box. Loosen the anchor screws on the end of the casing guides. On the back side of the shift box two holes are provided to permit inserting an Allen wrench to hold the underneath screw while the upper outside Allen screws are removed. After the screws have been loosened, pull the shift cable out of the cable casing.

4- To remove the cables at the engine end: Remove the self-locking nuts securing the cables to the engine. Remove the clip on the trunion. Slip the end of the cable out of the engine retainer, and then remove the cable from the boat. To remove the guide

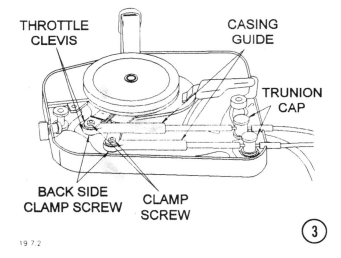

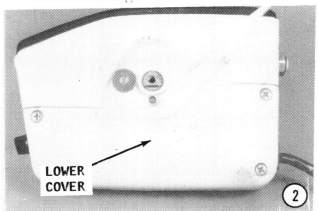

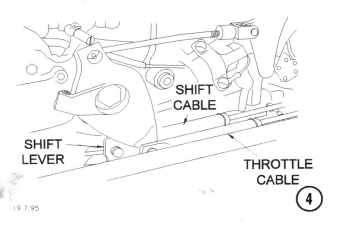

casings at the engine end: Loosen the Allen screws on both sides and pull the casing free.

Throttle Cable Lubrication

If the throttle cable is **NOT** to be replaced, now is an excellent time to lubricate the inner wire.

5- To lubricate the inner wire, remove the casing guide from the cable at both ends. Attach an electric drill to one end of the wire. Momentarily turn the drill on and off to rotate the wire and at the same time allow lubricant to flow into the cable, as shown.

ASSEMBLING

Shift or Throttle Cable Into Shift Box

The following procedures are to be followed to install either the shift cable or the throttle cable.

CRITICAL WORDS

Check the end of the cable to determine if the temper has been removed. If the end has a bluish appearance, it has been heated at an earlier date and the temper removed. The temper **MUST** be removed to permit the holding screw to make a crimp in the wire to hold an adjustment. If the wire has not been tempered, heat the end, but not enough to melt the wire.

It is very easy to shear the wire by applying **EXCESSIVE** force when tightening the screw to make the crimp. Therefore, play it cool. Tighten the screw; make one more complete turn to make the crimp; call it good; and then bring the other screw up just tight against the wire.

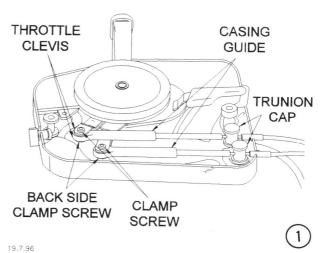

1- Place the shift control handle in the **NEUTRAL** position and the warmup lever in the **START** position. Coat the cable sleeve with anti-corrosive lubricant, and then slide the casing guide over the cable end. Thread one screw into the control cable clamp. Insert the casing guide into the shift control clevis and align the hole in the casing guide with the hole in the clevis. Lubricate and insert control wire clamp in the clevis hole with the screw toward the back of the control lever.

Align the wire holes in the clamp with wire holes in the casing guide. Feed the control wire through the clamp until the wire is flush to within 1/32" (0.8 mm) recessed with the end of the casing. Secure the cable wire in the casing guide. The control wire is held in place with two screws -- the clamp screw already in place and the anchor screw. The wire **MUST** be crimped as described in "Critical Words" at the be-

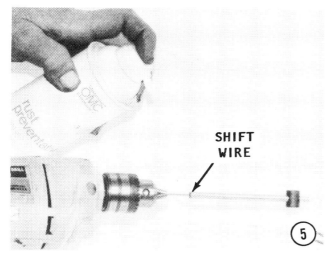

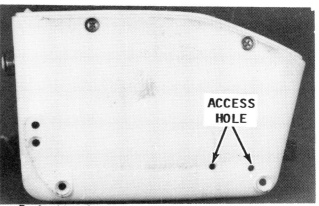

Back side of the shift box showing the two access holes through which an Allen wrench may be passed to tighten the inner cable retaining screws.

7-14 REMOTE CONTROLS

ginning of these Assembling procedures to hold the adjustment. This is accomplished by reaching through the access hole in the rear of the control box with a 3/32" Allen wrench and tightening the screw until it is up just snug against the wire. Now, tighten the screw **ONLY** one complete turn more to make the crimp. Install the second screw, from the front and bring it up just tight against the wire. The wire will now be held securely in the casing guide.

SPECIAL NOTE

The previous procedure is complete to install either the throttle or the shift cable. If both cables are to be replaced, the complete procedure must be followed again to install the second cable properly.

2- Snap the nylon trunnion caps onto the cable trunnion, and then position the trunnion in the remote control. If the cable has a spherical trunnion, use the anchor blocks included with the new cable instead of the nylon trunnion caps furnished with the remote control. The nylon caps may be discarded.

3- Insert the electric cable grommet into the remote control.

4- Mount the shift control box in the boat and secure it in place with the attaching screws. Install the shift control handle. Install the access cover with the throttle cable positioned in the machined recess in the cover and the nylon trunnion caps in place. Clamp the control cables to the boat along the run to the engine.

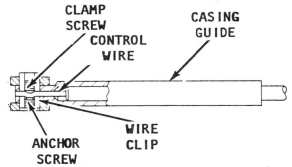

Cut-a-way section of the throttle or shift cable connection at the shift box. Note the crimp in the wire made by the clamp screws. The anchor screw is brought up just tight as explained in the text.

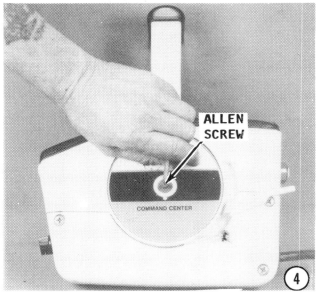

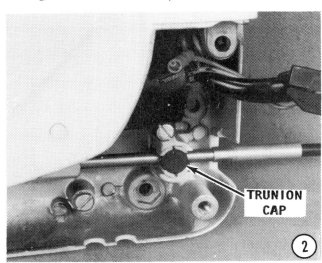

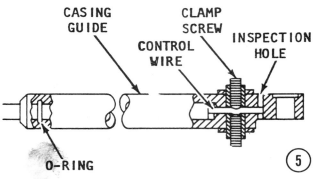

Shift or Throttle Cable at the Engine

5- Check to be sure the shift control lever is in the **NEUTRAL** position and the warmup lever is in the **RUN** position. Insert the clamp in the casing guide. Start the Allen screws into the clamp. Work the casing guide down over the cable (shift or throttle cable), until the wire protrudes out into the inspection hole of the guide.

CRITICAL WORD

The flat side of the casing guide **MUST** face toward the engine. This flat side is necessary to allow the guide to move as the throttle or shift lever is operated.

Tighten one of the Allen screws until it makes contact with the wire, and then give it **ONLY** one more complete turn to make the crimp in the wire. Tighten the other Allen screw just snug against the wire.

SPECIAL NOTE

The previous procedure is complete to install either the throttle or the shift cable at the engine. If both cables are to be replaced, the complete procedure must be followed again to install the second cable properly.

6- Place the shift or throttle cable onto the shift or throttle lever studs. Secure the cables in place with the washers and locknuts.

7- Move the throttle lever on the engine until the idle stop screw makes contact with the stop. Pull firmly on the throttle casing guide and trunnion nut to remove any backlash in the cable run and at the remote control at the shift box. If this backlash is not removed, the engine may not return to a consistent idle speed.

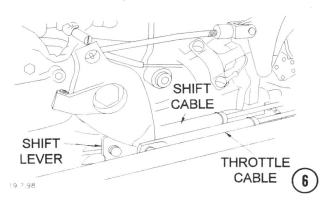

8- Adjust the trunnion adjustment nut on the throttle cable until the cable will slip into the trunnion. Install the throttle or shift cable into the trunnion and install the retaining cover over the top of the trunnion and tighten the screw in the center of the trunnion.

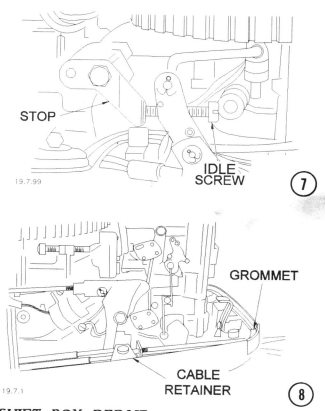

SHIFT BOX REPAIR

SAFETY WORD

Always disconnect the electrical leads at the battery terminals to prevent possible personal injury during the work.

1- Remove the throttle and shift handle by first removing the Allen screw in the center of the handle.

2- Remove the three Phillips screws on the lower side of the shift box, and then remove the cover.

3- Remove the screws from the top of the shift box, and then the arm rest. Remove the two screws from the back side of the shift box securing the upper panel to the outside of the shift box, and then lift off the panel.

7-16 REMOTE CONTROLS

4— Remove the screws from the inside of the shift box securing the shift box to the boat. Use an Allen wrench and working from the back side and another Allen wrench from the front side, remove the Allen screws in the control wire on the end of the casing guide. Remove these screws from the shift and the throttle wires.

GOOD WORDS

As the panel is lifted, notice how the arm and the mechanism for the throttle are mounted on one side of the panel. Notice the key switch, overheat horn, cam, and start-in-neutral switch installed on the other side of the panel. Also observe the spring and ball bearing under the plate installed under the shift cam.

5— Lift the cam assembly from the panel. **TAKE CARE** not to lose the spring and ball bearing installed under the plate.

6— Remove the countersunk screw, flat washer, shift lever, and bushing from the housing. Remove the screw and cover plate containing the spring and ball bearing on the right side of the shift box.

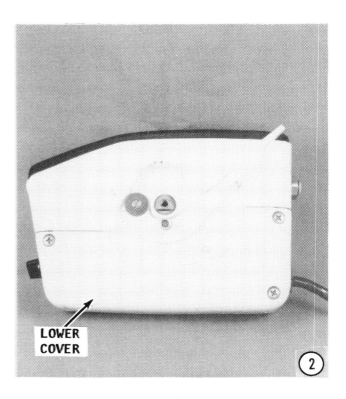

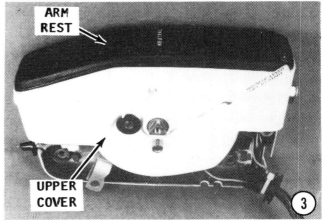

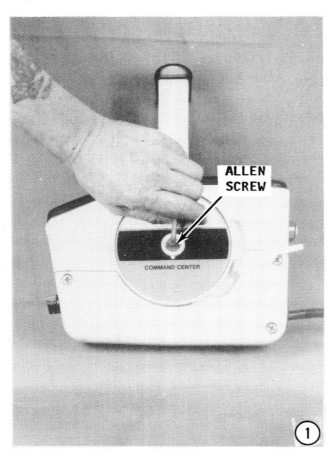

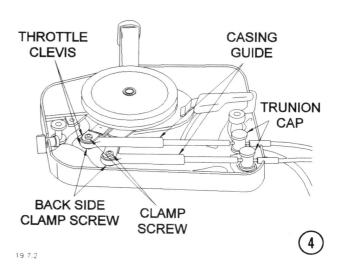

SHIFT BOX REPAIR 7-17

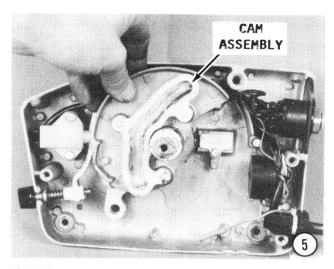

CLEANING AND INSPECTING

Disassemble and clean mechanical parts in solvent, and then blow them dry with compressed air. **NEVER** dip electrical parts in solvent. Check wiring and electrical parts for continuity with a test light or an ohmmeter. Faulty electrical parts **MUST** be replaced.

Inspect mechanical parts for wear, cracks, or other damage. Questionable parts should be replaced to ensure satisfactory service.

Pay special attention to the shift lever teeth. The teeth are made of a hard plastic material. Worn teeth will result in hard shifting.

ASSEMBLING

GOOD WORDS

During the assembling work, take time

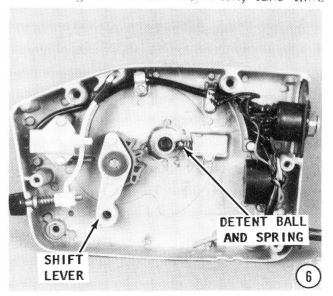

Worn (left) and new (right) shift levers. Note the worn teeth indicating this lever is no longer fit for service.

to coat the friction areas of mechanical moving parts with OMC Multi-purpose grease.

1- Slide the bushing onto the shift lever post in the housing, and then install the lever with the countersunk side of the washer facing **UP**. Tighten the screw to the specifications given in the Appendix. Before installing the shift lever, the detent spring and ball must be removed from the retainer.

2- Lower the shift lever down over the shift cam with the center tooth of the shift

7-18 REMOTE CONTROLS

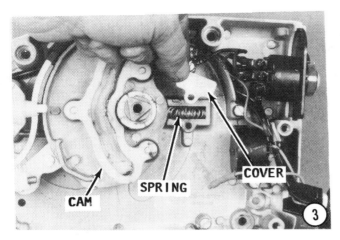

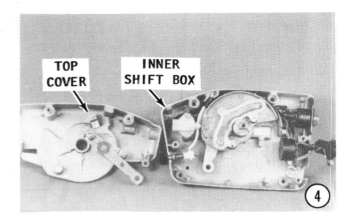

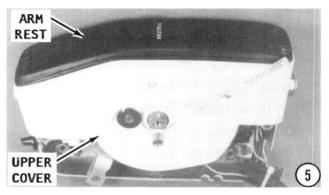

lever indexes with the center tooth on the cam.

3- Install the ball and detent spring into the recess, and then install the cover over the spring.

4- Place the upper housing onto the control box with the lever cam follower seats in shift lever cam channel. Tighten the screws on the back side of the shift box.

5- Install the arm rest to the top of the shift box and secure it in place with the retaining screws.

6- Slide the throttle and shift cables into the shift box and attach the casing guides in the shift and throttle levers. Tighten the Allen screws from the rear and side of the box. If difficulty is encountered during installation of the cables, see the more detailed instructions under Shift Cable Installation earlier in this section.

Install the shift box to the side of the boat. Check to be sure the cables are in their trunnions and the wiring harness is in the recess. Install the lower cover, and then install the shift handle and secure it with the Allen screw.

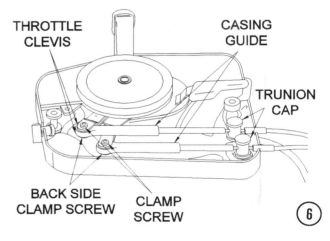

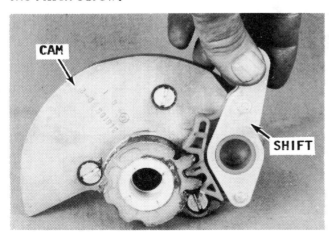

Shift lever and shift cam illustrating how the teeth of each indexes with the teeth of the other.

8
LOWER UNIT

8-1 DESCRIPTION

The lower unit is considered as that part of the outboard below the exhaust housing. The unit contains the propeller shaft, the driven and pinion gears, the driveshaft from the powerhead and the water pump. On models equipped with shifting capabilities, the forward and reverse gears, together with the clutch, shift assembly, and related linkage, are all housed within the lower unit.

The lower unit is removed by one of two methods depending on the model year and the engine horsepower.

1- Green and blue shift wires are disconnected on the port side of the engine.

2- The shift rod is disconnected at the linkage under and to the rear of the bottom carburetor.

The engine and model year is given in each section heading. Therefore, the Table of Contents may be used to determine which set of procedures to follow for the engine being serviced.

CHAPTER COVERAGE

Three different lower units are covered in this chapter with separate sections for each, as indicated:

Section 8-4 -- lower unit has propeller exhaust with electric shift. Two solenoids are used to affect the shift. One solenoid is used for neutral, and both solenoids are activated for the shift into reverse gear.

Section 8-5 -- lower unit has propeller exhaust and mechanical shift with a hydraulic assist pump.

Section 8-6 -- lower unit has propeller exhaust and mechanical shift with a sliding clutch dog. Movement of the clutch dog is mechanical utilizing a shift cradle and shift lever.

Each section is complete with detailed procedures. The lower units covered in Sections 8-4, 8-5, and 8-6 each contain their own troubleshooting procedures.

Check the Table of Contents and follow the procedures in the given section for the unit being serviced.

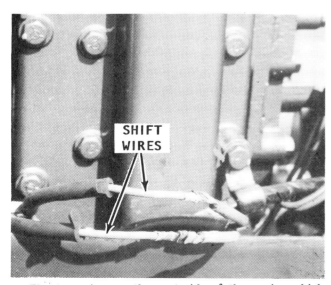

The two wires on the port side of the engine which must be disconnected before the lower unit is removed.

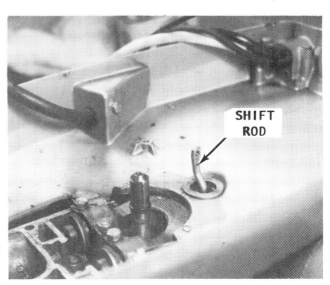

Shift rod extending up through the exhaust housing. The disconnection is made under the carburetor.

8-2 LOWER UNIT

Water Pump

Water pump service work is by far the most common reason for removal of the lower unit. Each lower unit service section contains complete detailed procedures to rebuild the water pump. The instructions given to prepare for the water pump work must be performed as listed. However, once the pump is ready for installation, if no other work is to be performed on the lower unit, the reader may jump to the pump assembling procedures and proceed with installation of the water pump.

Each section is presented with complete detailed instructions for removal, disassembly, cleaning and inspecting, assembling, adjusting, and installation of only one type unit.

ILLUSTRATIONS

Because this chapter covers such a wide range of models over an extended period of time, the illustrations included with the procedural steps are those of the most popular lower units. In some cases, the unit being serviced may not appear to be identical with the unit illustrated. However, the step-by-step work sequence will be valid in all cases. If there is a special procedure for a unique lower unit, the differences will be clearly indicated in the step.

SPECIAL WORDS

All threaded parts are right-hand unless otherwise indicated.

If there is any water in the lower unit or metal particles are discovered in the gear lubricant, the lower unit should be completely disassembled, cleaned, and inspected.

Actually, problems in the lower unit can be classified into three broad areas:

1- Lack of proper lubrication in the lower unit. Most often this is caused by failure of the operator to check the gear oil level frequently and to add lubricant when required.

2- A faulty seal allowing water to enter the lower unit. Water allowed to remain in the lower unit over a period of non-use time will separate from the oil and can be destructive.

3- Excessive clutch dog and clutch ear wear on the forward and reverse gears. This condition is caused by excessive wear in the bellcrank under the powerhead. A worn bellcrank will result in sloppy shifting of the lower unit and cause the clutch components to wear and develop shifting problems. Improper shifting techniques at the shift box will also result in excessive wear to the clutch dog and clutch ears of the forward and reverse gears.

Time will also take its toll. Continued service over a long period of time will cause parts to wear and require replacement.

8-2 PROPELLER SERVICE

EXHAUST PROPELLER

Propellers with the exhaust passing through the hub **MUST** be removed more frequently than the standard propeller. Removal after each weekend use or outing is not considered excessive. These propellers do not have a shear pin. The shaft and propeller have splines which **MUST** be coated with an anit-corrosion lubricant prior to installation as an aid to removal the next time the propeller is pulled. Even with the lubricant applied to the shaft splines, the propeller may be difficult to remove.

The propeller with the exhaust hub is more expensive than the standard propeller and therefore, the cost of rebuilding the unit, if the hub is damaged, is justified.

A replaceable diffuser ring on the backside of the propeller dispurses the exhaust away from the propeller blades as the boat moves through the water. If the ring becomes broken or damaged "ventilation"

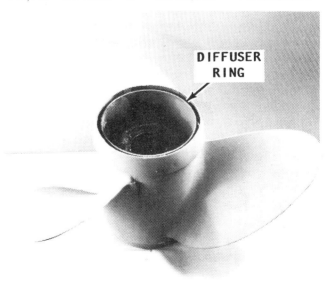

Propeller with exhaust hub. The defuser ring is clearly visible.

EXHAUST PROPELLER 8-3

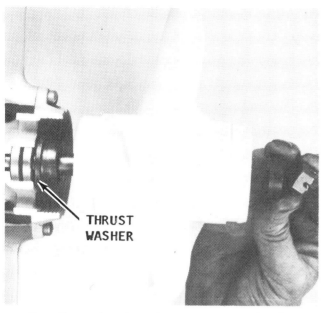

Propeller exhaust parts arrangement showing the thrust washer, propeller, splined washer, and propeller nut.

would be created pulling the exhaust gases back into the negative pressure area behind the propeller. This condition would create considerable air bubbles and reduce the effectiveness of the propeller.

PROPELLER WITH EXHAUST — REMOVAL

First, disconnect the high tension leads to the spark plugs to prevent accidental engine start. Next, pull the cotter pin from the propeller nut. Wedge a piece of wood between one of the propeller blades and the cavitation plate to prevent the propeller from rotating. Back off the castillated propeller nut and remove the splined wash-

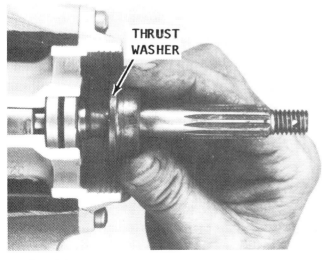

Installing the thrust washer onto the propeller shaft of a propeller exhaust unit.

Applying gasket sealer to the propeller shaft splines to prevent the propeller from becoming "frozen" to the shaft.

er. Pull the propeller straight off the shaft. It may be necessary to carefully tap on the front side of the propeller with a soft headed mallet to jar it loose. Remove the thrust washer from the propeller shaft.

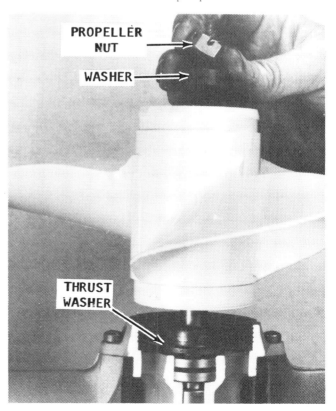

Installation of a propeller exhaust propeller with principle parts identified.

8-4 LOWER UNIT

"Frozen" Propeller

If the propeller appears to be "frozen" to the shaft, see Section 8-7 for special removal instructions. The thrust washer does not have to be removed unless it appears damaged.

EXHAUST PROPELLER INSTALLATION

Slide the thrust washer onto the propeller shaft. Coat the propeller shaft with Perfect Seal No. 4, Triple Guard Grease, or similar good grade of lubricant to prevent the propeller from becoming "frozen" to the shaft. Slide the propeller onto the shaft with the splines in the propeller indexing with the splines on the shaft. Slide the splined washer onto the shaft. Thread the castillated nut onto the shaft. Jamb a piece of board between one of the propeller blades and the cavitation plate to prevent the propeller from turning. Tighten the propeller nut securely and then a bit more to align the hole through the nut with the hole through the propeller shaft. Install the cotter pin through the nut and propeller shaft.

8-3 LOWER UNIT LUBRICATION

DRAINING LOWER UNIT

Position a suitable container under the lower unit, and then remove the **FILL** screw and the **VENT** screw.

Allow the gear lubricant to drain into the container. As the lubricant drains, catch some with your fingers, from time-to-

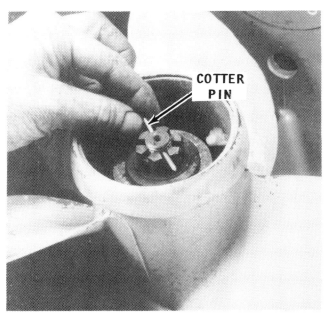

Installing the cotter pin through the castellated nut on a propeller exhaust unit.

time, and rub it between your thumb and finger to determine if any metal particles are present. If metal is detected in the lubricant, the unit must be completely disassembled, inspected, and the damaged parts replaced, illustration **"A"**.

Check the color of the lubricant as it drains. A whitish or creamy color indicates the presence of water in the lubricant. Check the drain pan for signs of water separation from the lubricant. The presence of any water in the gear lubricant is **BAD NEWS**. The unit must be completely disassembled, inspected, the cause of the problem determined, and then corrected.

Tightening the nut on a propeller exhaust unit. The cotter pin is installed after the nut is secure.

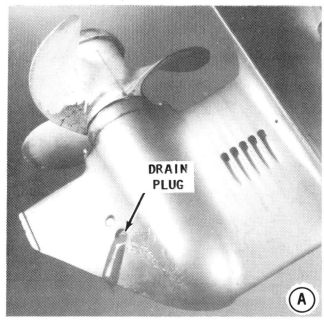

FILLING LOWER UNIT

Fill the lower unit with lubricant. Insert the lubricant tube into the bottom opening, and then fill the unit until lubricant is visible at the vent hole. The 2 hp model does not have a vent screw. Therefore, this unit must be laid in a horizontal position for filling and time taken to allow the lubricant to work into the lower unit cavity by raising the skeg slightly from time-to-time. Install the vent plug. Remove the gear lubricant tube and install the drain/fill plug.

After the lower plug has been installed, remove the vent plug again and using a squirt-type oil can, add lubricant through this vent hole. A squirt-type oil can must be used to allow the trapped air in the lower unit to escape at the same time the final lubricant is added. Once the unit is completely full, install and tighten the vent plug.

8-4 ELECTRIC SHIFT
TWO SOLENOIDS
65 HP — 1972

DESCRIPTION

The lower unit covered in this section is a three shift position, hydraulic activated, solenoid controlled, propeller exhaust unit.

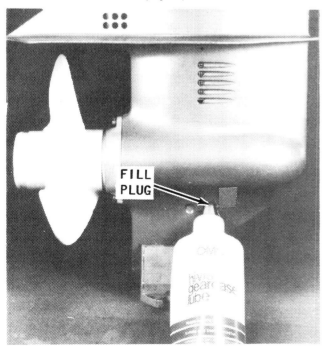

Filling a lower unit with OMC Gearcase Lubricant. Notice the vent plug has been removed to allow air to escape, as the unit fills with lubricant.

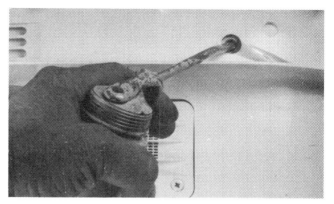

"Topping off" the lower unit using a squirt-type oil can through the vent hole, as described in the text.

A hydraulic pump mounted in the forward portion of the lower unit provides the force required to shift the unit. Two solenoids installed in the lower unit, above the pump, control and operate the pump valve. The pump valve directs the hydraulic force to place the clutch dog in the desired position for neutral, forward, or reverse gear position. One solenoid controls the valve for the neutral position. Both solenoids control the valve for the reverse position. When

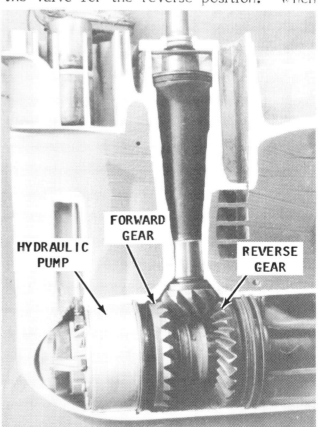

Cut-a-way view of a complete lower unit prior to disassembling. This type illustration is most helpful in gaining an appreciation of the internal parts and their relationship to one another.

neither solenoid is activated, the unit is at rest in the forward gear position.

In simple terms, something must be done (a solenoid activated and hydraulic pressure applied) to move the unit into neutral or reverse gear position. If no action is taken (shift mechanism at rest) the unit is in the forward gear position.

A full 12-volts is required to activate the solenoids. This means shifting is not possible if the battery should become low for any number of reasons. A potentially dangerous condition could exist because the unit could not be taken out of forward gear. Therefore, the only way to stop forward boat movement would be to shut the engine down. A low battery would also mean the electric starter motor would fail to crank the engine properly for engine start. Such a condition would require hand starting in an emergency, if a second battery were not available.

The lower unit houses the driveshaft and pinion gear, the forward and reverse driven gears, the propeller shaft, clutch dog, hydraulic pump, two solenoids, and the necessary shims, bearings, and associated parts to make it all work properly.

ONE MORE WORD

A useful piece of information to remember is that the green wire carries current for the neutral position; and the green and blue wires carry current for the reverse gear operation.

TROUBLESHOOTING

Preliminary Checks

Whenever the lower unit fails to shift properly the first place to check is the condition of the battery. Determine if the battery contains a full charge. Check the condition of the battery terminals, the battery leads to the engine, and the electrical connections.

The second area to check is the quantity and quality of the lubricant in the lower unit. If the lubricant level is low, contaminated with water, or is broken down because of overuse, the shift mechanism may be affected. Water in the lower unit is **VERY BAD NEWS** for a number of reasons, particularly when the lower unit contains electrical or hydraulic components. Electrical parts short out and hydraulic units will not function with water in the system.

BEFORE making any tests, remove the propeller, see Section 8-2. Check the propeller carefully to determine if the hub has been slipping and giving a false indication the unit is not in gear. If there is any doubt, the propeller should be taken to a shop properly equipped for testing, before the time and expense of disassembling the lower unit is undertaken. The expense of the propeller testing and possible rebuild is justified.

The following troubleshooting procedures are presented on the assumption the battery, including its connections, the lower unit lubricant, and the propeller have all been checked and found to be satisfactory.

Lower Unit Locked

Determine if the problem is in the powerhead or in the lower unit. Attempt to rotate the flywheel. If the flywheel can be moved even slightly in either direction, the problem is most likely in the lower unit. If it is not possible to rotate the flywheel, the problem is a "frozen" powerhead. To absolutely verify the powerhead is "frozen", separate the lower unit from the exhaust housing and then again attempt to rotate the flywheel. If the attempt is successful, the problem is definitely in the lower unit. If the attempt to rotate the flywheel, with the lower unit removed, still fails, a "frozen" powerhead is verified.

A "frozen" powerhead with burned pistons.

Unit Fails to Shift
Neutral, Forward, or Reverse
Disconnect the green and blue electrical wires from the lower unit at the engine.

Voltmeter Tests
Separate the green and blue wires at the engine by first sliding the sleeve back, and then making the disconnect. Connect one lead of the voltmeter to the green wire to the dash, and the other lead to a good ground on the engine. Turn the ignition key to the **ON** position. With the shift box handle in the forward position, the voltmeter should indicate **NO** voltage. Move the test lead from the green wire to the blue wire to the dash. With the shift lever still in the forward position, the voltmeter should indicate **NO** voltage.

Move the shift lever to the **NEUTRAL** position. With the voltmeter still connected to the blue wire, **NO** voltage should be indicated. Move the test lead to the green wire. Voltage **SHOULD** be indicated.

Move the shift lever to the **REVERSE** position. Voltage **SHOULD** be indicated on the green wire **AND** on the blue wire, reference illustration "A".

If the desired results are not obtained on any of these tests, the problem is in the shift box switch or the wiring under the dashboard. See Chapter 7.

Ohmmeter Tests
Set the ohmmeter to the low scale. Connect one lead to the green wire to the lower unit, and the other lead to a good ground. The meter should indicate 5 to 7 ohms. Connect the meter to the blue wire to the lower unit and ground. The meter should again indicate from 5 to 7 ohms, reference illustration "B".

BAD NEWS
If the unit fails the voltmeter and ohmmeter tests just outlined, the only course of action is to disassemble the lower unit to determine and correct the problem.

LOWER UNIT SERVICE

Propeller Removal
If the propeller was not removed, as directed for the troubleshooting, remove it now, according to the procedures outlined in Section 8-2.

Draining the Lower Unit
Drain the lower unit of lubricant, see Section 8-3.

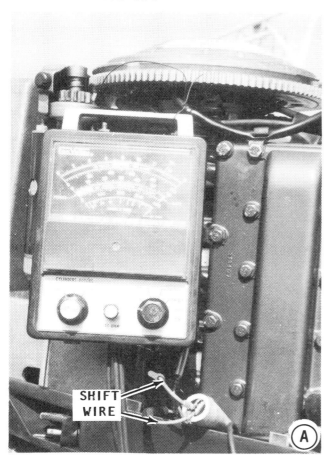

GOOD WORDS

If water is discovered in the lower unit and the propeller shaft seal is damaged and requires replacement, the lower unit does **NOT** have to be removed in order to accomplish the work.

The bearing carrier can be removed and the seal replaced without disassembling the lower unit. **HOWEVER** such a procedure is not considered good shop practice, but merely a quick-fix. If water has entered the lower unit, the unit should be disassembled and a detailed check made to determine if any other seals, bearings, bearing races, O-rings or other parts have been rendered unfit for further service by the water.

LOWER UNIT REMOVAL

1- Disconnect and ground the spark plug wires. Slide back the insulators on the shift wires at the engine. Disconnect the blue wire from the blue and the green wire from the green, at the engine.

HELPFUL WORD

Obtain a piece of electrical wire, about 5 ft. long. Connect one end to the green wire and the other end to the blue wire. Tape the connections. Now, when the lower unit is separated from the exhaust housing, the ends of the wire will feed down through the exhaust housing. When the lower unit is free, disconnect the wire ends from the blue and green wires and leave the wire loop in the exhaust housing. When it is time to

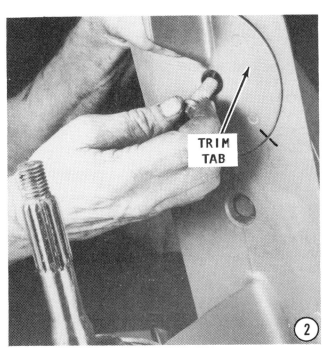

bring the lower unit together with the exhaust housing, the wire ends will be connected again and the blue and green wires easily pulled back up through the exhaust housing. No sweat! Alright, on with the work.

2- Scribe a mark on the trim tab and a matching mark on the lower unit to ensure the trim tab will installed in the same position from which it is removed. Remove the attaching hardware, and then remove the trim tab.

3- Use a 1/2" socket with a short extension and remove the bolt from inside the

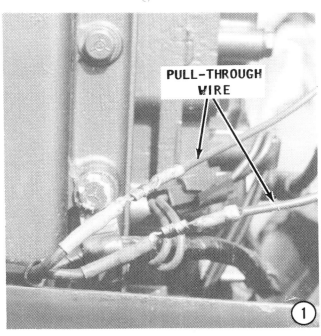

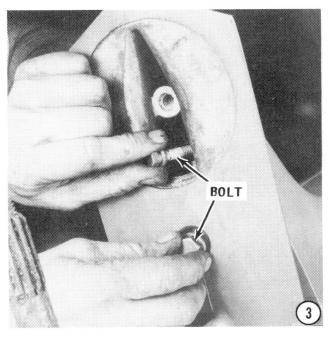

trim tab cavity. Remove the 5/8" countersunk bolt located just ahead of the trim tab position.

4- Remove the four 9/16" bolts, two on each side, securing the lower unit to the exhaust housing. Work the lower unit free of the exhaust housing. If the unit is still mounted on a boat, tilt the engine forward to gain clearance between the lower unit and the deck (floor, ground, whatever). **EXERCISE CARE** to withdraw the lower unit straight away from the exhaust housing to prevent bending the driveshaft. Once the lower unit is free of the exhaust housing, stop and disconnect the wires installed as described in the **"Helpful Word"** following Step 1. Leave the wire loop in the exhaust housing as an aid during installation.

WATER PUMP REMOVAL

5- Position the lower unit in the vertical position on the edge of the work bench resting on the cavitation plate. Secure the lower unit in this position with a C-clamp. The lower unit will then be held firmly in a favorable position during the service work. An alternate method is to cut a groove in a short piece of 2" x 6" wood to accommodate the lower unit with the cavitation plate resting on top of the wood. Clamp the wood in a vise and service work may then be performed with the lower unit erect (in its normal position), or inverted (upside down). In both positions, the cavitation plate is the supporting surface.

TAKE TIME

Take time to notice how the shift wires are routed and anchored in position with a clamp on top of the water pump. It is extremely important for the shift wires to be routed and secured in the same position during installation, reference illustration "C".

6- Remove the O-ring from the top of the driveshaft. Remove the bolts securing the water pump to the lower unit housing and the clamp securing the shift wires in place. Leave the clamp on the shift cable as an aid during installation. Pull the water pump housing up and free of the driveshaft.

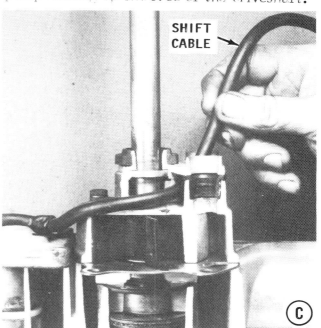

8-10 Lower Unit

7- Slide the water pump impeller up and free of the driveshaft. Pop the impeller Woodruff key out of the driveshaft keyway. Slide the water pump base plate up and off of the driveshaft.

GOOD WORDS

If the only work to be performed is service of the water pump, proceed directly to Page 8-28, Water Pump Installation.

Shift Solenoid — Removal

8- Remove the shift solenoid cover located just aft of the water pump position. Take care not to lose the wavy washer installed under the cover. Grasp the upper (green) shift solenoid and withdraw the solenoids and shift rod from the lower unit cavity as an assembly.

Bearing Carrier — Removal

9- Remove the four 5/16" bolts from inside the bearing carrier. Notice how each bolt has an O-ring seal. These O-rings should be replaced each time the bolts are removed. Also observe the word UP embossed into the metal rim of some bearing carriers. This word must face UP in relation to the lower unit during installation. Clean the surface and if the word "UP" does not show, the position of the carrier during installation is not important.

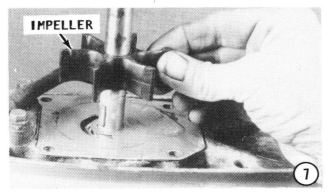

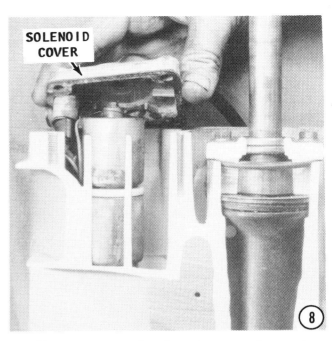

10- Remove the bearing carrier using one of the methods described in the following paragraphs, under Special Words.

SPECIAL WORDS

Several models of bearing carriers are used on the lower units covered in this section.

The bearing carriers are a very tight fit into the lower unit opening. Therefore, it is not uncommon to apply heat to the outside surface of the lower unit with a torch, at the same time the puller is being worked to remove the carrier. **TAKE CARE** not to overheat the lower unit.

One model carrier has two threaded holes on the end of the carrier. These threads permit the installation of two long bolts. These bolts will then allow the use of a flywheel puller to remove the bearing carrier, Illustration #10.

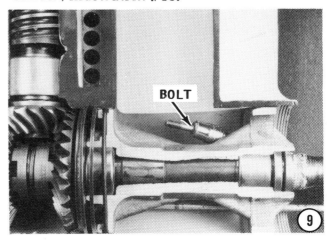

Another model does not have the threaded screw holes. To remove this type bearing carrier, a special puller with arms must be used. The arms are hooked onto the carrier web area, and then the carrier removed, reference illustration **"D"**.

WARNING

The next step involves a dangerous procedure and should be executed with care while wearing **SAFETY GLASSES**. The retaining rings are under tremendous tension in the groove and while they are being removed. If a ring should slip off the Truarc pliers, it will travel with incredible speed causing personal injury if it should strike a person. Therefore, continue to hold the ring and pliers firm after the ring is out of the groove and clear of the lower unit. Place the ring on the floor and hold it securely with one foot before releasing the grip on the pliers. An alternate method is to hold the ring inside a trash barrel, or other suitable container, before releasing the pliers.

11- Obtain a pair of Truarc pliers. Insert the tips of the pliers into the holes of the first retaining ring. Now, **CAREFULLY** remove the retaining ring from the groove and gear case without allowing the pliers to slip. Release the grip on the pliers in the manner described in the above **WARNING**. Remove the second retaining ring in the same manner. The rings are identical and either one may be installed first.

12- Remove the retainer plate. As the plate is removed, notice which surface is facing into the housing, as an aid during installation.

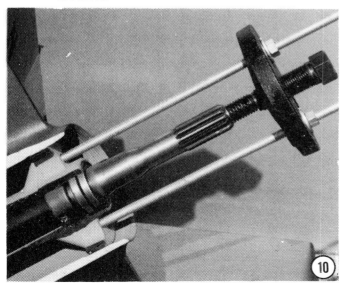

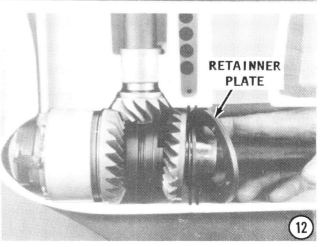

Propeller Shaft -- Removal

13- Grasp the propeller shaft firmly and withdraw it from the lower unit. The retainer plate, thrust washer, thrust bearing, reverse gear, reverse gear small thrust washer, and the clutch dog, will come out with the shaft.

"FROZEN" PROPELLER SHAFT

On rare occasions, especially if water has been allowed to enter the lower unit, it may not be possible to withdraw the propeller shaft as described in Step 13. The shaft may be "frozen" in the hydraulic pump due to corrosion.

If efforts to remove the propeller shaft after the bearing carrier has been removed, are unsuccessful, all is not lost. Thread an adaptor to the end of the propeller shaft. The adaptor has internal threads at both ends. Attach a slide hammer to the adaptor; operate the hammer; and remove the propeller shaft, reference illustration **"E"**.

BAD NEWS

Using the slide hammer, under these conditions, to remove the propeller shaft, will usually result in some internal part being damaged as the shaft is withdrawn. However, the cost of replacing the damaged part is reasonable considering the seriousness of a "frozen" shaft and getting it out.

Pinion Gear -- Removal

Special tool, OMC No. 316612 is required to turn the driveshaft in order to remove the pinion gear nut.

14- Obtain the special tool and slip it over the end of the driveshaft with the splines of the tool indexed with the splines on the driveshaft. Hold the pinion gear nut with the proper size wrench, and at the

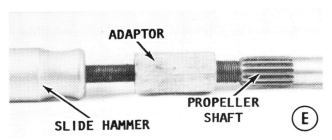

same time rotate the driveshaft, with the special tool and wrench **COUNTERCLOCKWISE** until the nut is free. If the special tool is not available, clamp the driveshaft in a vise equipped with soft jaws, in an area below the splines but not in the water pump impeller area. Now, with the proper size wrench on the pinion gear nut, rotate the complete lower unit **COUNTERCLOCKWISE** until the nut is free. This procedure will probably require the driveshaft to be loosened in the vise several times and reclamped in order to affect rotation of the lower unit and wrench. After the nut is free, proceed with the next step. The driveshaft will be withdrawn from the pinion gear.

Driveshaft -- Removal

15- Remove the four bolts from the top of the lower unit securing the bearing housing. **CAREFULLY** pry the bearing housing upward away from the lower unit, then slide it free of the driveshaft. An alternate method is to again clamp the driveshaft in a vise equipped with soft jaws. Use a soft-headed mallet and tap on the top side of the bearing housing. This action will jar the housing loose from the lower unit. Continue tapping with the mallet and the bearing housing, O-rings, shims, thrust washer, thrust bearing, and the driveshaft will all

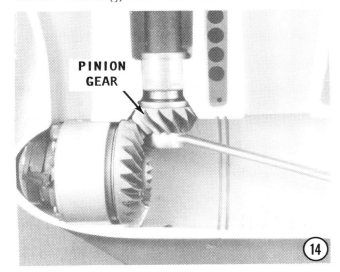

ELECTRIC SHIFT 8-13

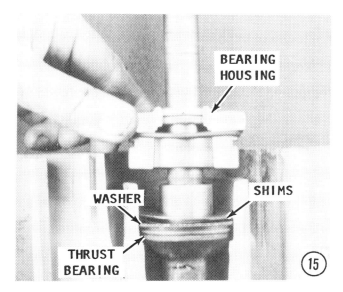

⑮

breakaway from the lower unit and may be removed as an assembly.

16- Remove the pinion gear from the lower unit cavity. Remove the forward gear from the hydraulic pump.

Hydraulic Pump — Removal

17- Obtain two long rods with 1/4" x 20 threads on both ends. Thread the two rods into the hydraulic pump housing. Attach a slide hammer to the rods and secure it with a nut on the end of each rod. Check to be sure the slide hammer is installed onto the rods **EVENLY** to allow an even pull on the pump. If the slide hammer is not installed to the rods properly, the pump may become tightly wedged in the lower unit. Operate the slide hammer and pull the hydraulic pump free. If the pump should happen to become lodged in the lower unit, stop operating the slide hammer **IMMEDIATELY**. Tap the hydraulic pump back into place in the lower unit and start the removal procedure over.

⑯

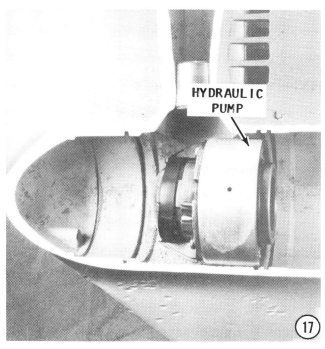

⑰

Lower Driveshaft Bearing — Removal

This bearing cannot be removed without the aid of a special tool. Therefore, **DO NOT** attempt to remove this bearing unless it is unfit for further service. To check the bearing, first use a flashlight and inspect it for corrosion or other damage. Insert a finger into the bearing, and then check for "rough" spots or binding while rotating it.

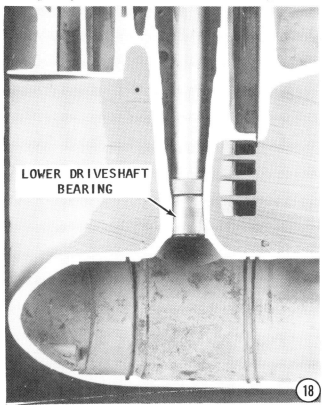

⑱

8-14 LOWER UNIT

18- Remove the **ALLEN** screw from the water pickup slots in the starboard side of the lower unit housing, reference illustration **"F"**. This screw secures the bearing in place and **MUST** be removed before an attempt is made to remove the bearing. The bearing must actually be **"PULLED"** upward to come free. **NEVER** make an attempt to "drive" it down and out or the lip in the lower unit holding the bearing will be broken off. **VERY BAD NEWS.** The lower unit housing would have to be replaced. Obtain special tool, OMC No. 385546. Use the special tool and "pull" the bearing from the lower unit, reference illustration **"G"**.

Propeller Shaft -- Disassembling

19- Notice the spring retainer on the outside surface of the clutch dog. Use a small screwdriver and work one end of the spring up onto the shoulder of the clutch dog. Continue working the spring out of the groove until it is free. **TAKE CARE** not to distort the spring. Place one end of the propeller shaft on the bench and push the pin free of the clutch dog. Raise the propeller end of the shaft upward and the piston, retainer, and spring, will come free of the shaft.

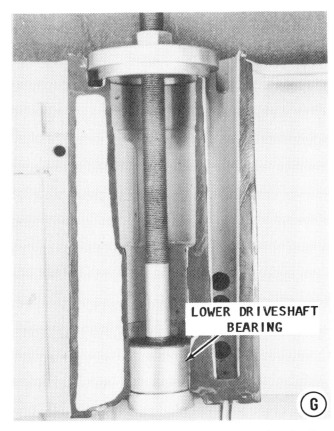

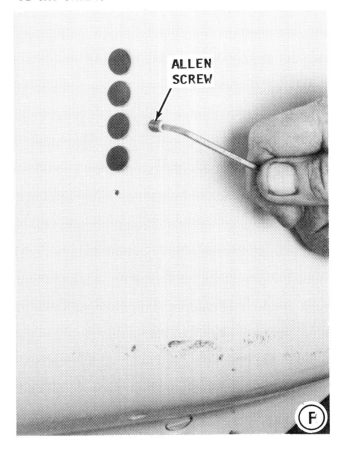

20- Notice how the small end of the retainer fits into the spring. Also notice the hole in the retainer. During installation, this hole must align with the hole in the propeller shaft and clutch dog to allow the pin to pass through. Slide the clutch dog free of the propeller shaft.

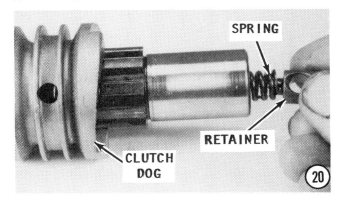

ELECTRIC SHIFT 8-15

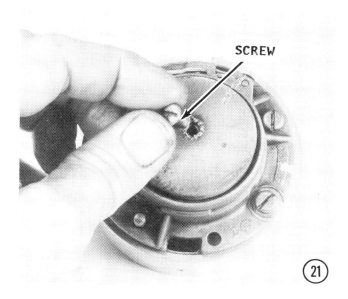

(21)

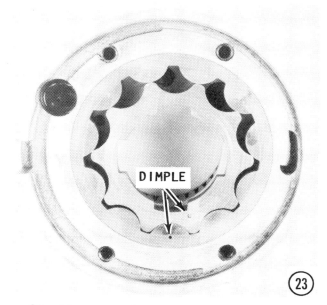

(23)

Hydraulic Pump — Disassembling

21- Remove the screw from the center of the screen on the back side of the pump. Remove the screen.

22- Remove the screws securing the valve housing to the pump, and then lift the housing free of the pump.

23- Lift the two gears out off the pump housing and **HOLD** them just as they were removed. Check the face of each gear for an indent mark (a dot, dimple, or similar identification). The identification mark will indicate how the gear **MUST** face in the housing. Make a note of how the mark faces, outward or inward, to **ENSURE** the gears will be installed properly in the same position from which they were removed.

Bearing Carrier — Disassembling

The bearings in the carrier need **NOT** be removed unless they are unfit for further service. Insert a finger and rotate the bearing. Check for "rough" spots or binding. Inspect the bearing for signs of corrosion or other types of damage. If the bearings must be replaced, proceed with the next step.

24- Use a seal remover to remove the two back-to-back seals or clamp the carrier in a vise and use a pry bar to pop each seal out.

25- Use a drift punch to drive the bearings free of the carrier. The bearings are being removed because they are unfit for service, therefore, additional damage is of no consequence.

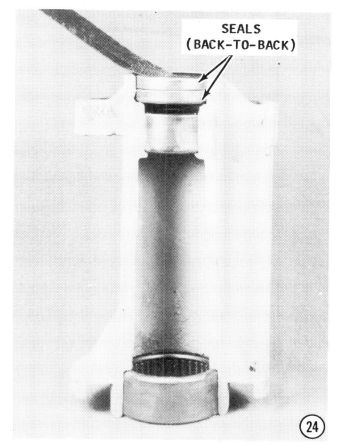

(24)

(22)

8-16 Lower Unit

Solenoid and Shift Assembly

It is not recommended to attempt service of this assembly. If troubleshooting has been performed and the determination made the unit or any part is faulty, the **ONLY** satisfactory solution is to purchase and install a new assembly.

CLEANING AND INSPECTING

Wash all, except **ELECTRICAL**, parts in solvent and dry them with compressed air. Discard all O-rings and seals that have been removed. A new seal kit for this lower unit is available from the local dealer. The kit will contain the necessary seals and O-rings to restore the lower unit to service.

Inspect all splines on shafts and in gears for wear, rounded edges, corrosion, and damage.

Carefully check the driveshaft and the propeller shaft to verify they are straight and true without any sign of damage. A complete check must be performed by turning the shaft in a lathe. This is only necessary if there is evidence to suspect the shaft is not true.

Check the water pump housing for corrosion on the inside and verify the impeller

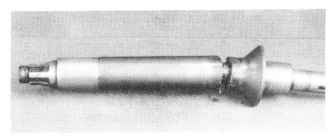

A two-section driveshaft with a weld section that has failed. The weld area of the driveshaft should be carefully checked anytime the lower unit is disassembled.

and base plate are in good condition. Actually, good shop practice dictates to rebuild or replace the water pump each time the lower unit is disassembled. The small cost is rewarded with "peace of mind" and satisfactory service.

Inspect the lower unit housing for nicks, dents, corrosion, or other signs of damage. Nicks may be removed with No. 120 and No. 180 emery cloth. Make a special effort to ensure all old gasket material has been removed and mating surfaces are clean and smooth.

Inspect the water passages in the lower unit to be sure they are clean. The screen may be removed and cleaned.

Check the gears and clutch dog to be sure the ears are not rounded. If doubt exists as to the part performing satisfactorily, it should be replaced.

Inspect the bearings for "rough" spots, binding, and signs of corrosion or damage.

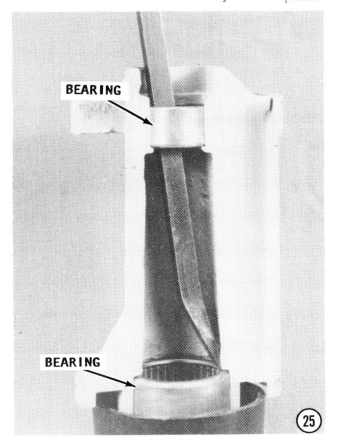

Damaged hydraulic pump. Water in the lower unit and a broken gear was the cause of this pump being destroyed.

ELECTRIC SHIFT 8-17

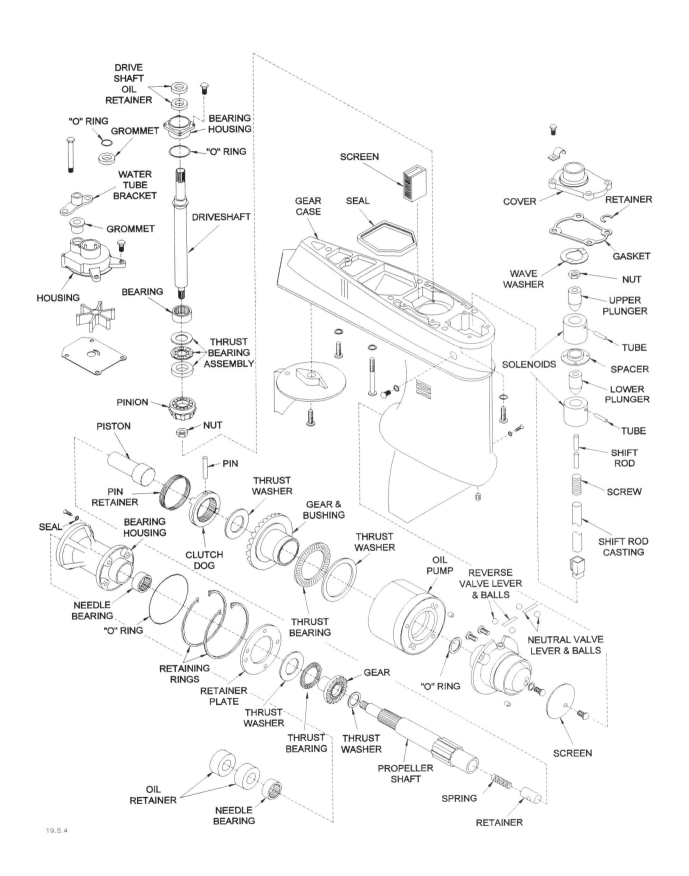

Exploded view of a 65 hp -- 1972 lower unit with major parts identified.

8-18 LOWER UNIT

Damaged reverse gear (left) and forward gear (right). This damage was caused from water entering the lower unit.

Test the neutral and reverse solenoids with an ohmmeter. A reading of 5 to 7 ohms is normal and indicates the solenoid is in satisfactory condition.

ASSEMBLING

READ AND BELIEVE

The lower unit should **NOT** be assembled in a dry condition. Coat all internal parts with OMC HI-VIS lube oil as they are assembled. All seals should be coated with OMC Gasket Seal Compound. When two

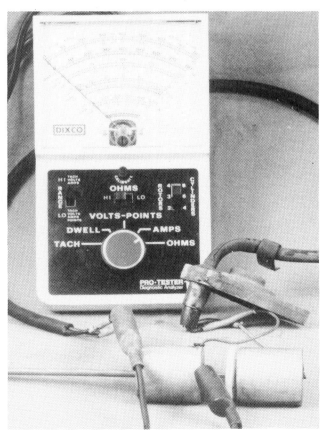

Using an ohmmeter to test the solenoids, as explained in the text.

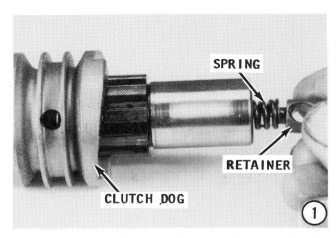

seals are installed back-to-back, use Triple Guard Grease between the seal surfaces.

AUTHORS APOLOGY

The accompanying illustrations show a rubber seal on the end of the hydraulic pump and a snap ring installed in front of the pump. During 1972, these two items were not used. Therefore, disregard the seal and snap ring, for this model year.

Propeller Shaft — Assembling

1- Slide the clutch dog onto the propeller shaft with the face of the dog marked **"PROP END"** facing toward the propeller end of the shaft, reference illustration **"A"**. Before the splines of the clutch dog engage the splines of the propeller shaft, rotate the dog until the hole for the pin appears to align with the hole through the propeller shaft. Slide the clutch dog onto the splines until the hole in the dog aligns with the hole in the shaft. If the hole is off just a bit, slide the clutch dog back off the splines, rotate it one spline in the required direction, and then slide it into place.

Insert the spring into the end of the propeller shaft. Secure the spring in place with the spring retainer. Install the retainer with the small end going into the propeller shaft **FIRST**.

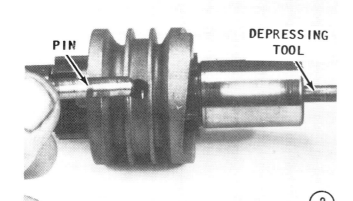

2- Depress the spring retainer and insert the pin through the clutch dog, the shaft, spring retainer, and out the other side of the shaft and clutch dog. Center the pin through the clutch dog.

3- Install the spring-type pin retainer around the clutch dog to secure the pin in place. **TAKE CARE** not to distort the pin retainer during the installation process.

Bearing Carrier Bearings & Seals Installation

4- Install the reverse gear bearing into the bearing carrier by pressing against the **LETTERED** side of the bearing with the proper size socket. Press the forward gear bearing into the bearing carrier in the same manner. Press against the **LETTERED** side of the bearing.

5- Coat the outside surfaces of the seals with HI-VIS oil. Install the first seal with the flat side facing **OUT**. Coat the flat surface of both seals with Triple Guard Grease, and then install the second seal with the flat side going in **FIRST**. The seals are

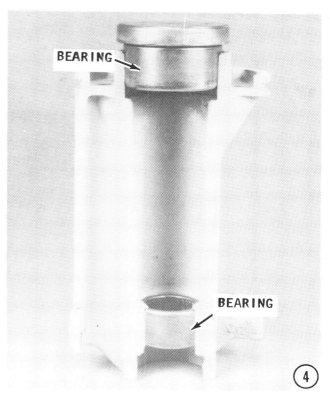

then back-to-back with the grease between the two surfaces. The outside seal prevents water from entering the lower unit and the inside seal prevents the lubricant in the lower unit from escaping.

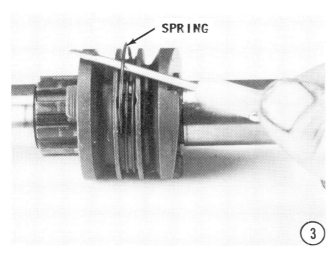

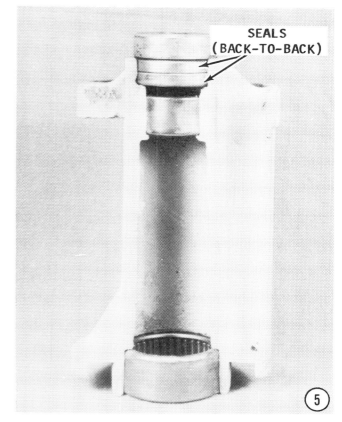

A set of double seals showing the back side (left) and the front side (right). These seals are installed back-to-back (flat side-to-flat side) with Triple Guard Grease between the surfaces. This arrangement prevents fluid from passing in either direction.

Hydraulic Pump -- Assembling

6- Check the note made during disassembling, per Step 23, to determine how the identifying marks (dots, dimples, whatever) on the gears must face -- inward or outward. The gears **MUST** be installed in the same position from which they were removed.

7- Install the rear valve housing with the tang on the outside edge of the housing indexed with the small slot in the pump housing. Secure the valve housing in place with the attaching screws tightened securely.

8- Place the screen in position on the back side of the valve housing, and then secure it in place with the screw.

9- Install the forward gear into the pump housing. It may be necessary to work the gears around in the pump to permit the tangs on the forward gear shank to index in the slots in the housing. Set the assembly aside for later installation.

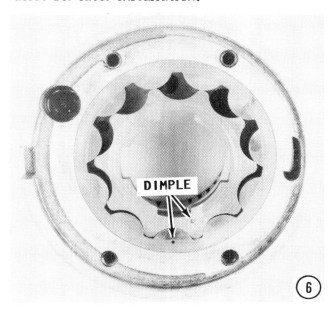

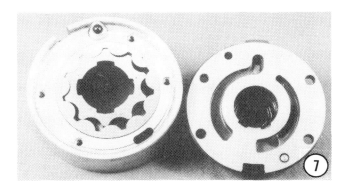

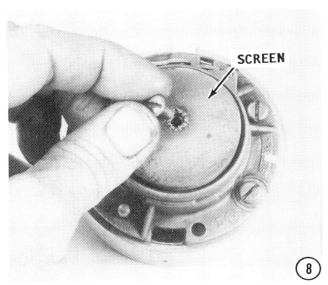

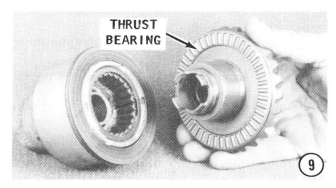

Two types of hydraulic pumps. The pump on the left is the most common with the shift rod passing through a hole in two levers on top of the pump. The pump on the right has the shift rod passing directly through a hole in the pump top.

Lower Driveshaft Bearing — Installation

10- Obtain tool, OMC No. 385546. Assemble the tool with the washer, guide sleeve, and remover portion of the tool, in the order given. The shoulder of the tool must face **DOWN**. Place the bearing onto the end of the tool, with the lettered side of the bearing facing the tool. Drive the bearing down until the large washer on the tool makes contact with the surface of the lower unit. The bearing is then seated to the proper depth.

11- If an Allen screw is used to secure the bearing in place, apply Loctite to the threads, amd then install the screw through the lower unit, as shown.

Hyraulic Pump — Installation
First, These Words

Observe the tang on the backside of the pump. This tang **MUST** face directly up in relation to the lower unit housing to permit installation of the shift rod into the pump. Also notice the pin on the backside of the pump. This pin **MUST** index into a matching hole in the housing to restrain the pump from rotating.

12- Secure the lower unit housing in the horizontal position with the bearing carrier opening facing up. Remove the forward gear from the pump. Obtain two long 1/4 x

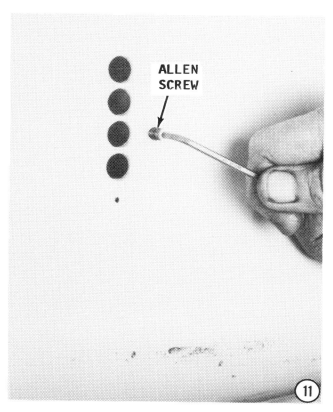

20 rods with threads on both ends. Thread the rods into the pump and then lower the pump into the lower unit housing. To index the pin on the back of the pump housing into the hole in the lower unit is not an easy task. However, exercise patience and rotate the pump ever so slowly. A helpful hint at this point: As the pump is being lowered into the cavity, align the opening and tang on top of the pump in the approximate position your eye indicates the shift rod may be installed. When the pin indexes, it will

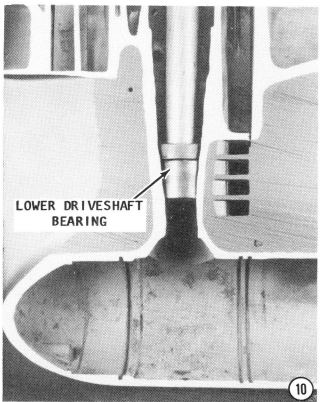

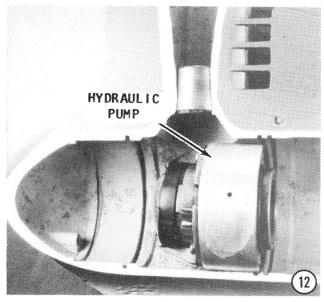

not be possible to rotate the pump. The pump **MUST** be properly seated to permit installation of the shift rod and the pinion gear. After the pump is in place, remove the two rods used during installation.

13- Coat the plunger with oil, and then lower the large end of the plunger into the hydraulic pump. Check to be sure it seats all the way into place.

14- Install the thrust washer and thrust bearing into the pump with the flat side of the bearing facing **OUTWARD**. Lower the forward gear into the pump. Work the gear slowly until the teeth index with the teeth of the pump gear. This should not be too difficult because the forward gear was installed once, and then removed in the previous step. However, it is entirely possible the gears moved when the pump was installed. Therefore, use a flashlight and check the position of the gears. If necessary, use a long shank screwdriver and rotate the gears until they are close to center, then install the forward gear.

Driveshaft and Pinion Gear -- Installation

CRITICAL WORDS

The driveshaft and pinion gear must be assembled prior to installation, and then checked with a special shimming gauge. This shimming must be accomplished properly, the unit disassembled, and then installed into the lower unit. Use of the shimming gauge is the **ONLY** way to determine the proper amount of shimming required at the upper end of the driveshaft. The following detailed step outlines the procedure.

15- Clamp the driveshaft in a vise equipped with soft jaws and in such a manner that the splines, water pump area, or other critical portions of the shaft cannot

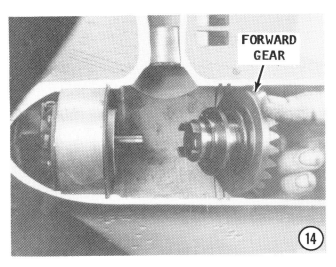

be damaged. Slide the pinion gear onto the driveshaft with the bevel of the gear teeth facing toward the lower end of the shaft. Install the pinion gear nut and tighten it to a torque value of 40 to 45 ft-lbs. No parts should be installed on the upper end of the driveshaft at this point. Slide the same amount of shim material removed during disassembling, onto the driveshaft and seat it against the driveshaft shoulder.

Obtain special shimming tool, OMC No. 315767. Slip the special tool down over the driveshaft and onto the upper surface of the top shim. Measure the distance between the

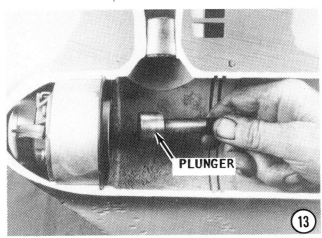

top of the pinion gear and the bottom of the tool. The tool should just barely make contact with the pinion gear surface for **ZERO** clearance. Add or remove shims from the upper end of the driveshaft to obtain the required **ZERO** clearance. Remove the tool and set the shims aside for installation later. Back off the pinion gear nut and remove the pinion gear. Remove the driveshaft from the vise.

16- Insert the pinion gear into the cavity in the lower unit with the flat side of the bearing facing **UPWARD**. Hold the pinion gear in place and at the same time lower the driveshaft down into the lower unit. As the driveshaft begins to make contact with the pinion gear, rotate the shaft slightly to permit the splines on the shaft to index with the splines of the pinion gear. After the shaft has indexed with the pinion gear, thread the pinion gear nut onto the end of the shaft. Obtain special tool, OMC No. - 312752. Slide the special tool over the upper end of the driveshaft with the splines of the tool indexed with the splines on the shaft. Attach a torque wrench to the special tool. Now, hold the pinion gear nut with the proper size wrench and rotate the driveshaft **CLOCKWISE** with the special tool until the pinion gear nut is tightened to a torque value of 40 to 45 ft-lbs. Remove the special tool.

17- Slide the thrust bearing, thrust washer, and the shims, set aside after the shim gauge procedure in Step 15, onto the driveshaft.

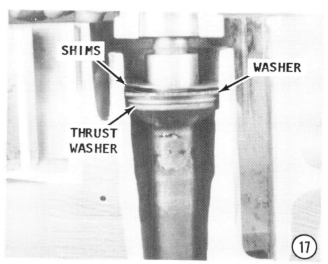

18- Install the two seals back-to-back into the opening on top of the bearing housing. Coat the outside surface of the **NEW** seals with OMC Lubricant. Press the first seal into the housing with the flat side of the seal facing **OUTWARD**. After the seal is in place, apply a coating of Triple Guard Grease to the flat side of the installed seal and the flat side of the second seal. Press the second seal into the bearing housing with the flat side of the seal facing **INWARD**. Insert a **NEW** O-ring into the bottom opening of the bearing housing.

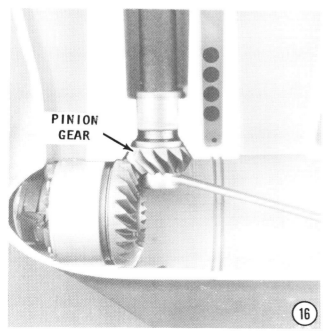

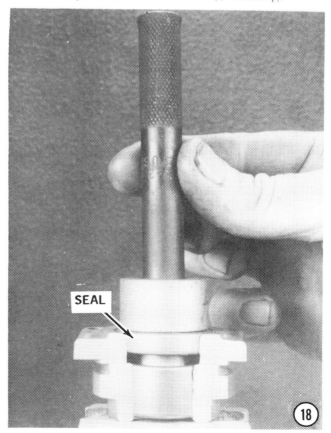

8-24 LOWER UNIT

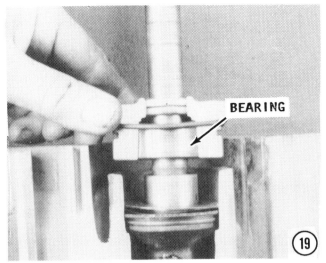

19- Wrap friction tape around the splines of the driveshaft to protect the seals in the bearing housing as the housing is installed. Now, slide the assembled bearing housing down the driveshaft and seat it in the lower unit housing. Secure the housing in place with the four bolts. Tighten the bolts **ALTERNATELY** and **EVENLY**.

Propeller Shaft -- Installation

20- Secure the lower unit in the horizontal position with the bearing carrier opening facing **UPWARD**. Check the inside diameter of the forward gear to be sure the small thrust washer in the gear is still in place. Lower the propeller shaft assembly down into the lower unit with the inside diameter of the shaft indexing over the shaft of the piston.

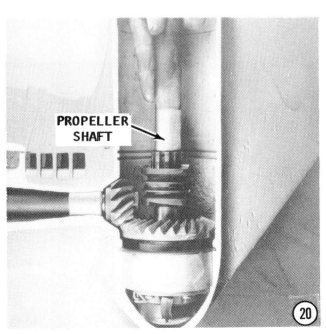

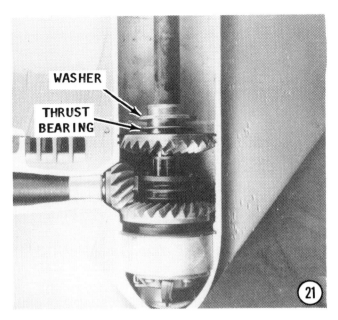

21- Apply a light coating of grease to the inside surface of the reverse gear to hold the thrust washer in place. Insert the thrust washer into the reverse gear. Lower the thrust bearing and thrust washer down onto the shank of the reverse gear. Slide the reverse gear down the propeller shaft with the splines of the reverse gear indexing with the splines of the shaft.

22- Insert the retainer plate into the lower unit against the reverse gear.

WARNING

This next step can be dangerous. Each snap ring is placed under tremendous tension with the Truarc pliers while it is being

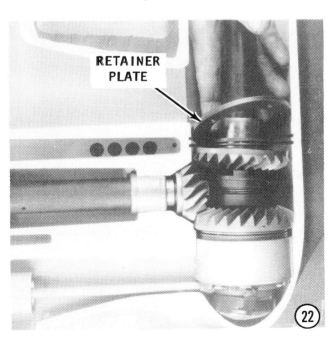

placed into the groove. Therefore, wear **SAFETY GLASSES** and exercise care to prevent the snap ring from slipping out of the pliers. If the snap ring should slip out, it would travel with incredible speed and cause personal injury if it struck a person.

23- Install the Truarc snap rings one at-a-time following the precautions given in the **WARNING** and the **ADVICE** given in the following paragraph.

WORDS OF ADVICE

The two snap rings index into separate grooves in the lower unit housing. As the first ring is being installed, depth perception may play a trick on your eyes. It may appear that the first ring is properly indexed all the way around in the proper groove, when in reality, a portion may be in one groove and the remainder in the other groove. Should this happen, and the Truarc pliers be released from the ring, it is extremely difficult to get the pliers back into the ring to correct the condition. If necessary use a flashlight and carefully check to be sure the first ring is properly seated all the way around **BEFORE** releasing the grip on the pliers. Installation of the second ring is not so difficult because the one groove is filled with the first ring.

24- Obtain two long 1/4" rods with threads on one end. Thread the rods into the retainer plate opposite each other to act as guides for the bearing carrier.

25- Check the bearing carrier to be sure a **NEW** O-ring has been installed. Position

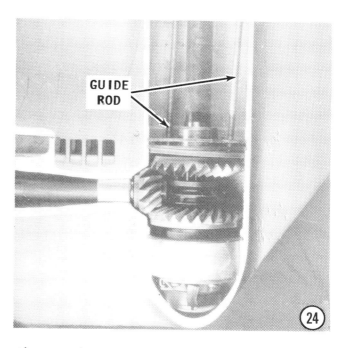

the carrier over the guide pins with the embossed word **UP** on the rim of the carrier facing **UP** in relation to the lower unit housing. Now, lower the bearing carrier down over the guide pins and into place in the lower unit housing.

26- Slide **NEW** little O-rings onto each bolt, and apply some OMC Sealer onto the threads. Install the bolts through the carrier and into the retaining plate. After a couple bolts are in place, remove the guide pins and install the remaining bolts. Tighten the bolts **EVENLY** and **ALTERNATELY** to the torque value given in the Appendix.

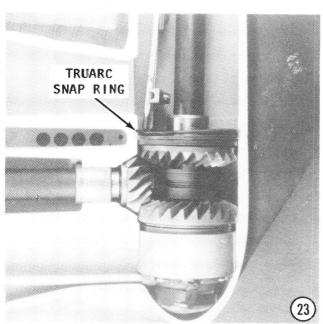

8-26 LOWER UNIT

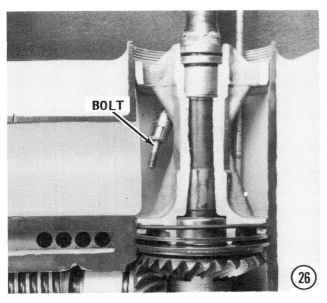

Shift Cable, Solenoids, and Rod — Assembling

CRITICAL WORDS

The following procedures **MUST** be performed exactly as given and in the order presented. Do not attempt any shortcuts or anticipate what will be done next. All parts must be installed and adjusted to the letter, for the unit to function properly. Separate the upper (green) solenoid from the lower (blue) solenoid, by pulling them apart. An inner shift rod will be released from the shift rod casing. Check to be sure the cap on the bottom of the shift rod casing is in place.

27- Lower the blue solenoid down into the lower unit housing. Continue lowering the solenoid until the cap on the end of the shift rod casing seats on top of the valve

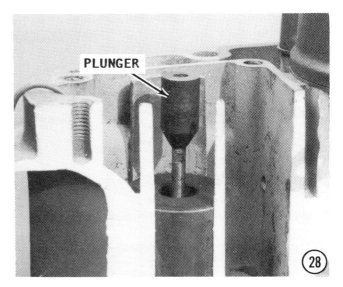

lever and check ball assembly. Check to be sure the solenoid is fully seated in the housing. The lower plunger should be flush with the top of the solenoid.

28- If the plunger is not flush with the solenoid, remove the solenoid and screw the lower plunger up or down on the shift rod casing, then install the solenoid again and check the plunger.

29- Install the spacer with the lip on the spacer facing **UPWARD**.

30- Lower the green solenoid into the lower unit housing. Insert the shift rod into the shift rod casing.

31- The upper plunger must be flush with the top surface of the solenoid. If it is not flush with the solenoid, remove the solenoid, loosen the nut, and make an adjustment. Install the solenoid again and check the upper plunger.

32- Check to be sure the inside plunger indexes into the hole in the hydraulic pump.

33- Install the wavy washer and a **NEW** gasket into the lower unit housing. Work the shift wires down into the lower unit cavity. Lower the cover into place in the lower unit housing. The wavy washer will give you a false pressure against the cap. Therefore, it takes a bit of patience to be sure the wavy washer indexes into the recess of the cover. Start the bolts securing the cover. Tighten the bolts **ALTERNATELY** and **EVENLY**. As the bolts are tighten-

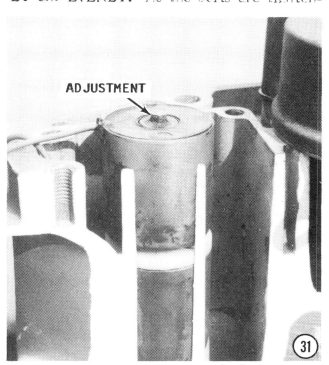

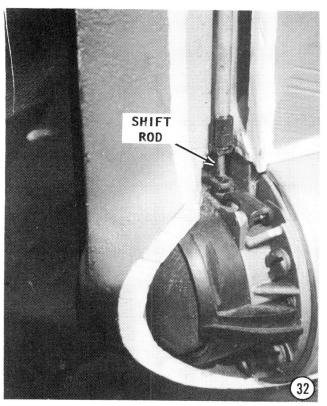

ed, make continuous checks to be sure the wavy washer and the green solenoid fit up into the cover as the bolts are tightened. This may not be accomplished on the first attempt, but keeping cool and working slowly will be rewarded with success.

34- Obtain an ohmmeter. Ground one lead of the meter, and then check the blue and green wires for continuity. The ohmmeter should indicate 5 to 7 ohms.

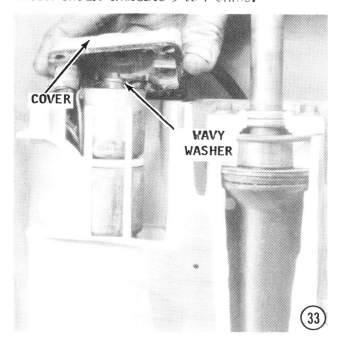

8-28 LOWER UNIT

WATER PUMP INSTALLATION

FIRST, THESE WORDS

An improved water pump is available as a replacement. If the old water pump housing is unfit for further service, only the new pump housing can be purchased. It is strongly recommended to replace the water pump with the improved model while the lower unit is disassembled. The accompanying illustration shows the original equipment (left) compared with the improved pump (right), reference illustration "A".

The new pump must be assembled before it is installed. Therefore, the following steps outline procedures for both pumps.

To assemble and install a replacement pump, perform steps 35 thru 43, then jump to Step 46.

To install an original equipment pump, proceed direectly to Step 44.

Assembling an Improved Pump Housing

35- Remove the water pump parts from the container. Insert the plate into the housing, as shown. The tang on the bottom side of the plate **MUST** index into the short slot in the pump housing.

36- Slide the pump liner into the housing with the two small tabs on the bottom side indexed into the two cutouts in the plate.

37- Coat the inside diameter of the liner with light-weight oil. Work the impeller into the housing with all of the blades bent back to the right, as shown. In this position,

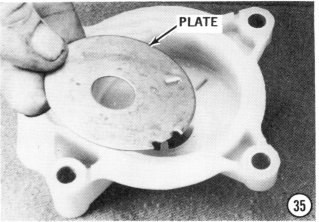

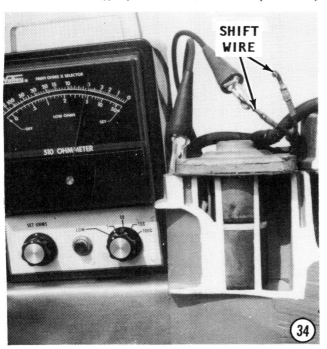

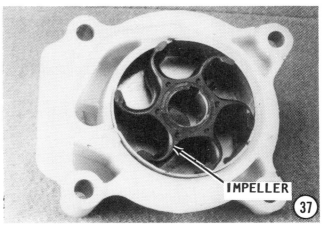

ELECTRIC SHIFT 8-29

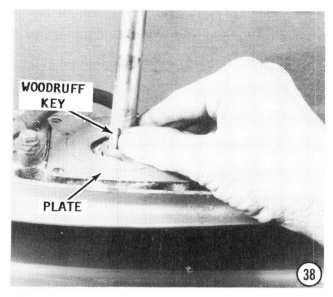

the blades will rotate properly when the pump housing is installed. Remember, the pump and the blades will be rotating **CLOCKWISE** when the housing is turned over and installed in place on the lower unit.

38- Coat the mating surface of the lower unit with 1000 Sealer. Slide the water pump base plate down the driveshaft and into place on the lower unit. Insert the Woodruff key into the key slot in the driveshaft.

39- Lay down a very thin bead of 1000 sealer into the irregular shaped groove in the housing. Insert the seal into the groove, and then coat the seal with the 1000 Sealer.

Water Pump Installation

40- Begin to slide the water pump down the driveshaft, and at the same time observe the position of the slot in the impeller. Continue to work the pump down the driveshaft, with the slot in the impeller indexed over the Woodruff key. The pump must be fairly well aligned before the key is covered because the slot in the impeller is not visible as the pump begins to come close to the base plate.

41- Install the short forward bolt through the pump and into the lower unit. **DO NOT** tighten this bolt at this time. Insert the grommet into the pump housing.

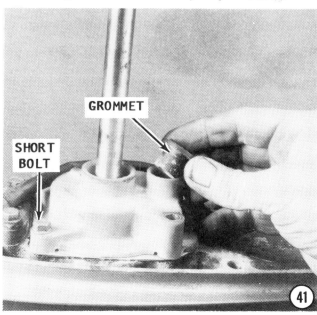

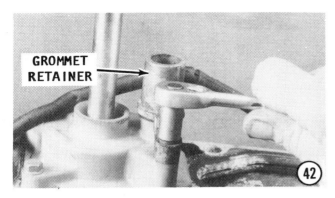

42- Install the grommet retainer and water tube guide onto the pump housing. Install the remaining pump attaching bolts. Tighten the bolts **ALTERNATELY** and **EVENLY**, and at the same time rotate the driveshaft **CLOCKWISE**. If the driveshaft is not rotated while the attaching bolts are being tightened, it is possible to pinch one of the impeller blades underneath the housing.

43- Slide the large grommet down the driveshaft and seat it over the pump collar. This grommet does not require sealer. Its function is to prevent exhaust gases from entering the water pump. Proceed directly to Lower Unit Installation, Step 47.

Water Pump Installation
Original Equipment

Perform the following two steps to install an original water pump.

44- Coat the water pump plate mating surface on the lower unit with 1000 Sealer. Slide the water pump plate down the driveshaft and **BEFORE** it makes contact with the sealer check to be sure the bolt holes in the plate will align with the holes in the housing. The plate will only fit one way. If the holes will not align, remove the plate, turn it over and again slide the plate down the driveshaft and into place on the housing.

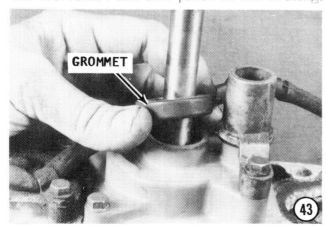

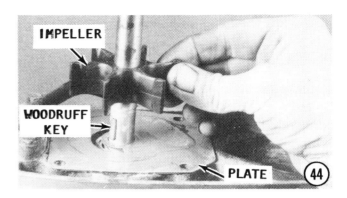

This checking will prevent accidently getting the sealer on both sides of the plate. If by chance sealer does get on the top surface it **MUST** be removed before the water pump impeller is installed.

Slide the water pump impeller down the driveshaft. Just before the impeller covers the cutout for the Woodruff key, install the key, and then work the impeller on down, with the slot in the impeller indexed over the Woodruff key. Continue working the impeller down until it is firmly in place on the surface of the pump plate.

45- Check to be sure **NEW** seals and O-rings have been installed in the water pump. Lubricate the inside surface of the water pump with light-weight oil. Lower the water pump housing down the driveshaft and over the impeller. **ALWAYS** rotate the driveshaft slowly **CLOCKWISE** as the housing is lowered over the impeller to allow the impeller blades to assume their natural and proper position inside the housing. Continue to rotate the driveshaft and work the water pump housing downward until it is seated on the plate. Coat the threads of the water pump attaching screws with sealer, and then secure the pump in place with the screws. Install the solenoid cable bracket with the same bolt and in the same position from which it was removed. On some units the

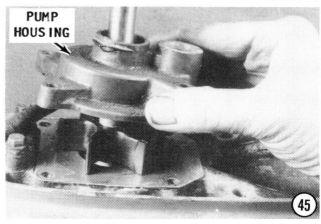

ELECTRIC SHIFT 8-31

solenoid cable fits into a recess of the water pump and is held in place in that manner. Tighten the screws **ALTERNATELY** and **EVENLY**.

LOWER UNIT INSTALLATION

46- Check to be sure the water tubes are clean, smooth, and free of any corrosion. Coat the water pickup tubes and grommets with lubricant as an aid to installation. Check to be sure the spark plug wires are disconnected from the spark plugs. Bring the lower unit housing together with the exhaust housing, and at the same time, guide the water tube into the rubber grommet of the water pump. Connect the ends of the wire left in the exhaust housing during removal to the blue and green wires from the lower unit. Tape the connections, and then oil the shift cable as an aid to slipping the cable through the exhaust housing. Continue to bring the two units together and at the same time: pull the wires through the exhaust housing; guide the water pickup tubes into the rubber grommet of the water pump; and, rotate the flywheel slowly to permit the splines of the driveshaft to index with the splines of the crankshaft. This may sound like it is necessary to do four things at the same time, and

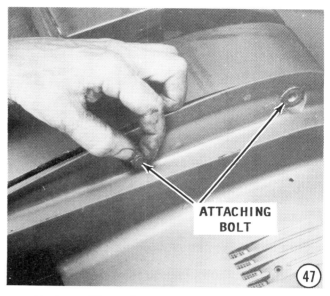

so it is. Therefore, make an earnest attempt to secure the services of an assistant for this task.

47- After the surfaces of the lower unit and exhaust housing are close, dip the attaching bolts in OMC Sealer and then start them in place. Two bolts are used on each side of the two housings.

48- Install the retaining bolt in the recess of the trim tab and another bolt in the cavitation plate. Tighten the bolts **ALTERNATELY** and **EVENLY** to the torque value given in the Appendix.

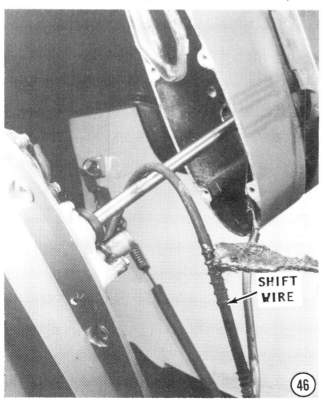

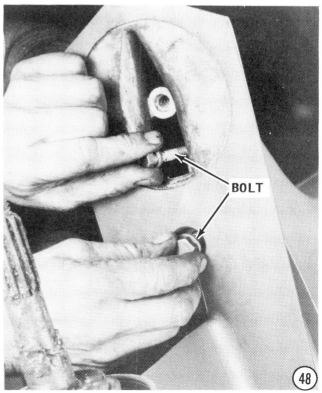

8-32 Lower Unit

49- Install the trim tab with the mark made on the tab during disassembling aligned with the mark made on the lower unit housing.

50- Disconnect and remove the "pull" wire used to feed the electrical wires up through the exhaust housing.

51- Obtain an ohmmeter. Ground the meter and again check the green and blue wires for continuity. The meter should indicate 5 to 7 ohms. Connect the green wire to the green wire and the blue wire to the blue wire. After the connections have been made, slide the sleeve down over the connections. A bit of oil on the wires will allow the sleeves to slide more easily.

Filling the Lower Unit

Fill the lower unit with lubricant according to the procedures in Section 8-3.

Propeller Installation

Install the propeller, see Section 8-2.

FUNCTIONAL CHECK

Perform a functional check of the completed work by mounting the engine in a test tank, in a body of water, or with a flush attachment connected to the lower unit. If the flush attachment is used, **NEVER** operate the engine above an idle speed, because the no-load condition on the propeller would allow the engine to **RUN-A-WAY** resulting in serious damage or destruction of the engine.

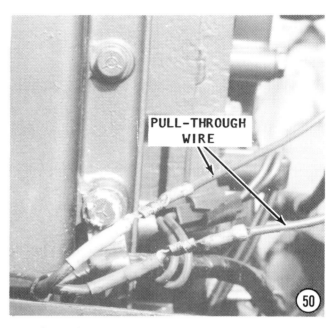

CAUTION: Water must circulate through the lower unit to the engine any time the engine is run to prevent damage to the water pump in the lower unit. Just five seconds without water will damage the water pump.

Start the engine and observe the tattle-tale flow of water from idle relief in the exhaust housing. The water pump installation work is verified. If a "Flushette" is connected to the lower unit, **VERY LITTLE** water will be visible from the idle relief port. Shift the engine into the three gears and check for smoothness of operation and satisfactory performance. Remember, when the unit is in forward gear, it is at rest with no current flow to either solenoid.

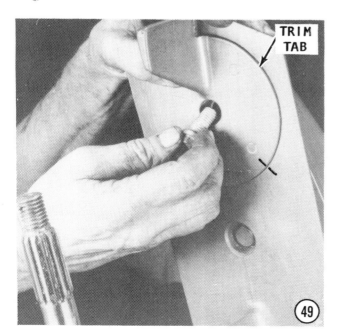

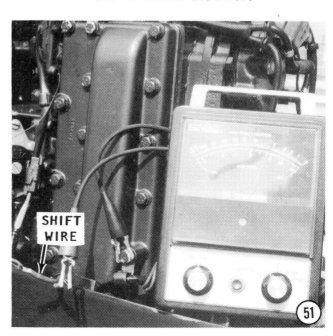

8-5 MECHANICAL SHIFT HYDRAULIC ASSIST SHIFT DISCONNECT UNDER LOWER CARBURETOR
65 HP TO 135 HP — 1973 THRU 1975
85 HP TO 200 HP — 1976 AND 1977

DESCRIPTION

The lower unit covered in this section is a three shift position, hydraulic activated, propeller exhaust unit. A hydraulic pump mounted in the forward portion of the lower unit provides the force required to shift the unit. A shift rod extends down through the exhaust housing and lower unit to an assist cylinder. A push rod connects the assist cylinder and the hydraulic pump. The pump valve directs the hydraulic force to place the clutch dog in the desired position for neutral, forward, or reverse gear position.

When the shift rod is in the upward position, the system is at rest and the lower unit is in forward gear.

In simple terms, something must be done to move the unit into neutral or reverse gear position. If no action is taken (shift mechanism at rest) the unit is in the forward gear position.

As the shift control lever is moved at the control box, the shift rod moves in the assist cylinder; the valve in the cylinder is activated; the push rod is moved, allowing oil from the pump to pass up through the hollow push rod into the assist cylinder; the shift mechanism is thereby "assisted" for easier and smoother shifting.

The lower unit houses the driveshaft and pinion gear, the forward and reverse driven gears, the propeller shaft, clutch dog, hydraulic pump, assist cylinder and the necessary shims, bearings, and associated parts to make it all work properly.

The water pump is considered a part of the lower unit.

TROUBLESHOOTING

Preliminary Checks

At rest, and without the engine running, the lower unit is in forward gear. Mount the engine in a test tank, in a body of water, or with a flush attachment connected to the lower unit. If the flush attachment is used, **NEVER** operate the engine above an idle speed, because the no-load condition on the propeller would allow the engine to **RUN-A-WAY** resulting in serious damage or destruction of the engine.

CAUTION: Water must circulate through the lower unit to the engine any time the engine is run to prevent damage to the water pump in the lower unit. Just five seconds without water will damage the water pump.

Cut-a-way view of a typical lower unit covered in this section. Major parts are exposed. This type of illustration will be most helpful in understanding the relationship of the various parts.

Helm-type shift box and steering wheel.

Attempt to shift the unit into **NEUTRAL** and **REVERSE**. It is possible the propeller may turn very slowly while the unit is in neutral, due to "drag" through the various gears and bearings. If difficult shifting is encountered, the problem is in the shift linkage or in the lower unit.

The second area to check is the quantity and quality of the lubricant in the lower unit. If the lubricant level is low, contaminated with water, or is broken down because of overuse, the shift mechanism may be affected. Water in the lower unit is **VERY BAD NEWS** for a number of reasons, particularly when the lower unit contains hydraulic components. Hydraulic units will not function with water in the system.

Remember, when the engine is not running, the unit should be in **FORWARD** gear.

BEFORE making any tests, remove the propeller, see Section 8-2. Check the propeller carefully to determine if the hub has been slipping and giving a false indication the unit is not in gear, reference illustration "**A**". If there is any doubt, the propeller should be taken to a shop properly equipped for testing, before the time and expense of disassembling the lower unit is undertaken. The expense of the propeller testing and possible rebuild is justified.

The following troubleshooting procedures are presented on the assumption the lower unit lubricant, and the propeller have been checked and found to be satisfactory.

Lower Unit Locked

Determine if the problem is in the powerhead or in the lower unit. Attempt to rotate the flywheel. If the flywheel can be moved even slightly in either direction, the problem is most likely in the lower unit. If it is not possible to rotate the flywheel, the problem is a "frozen" powerhead. To absolutely verify the powerhead is "frozen", separate the lower unit from the exhaust housing and then again attempt to rotate the flywheel. If the attempt is successful, the problem is definitely in the lower unit. If the attempt to rotate the flywheel, with the lower unit removed, still fails, a "frozen" powerhead is verified, reference illustration "**B**".

Unit Fails to Shift
Neutral, Forward, or Reverse

With the outboard mounted on a boat, in a test tank, or with a flush attachment connected, disconnect the shift lever at the engine. Attempt to manually shift the unit into **NEUTRAL, REVERSE, FORWARD**. At the same time move the shift handle at the shift box and determine that the linkage and shift lever are properly aligned for the shift positions. If the alignment is not correct, adjust the shift cable at the trunion. It is

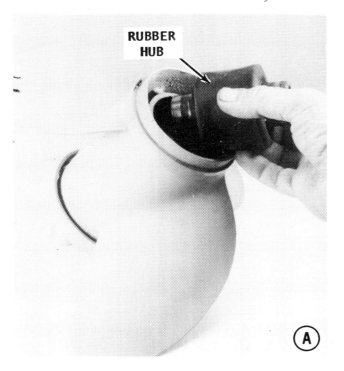

also possible the inner wire may have slipped in the connector at the shift box. This condition would result in a lack of inner cable to make a complete "throw" on the shift handle. Check to be sure the shift handle is moved to the full shift position and the linkage to the lower unit is moved to the full shift position.

If an adjustment is required at the shift box, see Chapter 7.

If it is not possible to shift the unit into gear by manually operating the shift rod while the engine is running, the lower unit requires service as described in this section.

LOWER UNIT SERVICE

Propeller Removal

If the propeller was not removed, as directed for the troubleshooting, remove it now, according to the procedures outlined in Section 8-2.

Draining the Lower Unit

Drain the lower unit of lubricant, see Section 8-3.

GOOD WORDS

If water is discovered in the lower unit and the propeller shaft seal is damaged and

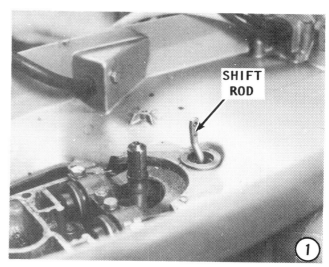

requires replacement, the lower unit does **NOT** have to be removed in order to accomplish the work.

The bearing carrier can be removed and the seal replaced without disassembling the lower unit. **HOWEVER** such a procedure is not considered good shop practice, but merely a quick-fix. If water has entered the lower unit, the unit should be disassembled and a detailed check made to determine if any other seals, bearings, bearing races, O-rings or other parts have been rendered unfit for further service by the water.

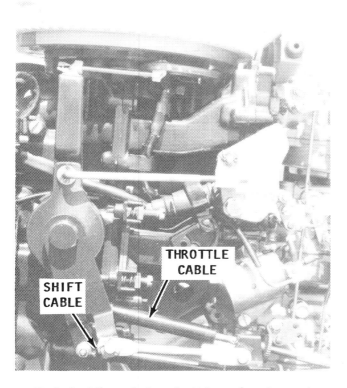

Typical shift and throttle linkage for the engines covered in this section.

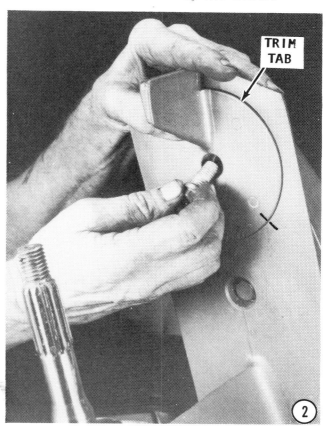

LOWER UNIT REMOVAL

1- Disconnect and ground the spark plug wires. Remove the bolt or bolts from the shift disconnect coupler under the lower carburetor.

2- Scribe a mark on the trim tab and a matching mark on the lower unit to ensure the trim tab will be installed in the same position from which it is removed. Use an Allen wrench and remove the trim tab.

3- Use a 1/2" socket with a short extension and remove the bolt from inside the trim tab cavity. Remove the 5/8" countersunk bolt located just ahead of the trim tab position.

4- Remove the four 9/16" bolts, two on each side, securing the lower unit to the exhaust housing.

5- Work the lower unit free of the exhaust housing. If the unit is still mounted on a boat, tilt the engine forward to gain clearance between the lower unit and the deck (floor, ground, whatever). **EXERCISE CARE** to withdraw the lower unit straight away from the exhaust housing to prevent bending the driveshaft.

WATER PUMP REMOVAL

6- Position the lower unit in the vertical position on the edge of the work bench resting on the cavitation plate. Secure the

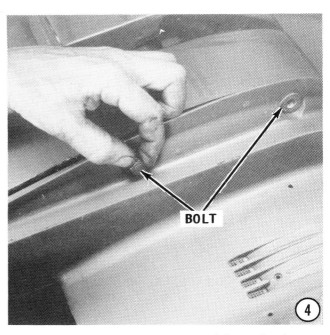

lower unit in this position with a C-clamp. The lower unit will then be held firmly in a favorable position during the service work. An alternate method is to cut a groove in a short piece of 2" x 6" wood to accommodate the lower unit with the cavitation plate resting on top of the wood. Clamp the wood in a vise and service work may then be performed with the lower unit erect (in its normal position), or inverted (upside down). In both positions, the cavitation plate is the supporting surface.

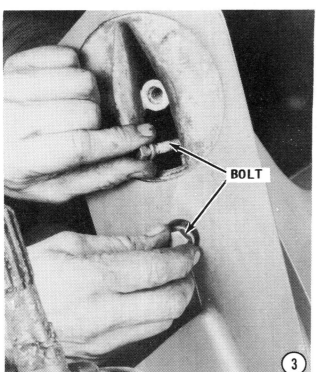

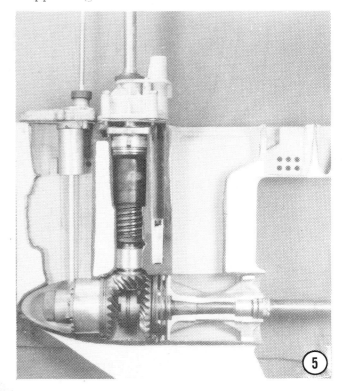

MECH. SHIFT HYDRAULIC ASSIST

7- Remove the O-ring from the top of the driveshaft. Remove the bolts securing the water pump to the lower unit housing. Pull the water pump housing up and free of the driveshaft. Slide the water pump impeller up and free of the driveshaft.

8- Pop the impeller Woodruff key out of the driveshaft keyway. Slide the water pump base plate up and off of the driveshaft.

GOOD WORDS

If the only work to be performed is service of the water pump, proceed directly to Page 8-59, Water Pump Installation.

Assist Cylinder — Removal

9- Remove the four screws at the front edge of the lower unit. Lift and slide the cover up and free of the shift rod.

Bearing Carrier — Removal

10- Remove the four 5/16" bolts from inside the bearing carrier. Notice how each bolt has an O-ring seal. These O-rings should be replaced each time the bolts are removed. In most cases, the word **UP** is embossed into the metal of the bearing carrier rim. This word must face **UP** in relation to the lower unit during installation.

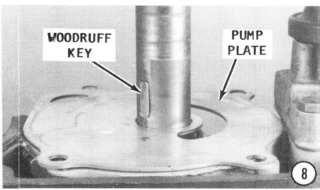
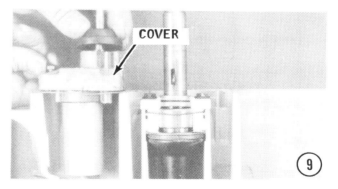

11- Remove the bearing carrier using one of the methods described in the following paragraphs, under Special Words.

SPECIAL WORDS

Several models of bearing carriers are used on the lower units covered in this section.

The bearing carriers are a very tight fit into the lower unit opening. Therefore, it is not uncommon to apply heat to the outside surface of the lower unit with a torch, at the same time the puller is being worked to remove the carrier. **TAKE CARE** not to overheat the lower unit.

One model carrier has two threaded holes on the end of the carrier. These threads permit the installation of two long bolts. These bolts will then allow the use of a flywheel puller to remove the bearing carrier, Illustruation #11.

Another model does not have the threaded screw holes. To remove this type bearing carrier, a special puller with arms must be used. The arms are hooked onto the carrier web area, and then the carrier removed, reference illustration **"C"**.

WARNING

The next step involves a dangerous procedure and should be executed with care while wearing **SAFETY GLASSES**. The retaining rings are under tremendous tension in the groove and while they are being removed. If a ring should slip off the Truarc pliers, it will travel with incredible speed

causing personal injury if it should strike a person. Therefore, continue to hold the ring and pliers firm after the ring is out of the groove and clear of the lower unit. Place the ring on the floor and hold it securely with one foot before releasing the grip on the pliers. An alternate method is to hold the ring inside a trash barrel, or other suitable container, before releasing the pliers.

12- Obtain a pair of Truarc pliers. Insert the tips of the pliers into the holes of the first retaining ring. Now, **CAREFULLY** remove the retaining ring from the groove and gear case without allowing the pliers to slip. Release the grip on the pliers in the manner described in the above **WARNING**. Remove the second retaining ring in the same manner. The rings are identical and either one may be installed first. Remove the retainer plate.

13- As the plate is removed, notice which surface is facing into the housing, as an aid during installation.

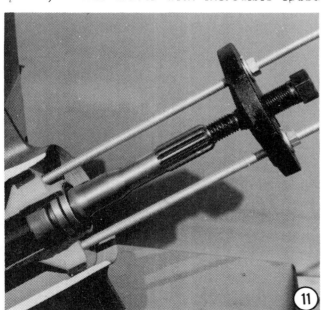

MECH. SHIFT HYDRAULIC ASSIST   8-39

Propeller Shaft -- Removal

14- Grasp the propeller shaft firmly with one hand and at the same time grasp the shift rod with your other hand. Now, pull the propeller shaft out and the shift rod upward at the same time. Remove both units from the lower unit. As the propeller shaft is withdrawn, the reverse gear, thrust bearing, washer, and retainer plate will all come out together with the propeller shaft.

"FROZEN" PROPELLER SHAFT

On rare occasions, especially if water has been allowed to enter the lower unit, it may not be possible to withdraw the propeller shaft as described in Step 14. The shaft may be "frozen" in the hydraulic pump due to corrosion.

If efforts to remove the propeller shaft after the bearing carrier has been removed fail, thread an adaptor to the end of the propeller shaft. Both ends of the adaptor have internal threads. Now, attach a slide hammer to the adaptor. Attach some type of slide hammer to the lower push rod or whatever is left in the lower unit of the assist cylinder mechanism. Obtain the help of an assistant. While the assistant operates one slide hammer, or whatever device is attached to the shift mechanism, operate the slide hammer on the propeller shaft at the same time, reference illustration **"D"**.

BAD NEWS

Using the slide hammer, under these conditions, to remove the propeller shaft,

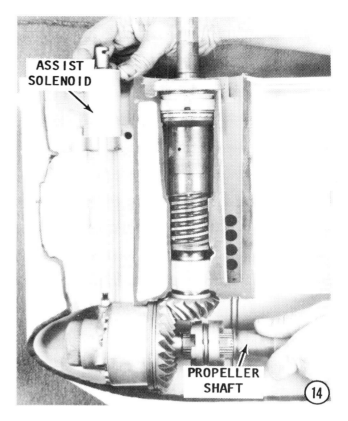

will usually result in some internal part being damaged as the propeller is withdrawn. Usually, the parts damaged will be the shift cylinder, piston, and/or the push rod. However, the cost of replacing the damaged parts is reasonable considering the seriousness of a "frozen" shaft and the problem of getting it out.

Pinion Gear -- Removal

Special tool, OMC No. 311875 is required to turn the driveshaft in order to remove the pinion gear nut.

15- Obtain the special tool and slip it over the end of the driveshaft with the splines of the tool indexed with the splines on the driveshaft. Hold the pinion gear nut with the proper size wrench, and at the same time rotate the driveshaft, with the special tool and wrench **COUNTERCLOCKWISE** until the nut is free. If the special tool is not available, clamp the driveshaft in a vise equipped with soft jaws, in an area below the splines but not in the water pump

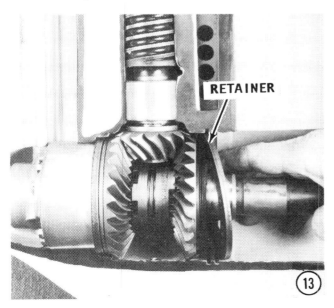

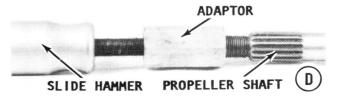

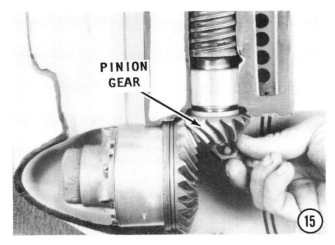

impeller area. Now, with the proper size wrench on the pinion gear nut, rotate the complete lower unit **COUNTERCLOCKWISE** until the nut is free. This procedure will require the lower unit to be rotated, then the wrench released, the lower unit turned back, the wrench again attached to the nut, and the unit rotated again. Continue with this little maneuver until the nut is free. After the nut is free, proceed with the next step. The driveshaft will be withdrawn from the pinion gear.

Driveshaft — Removal

16- Remove the four bolts from the top of the lower unit securing the bearing housing. **CAREFULLY** pry the bearing housing upward away from the lower unit, then slide it free of the driveshaft. An alternate method is to again clamp the driveshaft in a vise equipped with soft jaws. Use a soft-headed mallet and tap on the top side of the bearing housing. This action will jar the housing loose from the lower unit. Continue tapping with the mallet and the bearing housing, O-rings, shims, thrust washer, thrust bearing, and the driveshaft will all breakaway from the lower unit and may be removed as an assembly.

17- Remove the pinion gear from the lower unit cavity. Remove the forward gear, thrust washer, and thrust bearing, from the hydraulic pump.

Hydraulic Pump — Removal

18- Obtain two long rods with 1/4" x 20 threads on both ends. Thread the two rods into the hydraulic pump housing. Attach a slide hammer to the rods and secure it with a nut on the end of each rod. Check to be

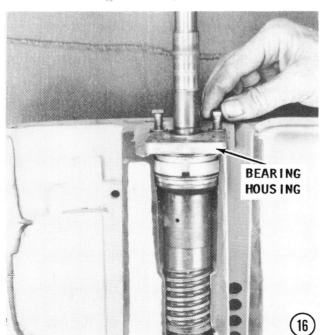

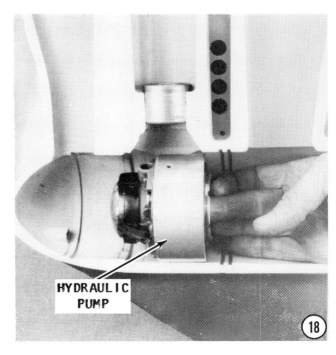

sure the slide hammer is installed onto the rods **EVENLY** to allow an even pull on the pump. If the slide hammer is not installed to the rods properly, the pump may become tightly wedged in the lower unit. Operate the slide hammer and pull the hydraulic pump free. If the pump should happen to become lodged in the lower unit, stop operating the slide hammer **IMMEDIATELY**. Tap the hydraulic pump back into place in the lower unit and start the removal procedure over.

Lower Driveshaft Bearing — Removal

This bearing cannot be removed without the aid of a special tool. Therefore, **DO NOT** attempt to remove this bearing unless it is unfit for further service. To check the bearing, first use a flashlight and inspect it for corrosion or other damage. Insert a finger into the bearing, and then check for "rough" spots or binding while rotating it.

19- Remove the Allen screw, if used, from the water pickup slots in the starboard side of the lower unit housing. This screw secures the bearing in place and **MUST** be removed before an attempt is made to remove the bearing. The bearing must be actually **"PULLED"** upward to come free. **NEVER** make an attempt to "drive" it down and out.

20- Obtain special tool, OMC No. 385547. Use the special tool and "pull" the bearing from the lower unit.

Propeller Shaft — Disassembling
SPECIAL WORDS

On the **1973** model, the plunger and shift rod bearing are removable. The plunger has a snap ring on the end to secure it in the end of the propeller shaft, exploded reference illustration "F".

On all other models covered in this section, a spring is installed into the end of the propeller shaft. Three detent balls are used on the shift rod and plunger assembling -- one ball between the plunger and the spring, and the other two ride partially indexed into holes on opposite sides of the plunger. The accompanying exploded reference illustration "G", will help clarify the relationship of these parts.

21- Notice the spring retainer on the outside surface of the clutch dog. Use a small screwdriver and work one end of the spring up onto the shoulder of the clutch dog. Continue working the spring out of the groove until it is free. **TAKE CARE** not to distort the spring.

22- Place one end of the propeller shaft on the bench and push the pin free of the clutch dog. Raise the propeller end of the shaft upward and the piston, plunger, detent balls, and spring, will come free of the shaft, all models, except 1973, reference illustration "E".

23- Slide the clutch dog free of the propeller shaft.

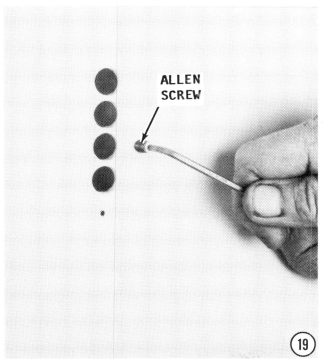

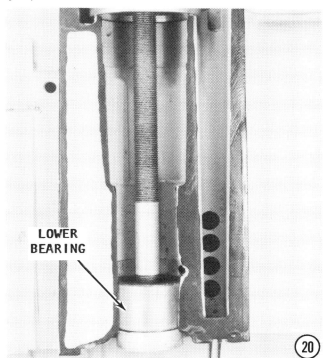

8-42 LOWER UNIT

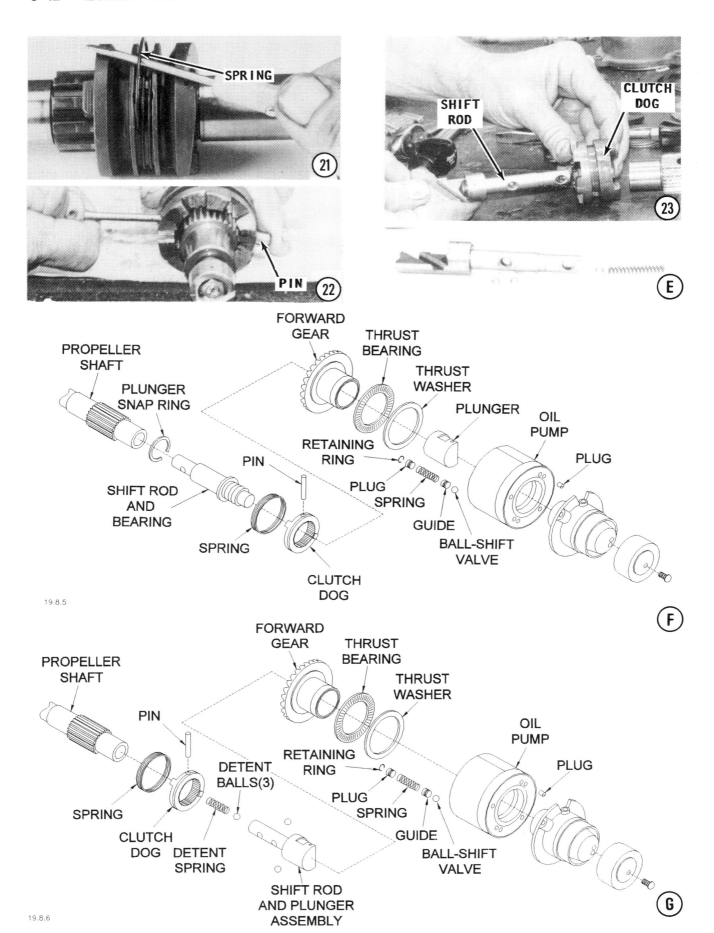

MECH. SHIFT HYDRAULIC ASSIST

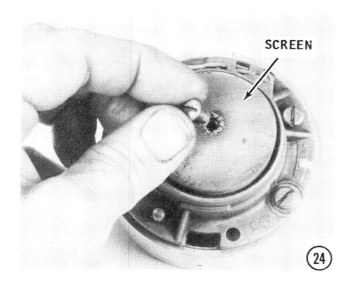

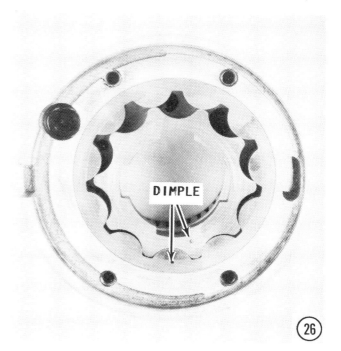

Hydraulic Pump — Disassembling

24- Remove the screw from the center of the screen on the back side of the pump. Remove the screen.

25- Remove the screws securing the valve housing to the pump, and then lift the housing free of the pump.

26- Lift the two gears out off the pump housing and **HOLD** them just as they were removed. Check the face of each gear for an indent mark (a dot, dimple, or similar identification). The identification mark will indicate how the gear **MUST** face in the housing. Make a note of how the mark faces, outward or inward, to **ENSURE** the gears will be installed properly in the same position from which they were removed.

Bearing Carrier — Disassembling

The bearings in the carrier need **NOT** be removed unless they are unfit for further service. Insert a finger and rotate the bearing. Check for "rough" spots or binding. Inspect the bearing for signs of corrosion or other types of damage. If the bearings must be replaced, proceed with the next step.

27- Use a seal remover to remove the two back-to-back seals or clamp the carrier in a vise and use a pry bar to pop each seal out.

28- Use a drift punch to drive the bearings free of the carrier. The bearings are being removed because they are unfit for service, therefore, additional damage is of no consequence.

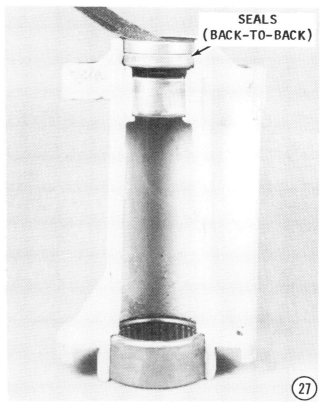

8-44   Lower Unit

Assist Cylinder — Disassembling

29- Back the shift rod out of the assist cylinder by rotating it **COUNTERCLOCKWISE** until it is free. The threads of the shift rod provide the adjustment, which will be made at the time of installation. Therefore, it is not necessary to count the number of turns at the time of removal.

Two special tools are required to disassemble the assist cylinder further. The assist cylinder can be purchased as a complete assembly. Therefore, it may be less expensive to purchase a new unit rather than buying the special tools and attempting an overhaul.

30- The two special tools required, are both OMC No. 386112. One is used on top of the cylinder and the other on the bottom. Obtain the two tools. Notice the hex head on both ends of the tool. Secure one end in a vise. Slide the lower push rod of the assist cylinder through the special tool and seat it onto the two prongs of the tool. Position the other tool onto the top of the cylinder and back off the cylinder cap.

31- Remove the valve and O-ring from the cylinder cap. Remove the two O-rings from the lower end of the cylinder.

32- Remove the retaining pin from the piston and the lower push rod will come free.

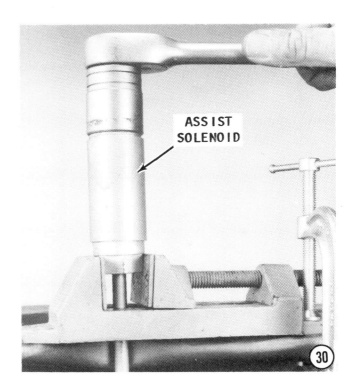

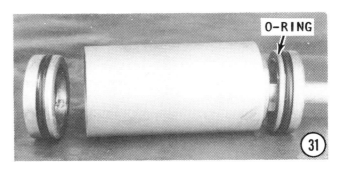

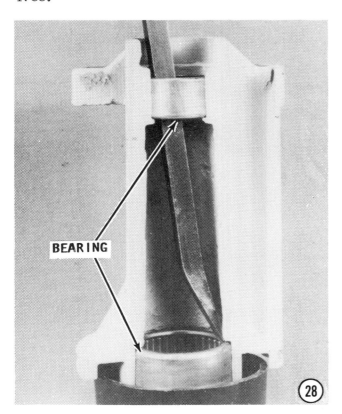

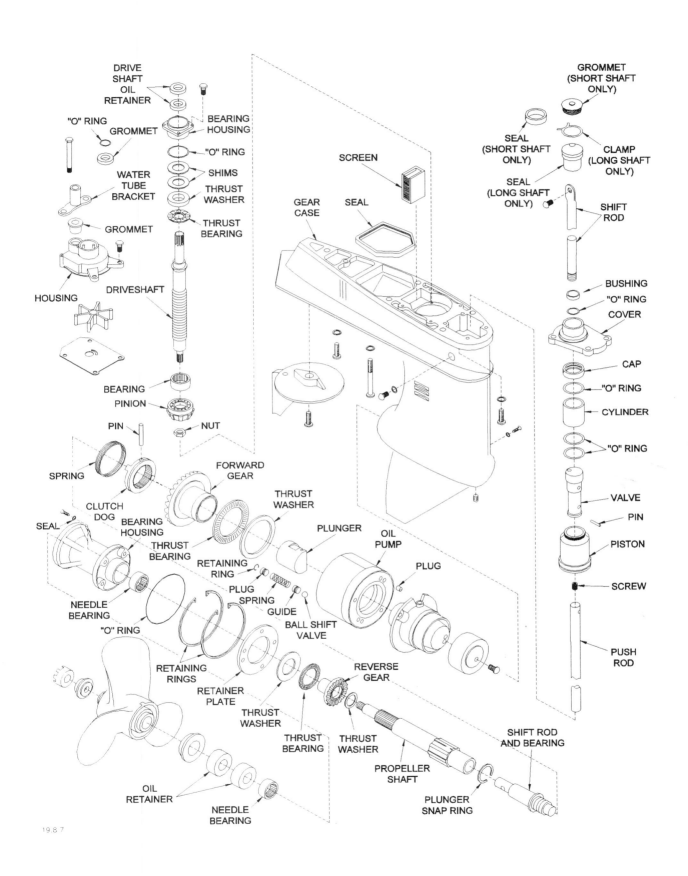

Exploded view of the lower unit used on the 65 hp to 135 hp -- 1973 engines.

8-46 LOWER UNIT

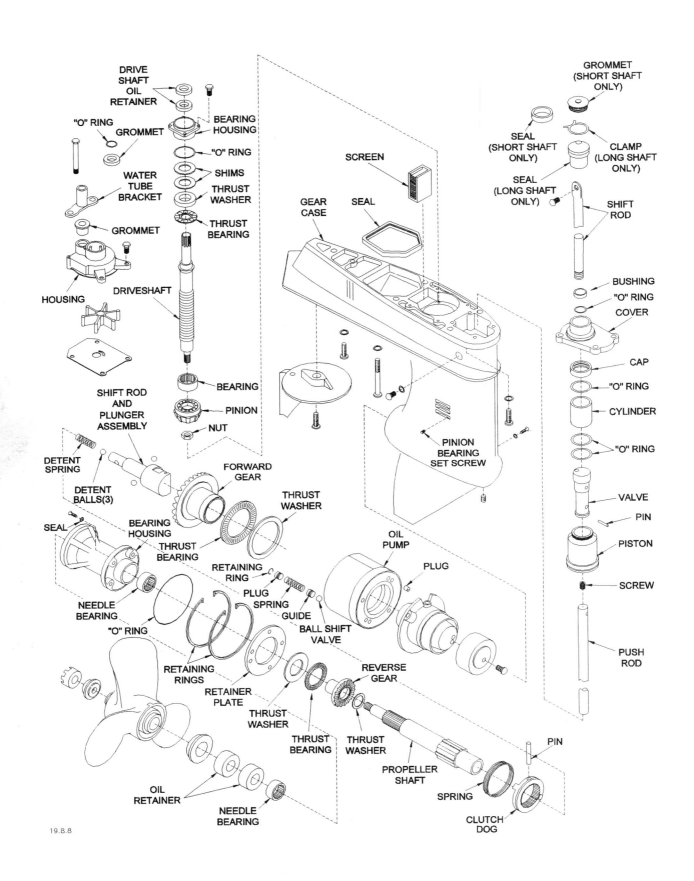

Exploded view of the lower unit for the 70 hp to 135 hp -- 1975, and the 85 hp to 140 hp -- 1976 and 1977 engines.

MECH. SHIFT HYDRAULIC ASSIST   8-47

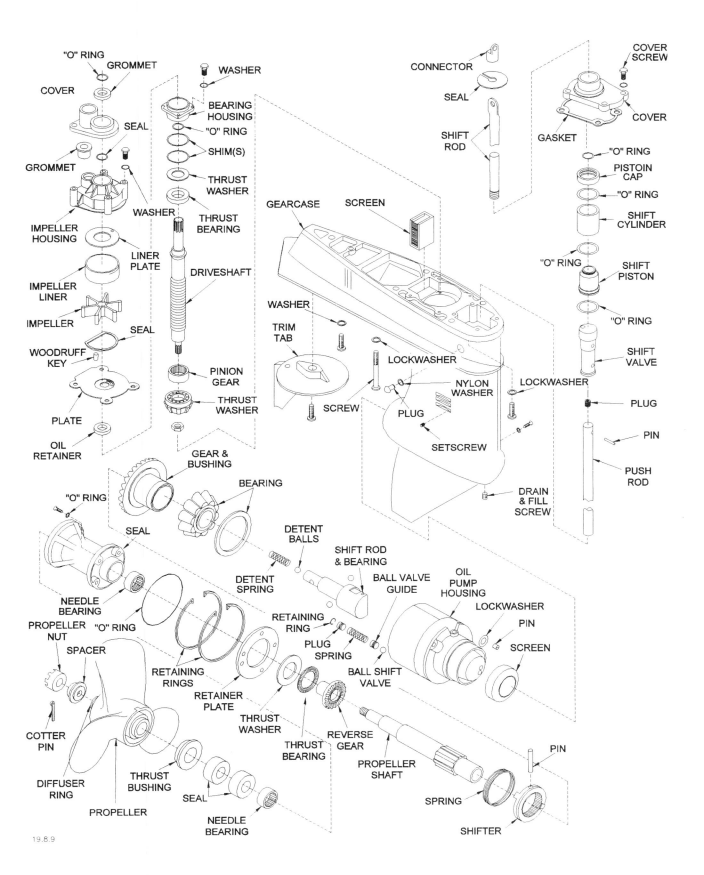

Exploded view of the lower unit for the 200 hp -- 1976 and the 175 hp and 200 hp -- 1977 engines.

CLEANING AND INSPECTING

Wash all parts in solvent and dry them with compressed air. Discard all O-rings and seals that have been removed. A new seal kit for this lower unit is available from the local dealer. The kit will contain the necessary seals and O-rings to restore the lower unit to service.

Inspect all splines on shafts and in gears for wear, rounded edges, corrosion, and damage.

Carefully check the driveshaft and the propeller shaft to verify they are straight and true without any sign of damage. A complete check must be performed by turning the shaft in a lathe. This is only necessary if there is evidence to suspect the shaft is not true.

Check the water pump housing for corrosion on the inside and verify the impeller and base plate are in good condition. Actually, good shop practice dictates to rebuild or replace the water pump each time the lower unit is disassembled. The small cost is rewarded with "peace of mind" and satisfactory service.

Inspect the lower unit housing for nicks, dents, corrosion, or other signs of damage. Nicks may be removed with No. 120 and No. 180 emery cloth. Make a special effort to ensure all old gasket material has been removed and mating surfaces are clean and smooth.

Inspect the water passages in the lower unit to be sure they are clean. The screen may be removed and cleaned.

Damaged hydraulic pump. Water in the lower unit and a broken gear was the cause of this pump being destroyed.

A two-section driveshaft with a weld section that has failed. This area of the driveshaft should be carefully checked anytime the lower unit is disassembled.

Check the gears and clutch dog to be sure the ears are not rounded. If doubt exists as to the part performing satisfactorily, it should be replaced.

Inspect the bearings for "rough" spots, binding, and signs of corrosion or damage.

ASSEMBLING

READ AND BELIEVE

The lower unit should **NOT** be assembled in a dry condition. Coat all internal parts with OMC HI-VIS lube oil as they are assembled. All seals should be coated with OMC Gasket Seal Compound. When two seals are installed back-to-back, use Triple Guard Grease between the seal surfaces.

Propeller Shaft -- Assembling
First, These Words

Assembling the propeller shaft for the 1973 model lower unit is somewhat different from all other models covered in this manual. Therefore, two sets of steps are presented -- one for each model.

For the 1973 model, follow Steps 1, 2, and 3, then jump to Step 7.

For all other models, begin by skipping to Step 3, and then carry on through.

For 1973 Model

1- Install the shift rod and bearing assembly into the end of the propeller shaft. Reference exploded illustration "A" on page 8-49 will be helpful during the assembling work. Slide the clutch dog onto the propeller shaft with the face of the dog marked **"PROP END"** facing toward the propeller end of the shaft. Before the splines of the clutch dog engage the splines of the propeller shaft, rotate the dog until the hole for the pin appears to align with the hole through the propeller shaft. Slide the clutch dog onto the splines until the hole in the dog aligns with the hole in the shaft. If the hole is off just a bit, slide the clutch dog back off the splines, rotate it one spline in the required direction, and then slide it into

MECH. SHIFT HYDRAULIC ASSIST

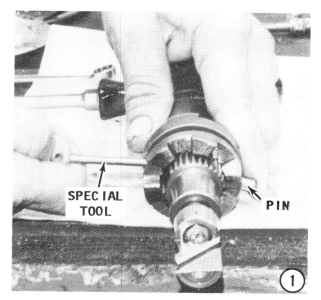

place. Insert the pin through the clutch dog, the shaft, and out the other side of the shaft and clutch dog. Center the pin through the clutch dog.

2- Install the spring-type pin retainer around the clutch dog to secure the pin in place. **TAKE CARE** not to distort the pin retainer during the installation process.

For All EXCEPT 1973 Model

3- In order to assemble the shift rod and clutch dog onto the propeller shaft, it will be necessary to make a simple detent spring depressing tool. This can be accomplished by cutting off a piece of 9/32" diameter round bar stock approximately 6-inches long. Flatten one end of the bar stock, as shown in the accompanying illustration.

4- Insert two of the detent balls into the detent ball hole in the side of the shift rod. Reference exploded illustration "B" on the next page will be helpful during the assembling work. Insert the third detent ball into the end of the shift rod, and then place the spring through the hole in the end of the

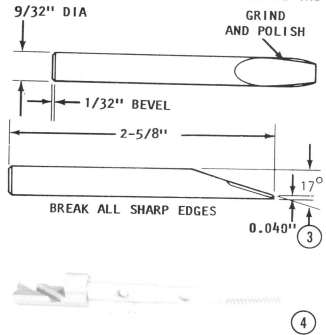

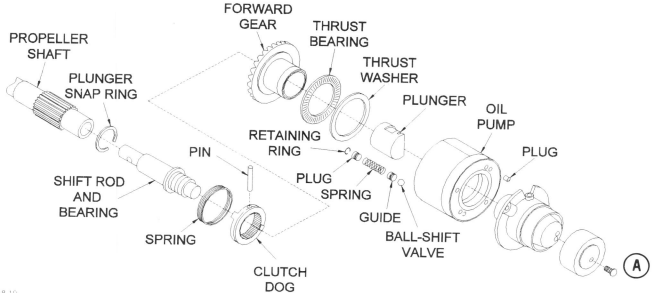

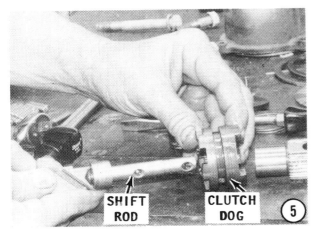

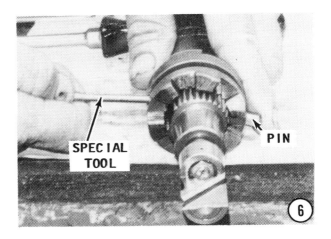

shift rod. Now, hold the balls in the shaft, and at the same time carefully align the holes in the shift rod and the clutch dog with the slot in the propeller shaft. Next, insert the shift rod into the propeller shaft until the detent balls slip into the grooves in the propeller shaft.

5- Cover the propeller shaft with a light coating of oil. Slide the clutch dog onto the shaft with the three-lug end facing the propeller end of the shaft. Align the holes in the clutch dog with the slot in the propeller shaft. The words **PROP END** are stamped on the side intended to face the propeller.

6- With the holes in the clutch dog still aligned with the holes in the shift rod, start the wedge end of the special tool made before Step 1, into the hole with the flat side facing the end of the detent spring. Continue pushing the tool into the hole until the end of the tool barely comes out the opposite side. Now, press the clutch dog retaining pin in through the clutch dog. As the retaining pin is inserted, the tool will be forced out.

7- Secure the pin in place with the retaining spring. **TAKE CARE** to be sure none of the spring coils overlap or the spring is not distorted in any way. Only through careful attention to installation of the retaining pin and spring can proper operation of the shift mechanism be expected. Set the completed assembly to one side until ready for installation into the lower unit.

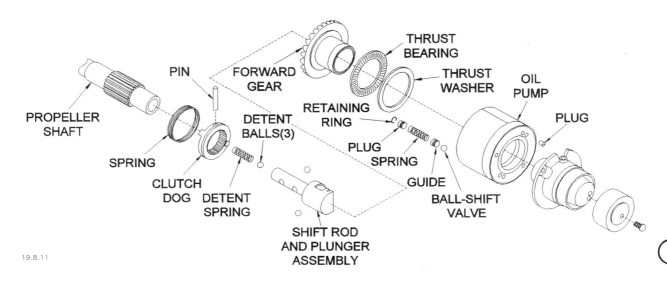

Bearing Carrier Bearings & Seals Installation

8- Install the reverse gear bearing into the bearing carrier by pressing against the **LETTERED** side of the bearing with the proper size socket. Press the forward gear bearing into the bearing carrier in the same manner. Press against the **LETTERED** side of the bearing.

9- Coat the outside surfaces of the seals with HI-VIS oil. Install the first seal with the flat side facing **OUT**. Coat the flat surface of both seals with Triple Guard Grease, and then install the second seal with the flat side going in **FIRST**. The seals are then back-to-back with the grease between the two surfaces. The outside seal prevents water from entering the lower unit and the inside seal prevents the lubricant in the lower unit from escaping.

Hydraulic Pump -- Assembling

10- Check the note made during disassembling, per Step 26, to determine how the identifying marks (dots, dimples, whatever) on the gears must face -- inward or outward. The gears **MUST** be installed in the same position from which they were removed.

11- After the gears have been installed into the pump, use a straightedge and check to be sure the gears are level in the pump. If the gears cannot be made level, either the gears or the pump must be replaced.

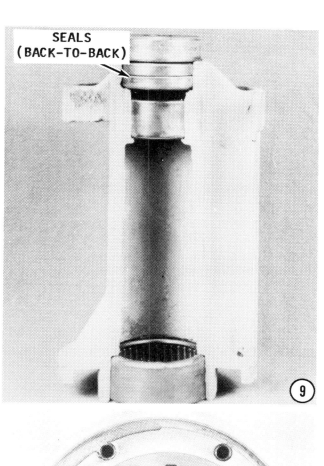

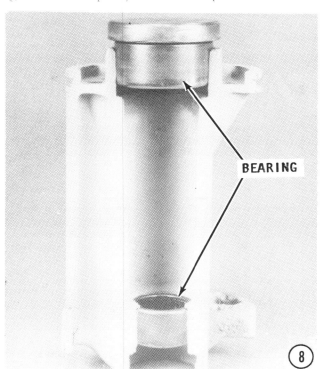

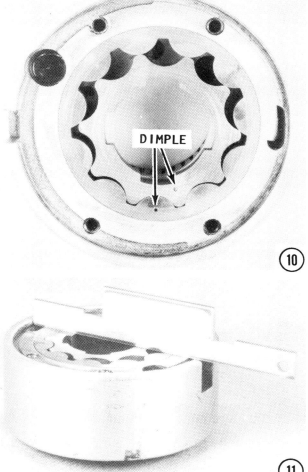

8-52 Lower Unit

12- Install the rear valve housing with the tang on the outside edge of the housing indexed with the small slot in the pump housing. Secure the valve housing in place with the attaching screws tightened securely.

13- Place the screen in position on the back side of the valve housing, and then secure it in place with the screw.

14- Install the forward gear into the pump housing. It may be necessary to work the gears around in the pump to permit the tangs on the forward gear shank to index in the slots in the housing. Set the assembly aside for later installation.

Lower Driveshaft Bearing — Installation

15- Obtain tool, OMC No. 385547. Assemble the tool with the washer, guide sleeve, and remover portion of the tool, in the order given. The shoulder of the tool must face **DOWN**. Place the bearing onto the end of the tool, with the lettered side of the bearing facing the tool. Drive the bearing down until the large washer on the tool makes contact with the surface of the lower unit. The bearing is then seated to the proper depth.

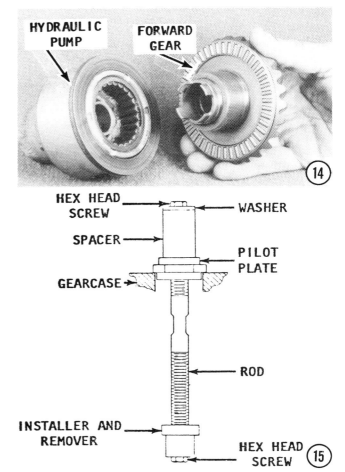

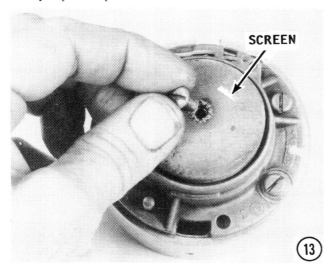

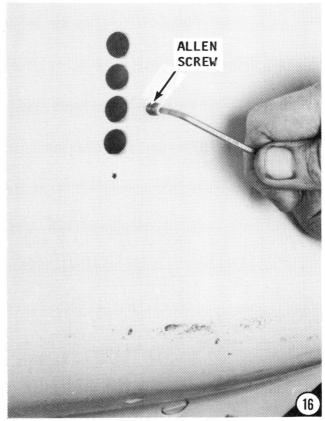

MECH. SHIFT HYDRAULIC ASSIST 8-53

16- After the bearing has been installed, install the Allen screw through the water pickup vent ramps to secure the bearing in place. Use Locktite on the screw threads before installation.

Hyraulic Pump — Installation
First, These Words

Observe the hole on the topside of the pump. This tang **MUST** face directly up in relation to the lower unit housing to permit installation of the shift rod into the pump. Also notice the pin on the backside of the pump. This pin **MUST** index into a matching hole in the housing to restrain the pump from rotating.

17- Secure the lower unit housing in the horizontal position with the bearing carrier opening facing up. Remove the forward gear from the pump. Obtain two long 1/4 x 20 rods with threads on both ends. Thread the rods into the pump and then lower the pump into the lower unit housing. To index the pin on the back of the pump housing into the hole in the lower unit is not an easy task. However, exercise patience and rotate the pump ever so slowly. A helpful hint at this point: As the pump is being lowered into the cavity, align the opening and tang on top of the pump in the approximate position your eye indicates the shift rod may be installed. When the pin indexes, it will not be possible to rotate the pump. The pump **MUST** be properly seated to permit installation of the shift rod and the pinion gear. After the pump is in place, remove the two rods used during installation.

18- Install the thrust washer and thrust bearing into the pump with the flat side of the bearing facing **OUTWARD**. Lower the forward gear into the pump. Work the gear slowly until the teeth index with the teeth of the pump gear. This should not be too difficult because the forward gear was installed once, and then removed in the previous step. However, it is entirely possible the gears moved when the pump was installed. Therefore, use a flashlight and check the position of the gears. If necessary, use a long shank screwdriver and rotate the gears until they are close to center, then install the forward gear.

Driveshaft and Pinion Gear — Installation

CRITICAL WORDS

The driveshaft and pinion gear must be assembled prior to installation, and then checked with a special shimming gauge. This shimming must be accomplished properly, the unit disassembled, and then installed into the lower unit. Use of the shimming gauge is the **ONLY** way to determine the proper amount of shimming required at the upper end of the driveshaft. The following detailed step outlines the procedure.

19- Clamp the driveshaft in a vise equipped with soft jaws and in such a manner that the splines, water pump area, or other critical portions of the shaft cannot be damaged. Slide the pinion gear onto the driveshaft with the bevel of the gear teeth facing toward the lower end of the shaft. Install the pinion gear nut and tighten it to a torque value of 40 to 45 ft-lbs. No parts should be installed on the upper end of the driveshaft at this point. Slide the shims removed during disassembly onto the driveshaft and seat them against the driveshaft shoulder. Obtain special shimming tool, OMC No. 315767. If servicing a 175 hp or 200 hp, 1976 or 1977, obtain special shimming tool OMC No. 321520. Slip the special tool down over the driveshaft and onto the upper surface of the top shim. Measure the distance between the top of the pinion gear and the bottom of the tool. The tool should just barely make contact with the pinion gear surface for ZERO clearance. Add or remove shims from the upper end of the driveshaft to obtain the required ZERO

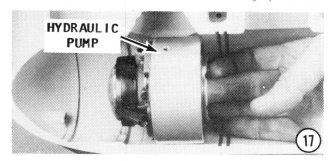

clearance. Remove the tool and set the shims aside for installation later. Back off the pinion gear nut and remove the pinion gear. Remove the driveshaft from the vise.

20- Insert the pinion gear into the cavity in the lower unit with the flat side of the bearing facing **UPWARD**. Hold the pinion gear in place and at the same time lower the driveshaft down into the lower unit. As the driveshaft begins to make contact with the pinion gear, rotate the shaft slightly to permit the splines on the shaft to index with the splines of the pinion gear. After the shaft has indexed with the pinion gear, thread the pinion gear nut onto the end of the shaft. Obtain special tool, OMC No.- 316612. Slide the special tool over the upper end of the driveshaft with the splines of the tool indexed with the splines on the shaft. Attach a torque wrench to the special tool. Now, hold the pinion gear nut with the proper size wrench and rotate the driveshaft **CLOCKWISE** with the special tool until the pinion gear nut is tightened to a torque value of 40 to 45 ft-lbs. Remove the special tool.

21- Slide the thrust bearing, thrust washer, and the shims, set aside after the shim gauge procedure in Step 19, onto the driveshaft.

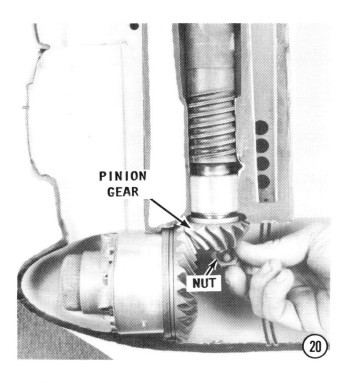

22- Install the two seals back-to-back into the opening on top of the bearing housing. Coat the outside surface of the **NEW** seals with OMC Lubricant. Press the first seal into the housing with the flat side of the seal facing **OUTWARD**. After the seal is in place, apply a coating of Triple

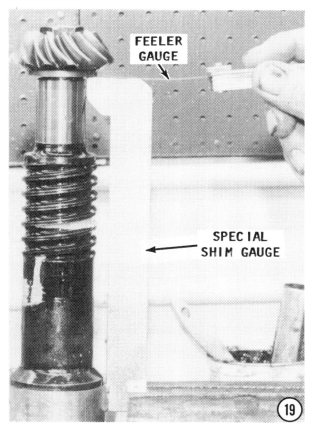

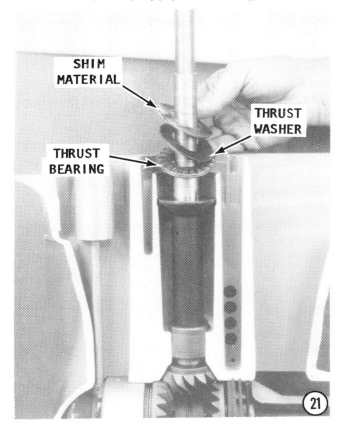

MECH. SHIFT HYDRAULIC ASSIST 8-55

A set of double seals showing the back side (left) and the front side (right). These seals are installed back-to-back (flat side-to-flat side) with Triple Guard Grease between the surfaces. This arrangement prevents fluid from passing in either direction.

Guard Grease to the flat side of the installed seal and the flat side of the second seal. Press the second seal into the bearing housing with the flat side of the seal facing **INWARD**. Insert a **NEW** O-ring into the bottom opening of the bearing housing.

23- Wrap friction tape around the splines of the driveshaft to protect the seals in the bearing housing as the housing is installed. Now, slide the assembled bearing housing down the driveshaft and seat it in the lower unit housing. Secure the housing in place with the four bolts. Tighten the bolts **ALTERNATELY** and **EVENLY**.

Propeller Shaft -- Installation
First, Some Good Words

Installation of the propeller shaft, assist valve, and plunger is not a simple task. However, with patience, attention to detail, and an understanding of the installation sequence, the work can proceed smoothly without serious problems or the need to disassemble and repeat certain steps. The best word of advice is to read through the following procedure at least twice, or until the various parts and their relationship to each other including the installation sequence is thoroughly familiar. Before these

units are installed in the lower unit, some assembling on the bench can be performed.

Assemble the propeller shaft and shifter as outlined at the beginning of this section, Steps 1 and 2 for the 1973 model; Steps 3 thru 6 for all other models. Observe how the boss on the plunger slides into the cut-a-way in the shift rod, reference illustration "C". This arrangement **MUST** be accomplished when these units are installed in the lower unit. Also observe that the shift rod on the end of the propeller shaft is heavy on the back side causing the rod to rotate. Therefore, when the propeller shaft is held horizontal, the heavy side will turn the shift rod downward.

24- Lay the lower unit on a flat surface with the port side facing up. Adjust the unit on the bench to offer a clear view into the assist cylinder bore and also into the propeller shaft opening. Now, hold the propeller shaft with the heavy side of the shift rod down and insert it through the oil pump and forward gear assembly.

25- Push the shaft in as far as possible. Use a flashlight to illuminate the interior of the shift assist cylinder bore. Observe the cut-a-way in the shifter rod. If you cannot see the cut-a-way, move the propeller shaft inward or outward until it is visible.

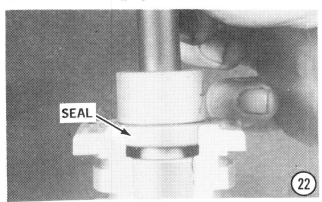

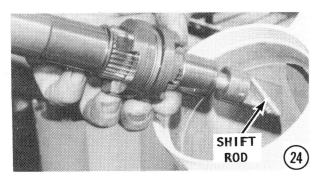

With the propeller shaft in this position, **CAREFULLY** insert the push rod and assist cylinder assembly into the gearcase with the boss on the cylinder facing down. Continue to slowly move the assembly into the gearcase until it makes contact with the plunger. When the valve and plunger are properly engaged, flat-to-flat, the valve and plunger cannot be rotated. Exert a **GENTLE** downward pressure on the assist cylinder and valve assembly and at the same time, slowly **EASE** the propeller shaft backward to engage the plunger keyway with the push rod key. The gentle downward pressure on the shift assist assembly will help the push rod key to slip into the plunger keyway. After the proper alignment has been made, as just described, push the propeller shaft and shift assist assembly inward until full engagement is reached. When the propeller shaft and the shift assist assembly are properly engaged, the assist valve and push rod **CANNOT** be rotated.

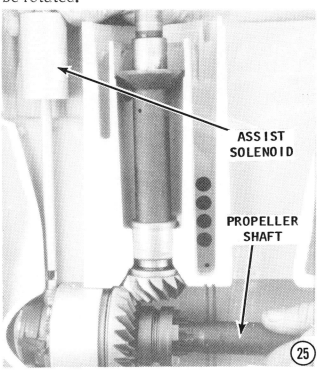

Solenoid Cover and Gasket — Installation

26- Slide the solenoid cover up onto the upper part of the shift rod, and then place a **NEW** gasket in place on the bottom side of the cover. Thread the shift rod into the valve assembly in the assist solenoid. Rotate the shift rod inward about three complete turns for a rough preliminary adjustment. Check to be sure the unit is in **NEUTRAL**. Now, measure the distance from the top of the housing to the center of the hole in the shift rod, reference illustration "D". This distance must be within 1/32" (0.79 mm) as follows for the horsepower and model year listed:

Long shaft is abbreviated l.s.

200 hp l.s. -- 1976 21.411" (54.38 cm)
200 hp extra l.s. -- 1976 26.411" (67.08 cm)
175 hp l.s. -- 1977 21.600" (54.86 cm)
175 hp extra l.s. -- 1977 26.00" (66.04 cm)
200 hp l.s. -- 1977 21.600" (54.86 cm)
200 hp extra l.s. -- 1977 26.000" (66.04 cm)

Rotate the shift rod inward or outward until the required measurement is obtained.

For all other engines covered in this section, simply thread the shift rod connector onto the shift rod until it seats, then back it out until the offset is facing forward.

If servicing a 70 hp or 75 hp -- 1975, obtain and use special tool OMC No. 320736.

Lower the cover and gasket down onto the lower unit housing. Secure the cover in place with the attaching hardware. Tighten the bolts **ALTERNATELY** and **EVENLY**.

Reverse Gear — Installation

27- Apply a light coating of grease to the inside surface of the reverse gear to hold the thrust washer in place. Insert the thrust washer into the reverse gear. Install the thrust bearing and thrust washer onto the shank on the backside of the reverse gear. Slide the reverse gear down the propeller shaft and index the teeth of the gear with the teeth of the pinion gear.

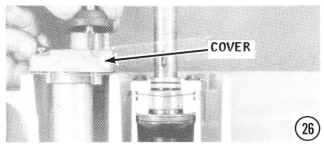

MECH. SHIFT HYDRAULIC ASSIST 8-57

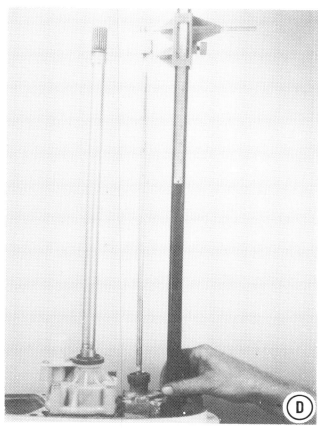

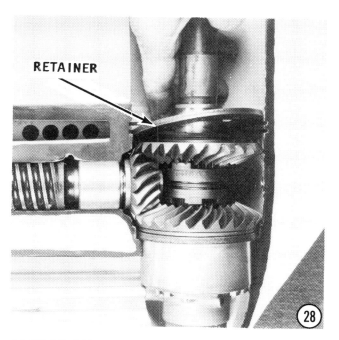

28- Insert the retainer plate into the lower unit against the reverse gear.

WARNING

This next step can be dangerous. Each snap ring is placed under tremendous tension with the Truarc pliers while it is being placed into the groove. Therefore, wear **SAFETY GLASSES** and exercise care to prevent the snap ring from slipping out of the pliers. If the snap ring should slip out, it would travel with incredible speed and cause personal injury if it struck a person.

29- Install the Truarc snap rings one at-a-time following the precautions given in the **WARNING** and the **ADVICE** given in the following paragraph.

WORDS OF ADVICE

The two snap rings index into separate grooves in the lower unit housing. As the first ring is being installed, depth perception may play a trick on your eyes. It may appear that the first ring is properly indexed all the way around in the proper groove, when in reality, a portion may be in one groove and the remainder in the other groove. Should this happen, and the Truarc pliers be released from the ring, it is extremely difficult to get the pliers back into the ring to correct the condition. If necessary use a flashlight and carefully check to be sure the first ring is properly seated all the way around **BEFORE** releasing the grip

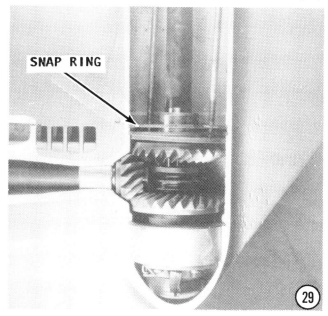

8-58　LOWER UNIT

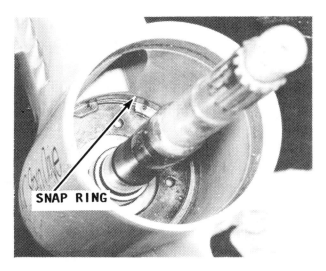

View into the lower unit showing the Truarc snap rings properly locked in the grooves.

on the pliers. Installation of the second ring is not so difficult because the one groove is filled with the first ring.

30- Obtain two long 1/4" rods with threads on one end. Thread the rods into the retainer plate opposite each other to act as guides for the bearing carrier.

31- Check the bearing carrier to be sure a **NEW** O-ring has been installed. Position the carrier over the guide pins with the embossed word **UP** on the rim of the carrier facing **UP** in relation to the lower unit housing, reference illustration **"E"**. Now, lower the bearing carrier down over the guide pins and into place in the lower unit housing.

32- Slide **NEW** little O-rings onto each bolt, and apply some OMC Sealer onto the threads. Install the bolts through the carrier and into the retaining plate. After a couple bolts are in place, remove the guide

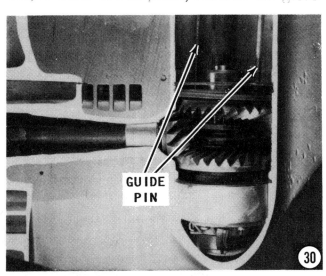

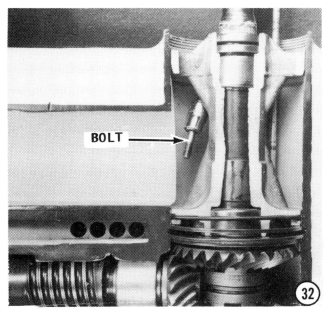

MECH. SHIFT HYDRAULIC ASSIST   8-59

pins and install the remaining bolts. Tighten the bolts **EVENLY** and **ALTERNATELY** to the torque value given in the Appendix.

WATER PUMP INSTALLATION

FIRST, THESE WORDS

An improved water pump is available as a replacement. If the old water pump housing is unfit for further service, only the new pump housing can be purchased. It is strongly recommended to replace the water pump with the improved model while the lower unit is disassembled. The accompanying illustration shows the original equipment (left) compared with the improved pump (right), reference illustration "F".

The new pump must be assembled before it is installed. Therefore, the following steps outline procedures for both pumps.

To assemble and install a replacement pump, perform Steps 33 thru 41, then jump to Step 46.

To install an original equipment pump, proceed directly to Step 42.

Assembling an Improved Pump Housing

33- Remove the water pump parts from the container. Insert the plate into the housing, as shown. The tang on the bottom side of the plate **MUST** index into the short slot in the pump housing.

34- Slide the pump liner into the housing with the two small tabs on the bottom side indexed into the two cutouts in the plate.

35- Coat the inside diameter of the liner with light-weight oil. Work the impeller into the housing with all of the blades bent back to the right, as shown. In this position, the blades will rotate properly when the pump housing is installed. Remember, the pump and the blades will be rotating **CLOCKWISE** when the housing is turned over and installed in place on the lower unit.

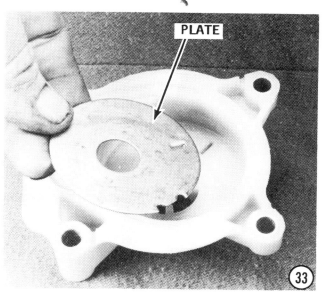
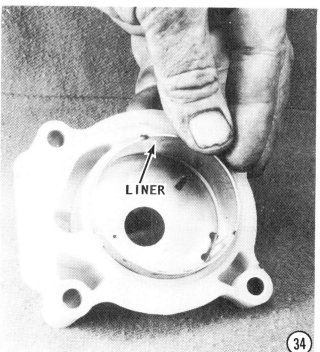
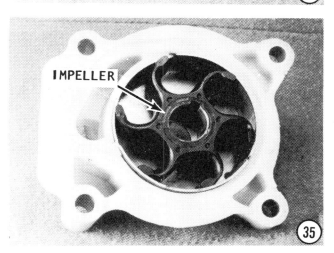

8-60 LOWER UNIT

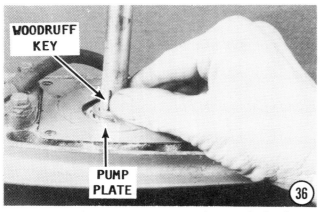

36- Coat the mating surface of the lower unit with 1000 Sealer. Slide the water pump base plate down the driveshaft and into place on the lower unit. Insert the Woodruff key into the key slot in the driveshaft.

37- Lay down a very thin bead of 1000 sealer into the irregular shaped groove in the housing. Insert the seal into the groove, and then coat the seal with the 1000 Sealer.

Water Pump Installation

38- Begin to slide the water pump down the driveshaft, and at the same time observe the position of the slot in the impeller. Continue to work the pump down the

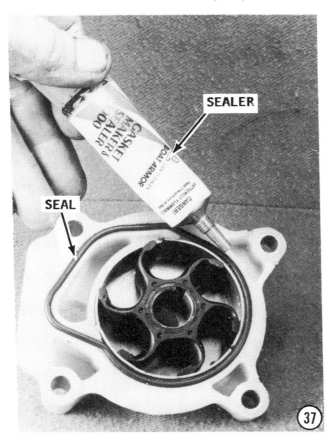

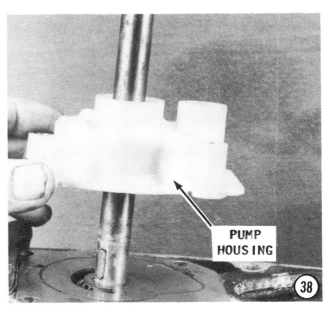

driveshaft, with the slot in the impeller indexed over the Woodruff key. The pump must be fairly well aligned before the key is covered because the slot in the impeller is not visible as the pump begins to come close to the base plate.

39- Install the short forward bolt through the pump and into the lower unit. **DO NOT** tighten this bolt at this time. Insert the grommet into the pump housing.

40- Install the grommet retainer and water tube guide onto the pump housing. Install the remaining pump attaching bolts. Tighten the bolts **ALTERNATELY** and **EVENLY,** and at the same time rotate the driveshaft **CLOCKWISE.** If the driveshaft is not rotated while the attaching bolts are being tightened, it is possible to pinch one of the impeller blades underneath the housing.

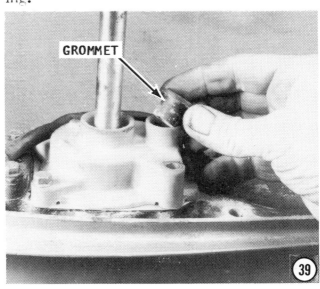

MECH. SHIFT HYDRAULIC ASSIST 8-61

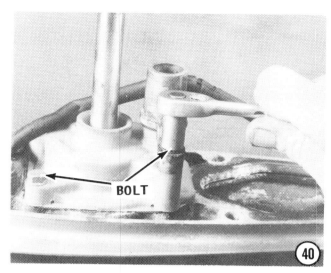

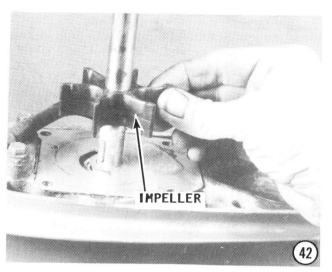

41- Slide the large grommet down the driveshaft and seat it over the pump collar. This grommet does not require sealer. Its function is to prevent exhaust gases from entering the water pump. Proceed directly to Lower Unit Installation, Step 44.

Water Pump Installation
Original Equipment

Perform the following two steps to install an original water pump.

42- Coat the water pump plate mating surface on the lower unit with 1000 Sealer. Slide the water pump plate down the driveshaft and **BEFORE** it makes contact with the sealer check to be sure the bolt holes in the plate will align with the holes in the housing. The plate will only fit one way. If the holes will not align, remove the plate, turn it over and again slide the plate down the driveshaft and into place on the housing. This checking will prevent accidently getting the sealer on both sides of the plate. If by chance sealer does get on the top surface it **MUST** be removed before the water pump impeller is installed.

Slide the water pump impeller down the driveshaft. Just before the impeller covers the cutout for the Woodruff key, install the key, and then work the impeller on down, with the slot in the impeller indexed over the Woodruff key. Continue working the impeller down until it is firmly in place on the surface of the pump plate.

43- Check to be sure **NEW** seals and O-rings have been installed in the water pump. Lubricate the inside surface of the water pump with light-weight oil. Lower the water pump housing down the driveshaft and over the impeller. **ALWAYS** rotate the driveshaft slowly **CLOCKWISE** as the housing is lowered over the impeller to allow the impeller blades to assume their natural and

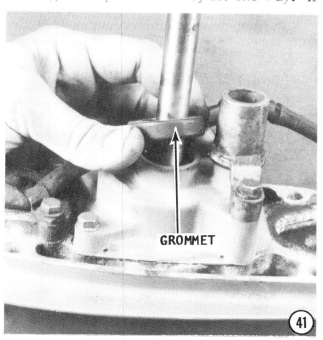

8-62 LOWER UNIT

proper position inside the housing. Continue to rotate the driveshaft and work the water pump housing downward until it is seated on the plate. Coat the threads of the water pump attaching screws with sealer, and then secure the pump in place with the screws. Install the solenoid cable bracket with the same bolt and in the same position from which it was removed. On some units the solenoid cable fits into a recess of the water pump and is held in place in that manner. Tighten the screws **ALTERNATELY** and **EVENLY**.

LOWER UNIT INSTALLATION

44- Check to be sure the water tubes are clean, smooth, and free of any corrosion. Coat the water pickup tubes and grommets with lubricant as an aid to installation. Check to be sure the spark plug wires are disconnected from the spark plugs. Bring the lower unit housing together with the exhaust housing, and at the same time, guide the water tube into the rubber grommet of the water pump. Continue to bring the two units together, at the same time guiding the water pickup tubes into the rubber grommet of the water pump, and, simultaneously rotating the flywheel slowly to permit the splines of the driveshaft to index with the splines of the crankshaft. This may sound like it is necessary to do four things at the same time, and so it is. Therefore, make an earnest attempt to secure the services of an assistant for this task.

45- After the surfaces of the lower unit and exhaust housing are close, dip the attaching bolts in OMC Sealer and then start them in place. Install the two bolts on each side of the two housings. Install the bolt in

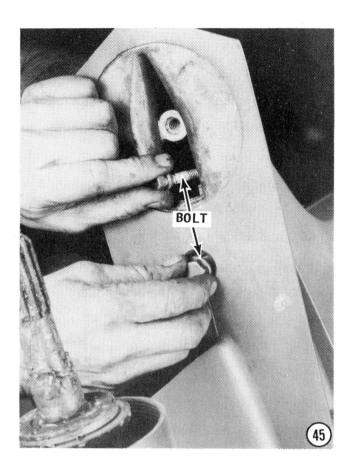

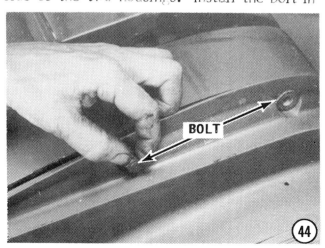

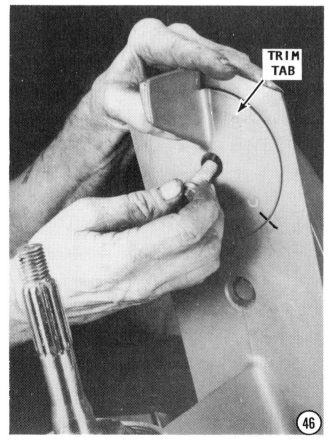

the recess of the trim tab. Install the bolt just forward of the trim tab that extends up through the cavitation plate. Tighten the bolts **ALTERNATELY** and **EVENLY** to the torque value given in the Appendix.

46- Install the trim tab with the mark made on the tab during disassembling aligned with the mark made on the lower unit housing.

47- The shift rod is visible beneath and to the rear of the lower carburetor. Install the shift connector with a bolt or pin, depending on the model being serviced.

Filling the Lower Unit

Fill the lower unit with lubricant according to the procedures in Section 8-3.

Propeller Installation

Install the propeller, see Section 8-2.

FUNCTIONAL CHECK

Perform a functional check of the completed work by mounting the engine in a test tank, in a body of water, or with a flush attachment connected to the lower unit. If the flush attachment is used, **NEVER** operate the engine above an idle speed, because the no-load condition on the propeller would allow the engine to **RUNAWAY** resulting in serious damage or destruction of the engine.

CAUTION: Water must circulate through the lower unit to the engine any time the engine is run to prevent damage to the water pump in the lower unit. Just five seconds without water will damage the water pump.

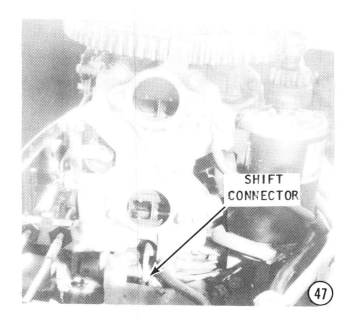

Start the engine and observe the tattle-tale flow of water from idle relief in the exhaust housing. The water pump installation work is verified. If a "Flushette" is connected to the lower unit, **VERY LITTLE** water will be visible from the idle relief port. Shift the engine into the three gears and check for smoothness of operation and satisfactory performance. Remember, when the unit is in forward gear, it is at rest with no current flow to either solenoid.

8-6 MECHANICAL SHIFT
SHIFT DISCONNECT UNDER
LOWER CARBURETOR
70 HP -- 1976 AND 1977
75 HP -- 1975 THRU 1977
ALL UNITS 1978 AND ON

DESCRIPTION

The lower unit covered in this section is a complete mechanical shift unit. A shift cable connects the shift box to the shift linkage at the engine. A shift rod extends from the engine down through the exhaust housing to the lower unit.

Shifting into forward, neutral, and reverse, is accomplished directly through mechanical means from the shift control handle through the cable and linkage to the clutch dog in the lower unit.

The lower unit houses the driveshaft and pinion gear, the forward and reverse driven gears, the propeller shaft, shift lever, cradle, shift shaft, clutch dog, shift rod, and the necessary shims, bearings, and associated parts to make it all work properly. A detent ball and spring is installed on some models.

The water pump is considered a part of the lower unit.

Two different lower units are covered in this section. One unit is used with the electric start model and the other with manual start. The model years actually overlap. The upper driveshaft bearing and the shift mechanism differ between the two units. These differences are clearly indicated in the procedural steps and illustrations.

TROUBLESHOOTING

Preliminary Checks

At rest, and without the engine running, the lower unit is in forward gear. Mount the engine in a test tank, in a body of water, or

8-64 Lower Unit

with a flush attachment connected to the lower unit. If the flush attachment is used, **NEVER** operate the engine above an idle speed, because the no-load condition on the propeller would allow the engine to **RUN-A-WAY**, resulting in serious damage or destruction of the engine.

CAUTION: Water must circulate through the lower unit to the engine any time the engine is run to prevent damage to the water pump in the lower unit. Just five seconds without water will damage the water pump.

Attempt to shift the unit into **NEUTRAL** and **REVERSE**. It is possible the propeller may turn very slowly while the unit is in neutral, due to "drag" through the various gears and bearings. If difficult shifting is encountered, the problem is in the shift linkage or in the lower unit.

The second area to check is the quantity and quality of the lubricant in the lower unit. If the lubricant level is low, contaminated with water, or is broken down because of overuse, the shift mechanism may be affected. Water in the lower unit is **VERY BAD NEWS** for a number of reasons, particularly when the lower unit contains hydraulic components. Hydraulic units will not function with water in the system.

Remember, when the engine is not running, the unit should be in **FORWARD** gear.

BEFORE making any tests, remove the propeller, see Section 8-2. Check the propeller carefully to determine if the hub has

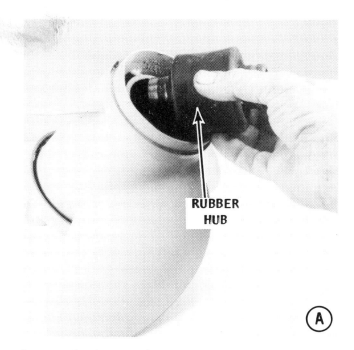

been slipping and giving a false indication the unit is not in gear, reference illustration "A". If there is any doubt, the propeller should be taken to a shop properly equipped for testing, before the time and expense of disassembling the lower unit is undertaken. The expense of the propeller testing and possible rebuild is justified.

The following troubleshooting procedures are presented on the assumption the lower unit lubricant, and the propeller have been checked and found to be satisfactory.

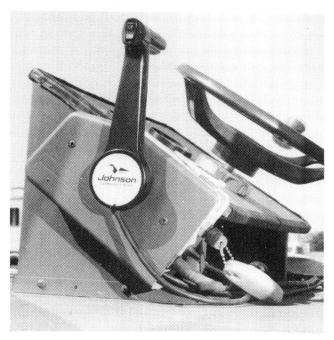

Helm-type shift box and steering wheel.

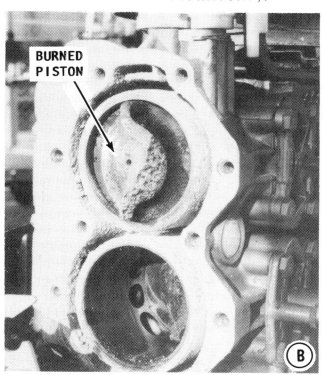

MECHANICAL SHIFT 8-65

Lower Unit Locked

Determine if the problem is in the powerhead or in the lower unit. Attempt to rotate the flywheel. If the flywheel can be moved even slightly in either direction, the problem is most likely in the lower unit. If it is not possible to rotate the flywheel, the problem is a "frozen" powerhead. To absolutely verify the powerhead is "frozen", separate the lower unit from the exhaust housing and then again attempt to rotate the flywheel. If the attempt is successful, the problem is definitely in the lower unit. If the attempt to rotate the flywheel, with the lower unit removed, still fails, a "frozen" powerhead is verified, reference illustration "B".

Unit Fails to Shift
Neutral, Forward, or Reverse

With the outboard mounted on a boat, in a test tank, or with a flush attachment connected, disconnect the shift lever at the engine. Attempt to manually shift the unit into **NEUTRAL, REVERSE, FORWARD**. At the same time move the shift handle at the shift box and determine that the linkage and shift lever are properly aligned for the shift positions. If the alignment is not correct, adjust the shift cable at the trunnion. It is also possible the inner wire may have slipped in the connector at the shift box. This condition would result in a lack of inner cable to make a complete "throw" on the shift handle. Check to be sure the shift handle is moved to the full shift position and the linkage to the lower unit is moved to the full shift position.

If an adjustment is required at the shift box, see Chapter 7.

If it is not possible to shift the unit into gear by manually operating the shift rod while the engine is running, the lower unit requires service as described in this section.

LOWER UNIT SERVICE

Propeller Removal

If the propeller was not removed, as directed for the troubleshooting, remove it now, according to the procedures outlined in Section 8-2.

Draining the Lower Unit

Drain the lower unit of lubricant, see Section 8-3.

GOOD WORDS

If water is discovered in the lower unit and the propeller shaft seal is damaged and requires replacement, the lower unit does **NOT** have to be removed in order to accomplish the work.

The bearing carrier can be removed and the seal replaced without disassembling the lower unit. **HOWEVER**, such a procedure is not considered good shop practice, but merely a quick-fix. If water has entered the lower unit, the unit should be disassembled and a detailed check made to determine if any other seals, bearings, bearing races, O-rings or other parts have been rendered unfit for further service by the water.

LOWER UNIT REMOVAL

1- Disconnect and ground the spark plug wires. Remove the starter motor. Remove the bolt or bolts from the shift disconnect coupler under the lower carburetor.

2- Scribe a mark on the trim tab and a matching mark on the lower unit to ensure the trim tab will be installed in the same position from which it is removed. Use an Allen wrench and remove the trim tab.

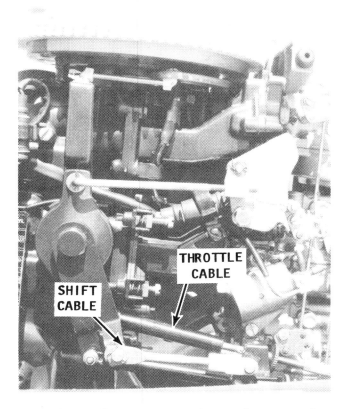

Typical shift and throttle linkage for the engines covered in this section.

8-66 LOWER UNIT

3- On some units, an attaching bolt is installed inside the trim tab cavity. Use a 1/2" socket with a short extension and remove this bolt. Failure to remove this bolt from inside the trim tab cavity may result in an expensive part being broken in an attempt to separate the lower unit from the exhaust housing. Remove the 5/8" countersunk bolt located just ahead of the trim tab position.

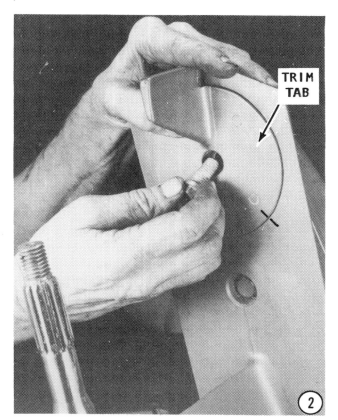

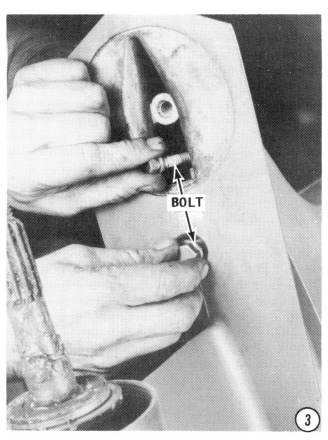

4- Remove the four 9/16" bolts, two on each side, securing the lower unit to the exhaust housing. Work the lower unit free of the exhaust housing. If the unit is still mounted on a boat, tilt the engine forward to gain clearance between the lower unit and the deck (floor, ground, whatever). **EXERCISE CARE** to withdraw the lower unit straight away from the exhaust housing to prevent bending the driveshaft.

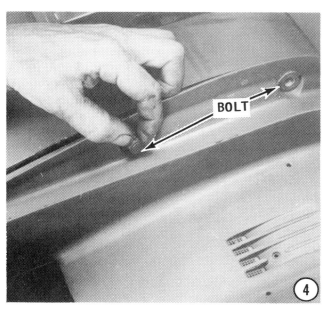

WATER PUMP REMOVAL

5- Position the lower unit in the vertical position on the edge of the work bench resting on the cavitation plate. Secure the lower unit in this position with a C-clamp. The lower unit will then be held firmly in a favorable position during the service work. An alternate method is to cut a groove in a short piece of 2" x 6" wood to accommodate the lower unit with the cavitation plate resting on top of the wood. Clamp the wood in a vise and service work may then be performed with the lower unit erect (in its normal position), or inverted (upside down). In both positions, the cavitation plate is the supporting surface.

6- Remove the O-ring from the top of the driveshaft. Remove the bolts securing the water pump to the lower unit housing. Pull the water pump housing up and free of the driveshaft. Slide the water pump impeller up and free of the driveshaft.

7- Pop the impeller Woodruff key out of the driveshaft keyway. Slide the water pump base plate up and off of the driveshaft.

GOOD WORDS

If the only work to be performed is service of the water pump, proceed directly to Page 8-85, Water Pump Installation.

Bearing Carrier — Removal

8- Remove the four 5/16" bolts from inside the bearing carrier. Notice how each bolt has an O-ring seal. These O-rings should be replaced each time the bolts are removed. In most cases, the word **UP** is

embossed into the metal of the bearing carrier rim. This word must face **UP** in relation to the lower unit during installation.

9- Remove the bearing carrier using one of the methods described in the following paragraphs, under Special Words.

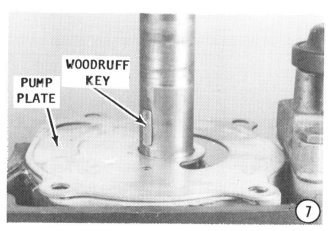

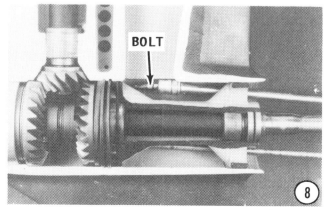

8-68　LOWER UNIT

SPECIAL WORDS

Several models of bearing carriers are used on the lower units covered in this section.

The bearing carriers fit very tightly into the lower unit opening. Therefore, it is not uncommon to apply heat to the outside surface of the lower unit with a torch, at the same time the puller is being worked to remove the carrier. **TAKE CARE** not to overheat the lower unit.

One model carrier has two threaded holes on the end of the carrier. These threads permit the installation of two long bolts. These bolts will then allow the use of a flywheel puller to remove the bearing carrier, Illustration #9.

Another model does not have the threaded screw holes. To remove this type bearing carrier, a special puller with arms must be used. The arms are hooked onto the carrier web area, and then the carrier removed, reference illustration "C".

WARNING

The next step involves a dangerous procedure and should be executed with care while wearing **SAFETY GLASSES**. The retaining rings are under tremendous tension in the groove and while they are being removed. If a ring should slip off the Truarc pliers, it will travel with incredible speed causing personal injury if it should strike a person. Therefore, continue to hold the ring and pliers firm after the ring is out of the groove and clear of the lower unit. Place

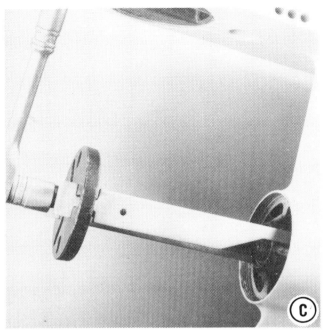

the ring on the floor and hold it securely with one foot before releasing the grip on the pliers. An alternate method is to hold the ring inside a trash barrel, or other suitable container, before releasing the pliers.

10- Obtain a pair of Truarc pliers. Insert the tips of the pliers into the holes of the first retaining ring. Now, **CAREFULLY** remove the retaining ring from the groove and gear case without allowing the pliers to slip. Release the grip on the pliers in the manner described in the above **WARNING**. Remove the second retaining ring in the same manner. The rings are identical and either one may be installed first.

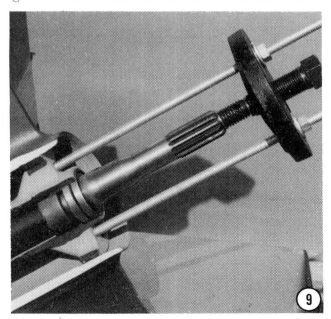

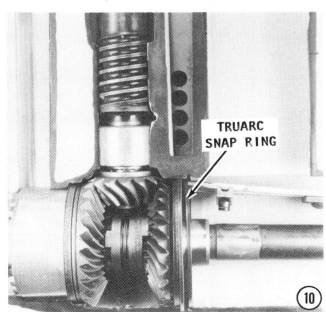

TRUARC SNAP RING

MECHANICAL SHIFT 8-69

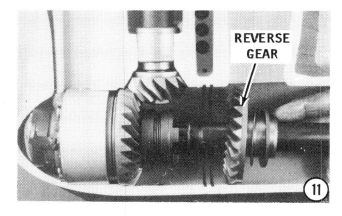

11- Remove the retainer plate, thrust washer, thrust bearing, and reverse gear from the propeller shaft.

12- Remove the four bolts from the driveshaft bearing housing. Pull up on the shift rod. This action will place shift dog into the forward gear position. Rotate the propeller shaft a bit as a check to be sure the unit is in forward gear.

Pinion Gear — Removal

For the 70 hp and 75 hp special tool OMC No. 312752 is required. For all other units covered in this section, special tool OMC No. 311875 is required to turn the driveshaft in order to remove the pinion gear nut.

13- Obtain the special tool and slip it over the end of the driveshaft with the splines of the tool indexed with the splines on the driveshaft. Hold the pinion gear nut with the proper size wrench, and at the same time rotate the driveshaft, with the special tool and wrench COUNTERCLOCKWISE until the nut is free. If the special tool is not available, clamp the driveshaft in a vise equipped with soft jaws, in an area below the splines but not in the water pump impeller area. Now, with the proper size wrench on the pinion gear nut, rotate the complete lower unit COUNTERCLOCKWISE until the nut is free. This procedure will require the lower unit to be rotated, then

the wrench released, the lower unit turned back, the wrench again attached to the nut, and the unit rotated again. Continue with this little maneuver until the nut is free. After the nut is free, proceed with the next step. The driveshaft will be withdrawn from the pinion gear.

Driveshaft — Removal

14- CAREFULLY pry the bearing housing upward away from the lower unit, then slide it free of the driveshaft. An alternate method is to again clamp the driveshaft in a vise equipped with soft jaws. Use a soft-headed mallet and tap on the top side of the bearing housing. This action will jar the housing loose from the lower unit. Continue tapping with the mallet and the bearing housing, O-rings, shims, thrust washer, thrust bearing, and the driveshaft will all breakaway from the lower unit and may be removed as an assembly. As the driveshaft is removed, reach into the lower unit, catch, and remove the pinion gear.

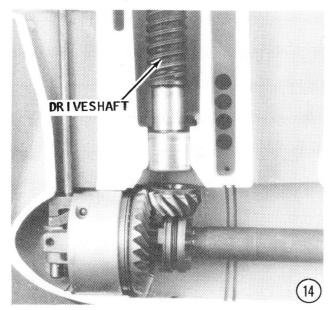

15- Push on the shift rod to move the unit into the reverse gear position. Remove the four bolts securing the shift rod cover. Rotate the shift rod **COUNTERCLOCKWISE** until it is free of the shifter detent in the lower unit. Remove the shift rod and cover as an assembly. As the plate is removed, notice which surface is facing into the housing, as an aid during installation.

Propeller Shaft -- Removal

16- Grasp the propeller shaft firmly with one hand and remove the propeller shaft from the lower unit. The forward gear, and bearing housing will come out with the shaft as an assembly.

"FROZEN" PROPELLER SHAFT

On rare occasions, especially if water has been allowed to enter the lower unit, it may not be possible to withdraw the propeller shaft as described in Step 16. The shaft may be "frozen" in the hydraulic pump due to corrosion.

If efforts to remove the propeller shaft after the bearing carrier has been removed fail, thread an adaptor to the end of the propeller shaft. Both ends of the adaptor have internal threads. Now, attach a slide hammer to the adaptor. Attach some type of slide hammer to the lower push rod or whatever is left in the lower unit of the assist cylinder mechanism. Obtain the help of an assistant. While the assistant operates

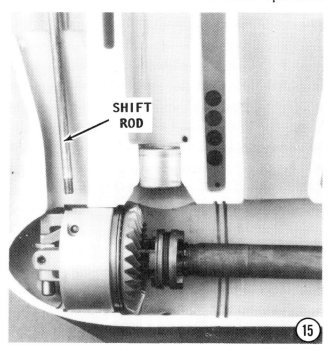

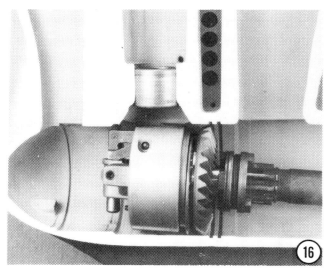

one slide hammer, or whatever device is attached to the shift mechanism, operate the slide hammer on the propeller shaft at the same time, reference illustration **"D"**.

BAD NEWS

Using the slide hammer, under these conditions, to remove the propeller shaft, will usually result in some internal part being damaged as the propeller is withdrawn. Usually, the parts damaged will be the shift cylinder, piston, and/or the push rod. However, the cost of replacing the damaged parts is reasonable considering the seriousness of a "frozen" shaft and the problem of getting it out.

Driveshaft Disassembling
For Unit On Page 8-76 ONLY

17- The driveshaft upper bearing and cone are replaced as an assembly. If replacement is required, obtain special tool OMC No. 387131. Clamp the tool below the bearing, and then place the unit in an arbor press equipped with a deep throat pedestal. Press the shaft from the bearing. An alternate method is to clamp special tool OMC No. 387206 to the driveshaft just above the shoulder. Now, invert the driveshaft and

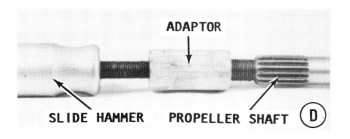

MECHANICAL SHIFT 8-71

place it in an arbor press with special tool OMC No. 387206 seated on the press. Place a piece of pipe over the driveshaft, and then press against tool OMC No. 387131 to remove the bearing.

Bearing Carrier -- Disassembling
INSPECTION FIRST

The bearings in the carrier need **NOT** be removed unless they are unfit for further service. Insert a finger and rotate the bearing. Check for "rough" spots or binding. Inspect the bearing for signs of corrosion or other types of damage. If the bearings must be replaced, proceed with the next step.

18- Use a seal remover to remove the two back-to-back seals or clamp the carrier in a vise and use a pry bar to pop each seal out.

19- Use a drift punch to drive the bearings free of the carrier. The bearings are being removed because they are unfit for service, therefore, additional damage is of no consequence.

Propeller Shaft -- Disassembling

20- Remove the coil spring from the outside groove of the clutch dog. **EXERCISE CARE** not to distort the spring as it is removed. Remove the clutch dog pin by

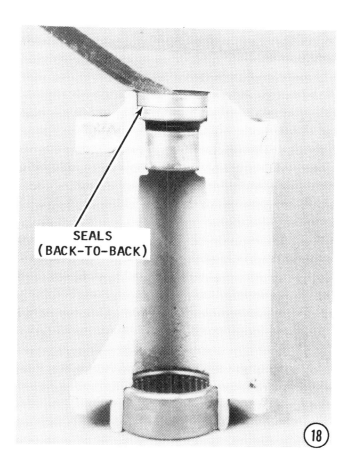

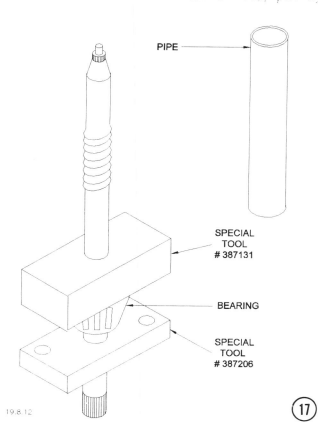

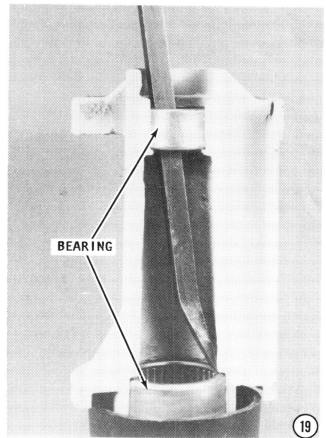

8-72 LOWER UNIT

pushing it through the clutch dog and propeller shaft. This pin is not a tight fit, therefore, it is not difficult to remove. Grasp the propeller shaft and pull it free of the bearing housing.

SPECIAL WORDS

Two different shifter arrangements on the propeller shaft are used on the units covered in this section. An exploded drawing of one type is shown on Page 8-75 together with reference illustration "D" on Page 8-72. An exploded drawing of the other lower unit will be found on Page 8-76 together with reference illustration "E" on Page 8-73. Take special notice of how one type has a shifter shaft, bearing, one detent ball, and one spring. The other model has a clutch dog shaft, thrust washer, and two detent balls and springs. Compare the propeller shaft from the unit being serviced with the two drawings.

Now, if the unit being serviced is the one shown on Page 8-75, perform Steps 21 and 22, then skip to Step 26.

If the unit being serviced is the one shown on Page 8-76, skip to Step 23 and perform the work thru Step 25, then continue with Step 26.

NOTE: The following two steps are to be performed if servicing a unit as illustrated on Page 8-75 and reference illustration "D".

21- Remove the two Allen screws from the back side of the bearing housing and at the same time be prepared to catch the two detent balls and springs.

22- Remove the shift lever pin. Remove the shift lever and shift detent, from the bearing housing.

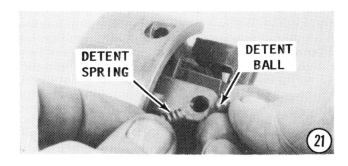

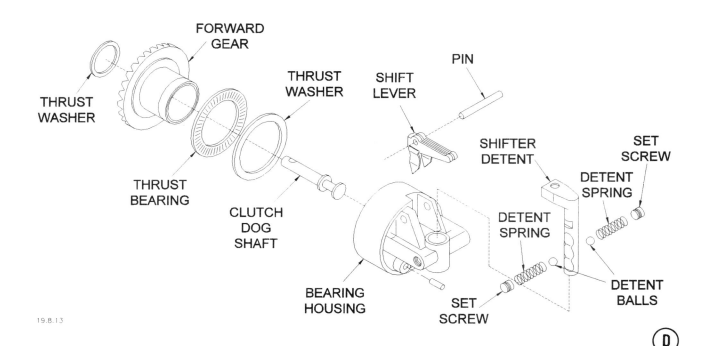

MECHANICAL SHIFT 8-73

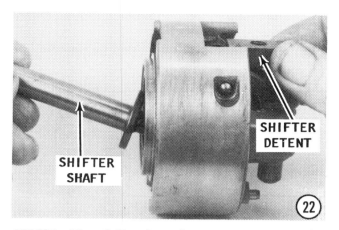

NOTE: The following three steps are to be performed if servicing a unit as illustrated on Page 8-76 and reference illustration "E".

23- Remove the shift lever pin and disengage the shift lever from the cradle in the shift shaft.

24- Remove the shifter shaft and cradle. Remove the shift lever.

SAFETY WORD

The next step is dangerous. The detent balls are under tension from the springs. The balls are released with pressure. Therefore, **SAFETY GLASSES** should be worn as personal protection for the eyes.

25- Observe the short arm at the upper end of the shifter detent. Now, rotate the shifter detent until this arm is 180° from its original position (facing toward the opposite direction). Rotating the shifter detent 180° will depress the detent ball and spring. From this position, work the shifter detent up out of the housing.

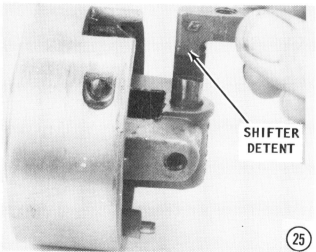

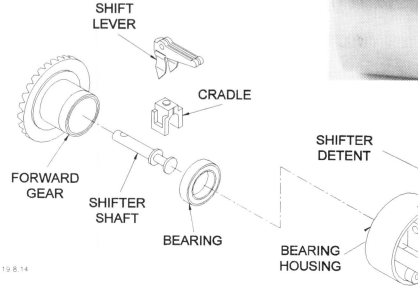

8-74 LOWER UNIT

Lower Driveshaft Bearing — Removal
SPECIAL WORDS

Two different type bearings are used on the lower unit models covered in this section. Exploded drawings of both type bearings are given on Page 8-75 and Page 8-76. To determine which type bearing is used on the unit being serviced, check the shifter arrangement on the propeller shaft and compare it with the two illustrations.

A special tool **MUST** be used to remove either type bearing. Therefore, **DO NOT** attempt to remove this bearing unless it is unfit for further service. To check the bearing, first use a flashlight and inspect it for corrosion or other damage. Insert a finger into the bearing, and then check for "rough" spots or binding while rotating it.

26- Remove the Allen screw, from the water pickup slots in the starboard side of the lower unit housing. This screw secures the bearing in place and **MUST** be removed before an attempt is made to remove the bearing. The bearing must be actually **"PULLED"** upward to come free. **NEVER** make an attempt to "drive" it down and out.

27- If servicing a unit shown in the exploded illustration on Page 8-75, obtain special tool, OMC No. 320672. If servicing a unit shown in the exploded drawing on Page 8-76, obtain special tool, OMC No. 391257. Use the special tool and "pull" the bearing from the lower unit.

CLEANING AND INSPECTING

Wash all parts in solvent and dry them with compressed air. Discard all O-rings and seals that have been removed. A new seal kit for this lower unit is available from the local dealer. The kit will contain the necessary seals and O-rings to restore the lower unit to service.

Inspect all splines on shafts and in gears for wear, rounded edges, corrosion, and damage.

Carefully check the driveshaft and the propeller shaft to verify they are straight and true without any sign of damage. A complete check must be performed by turning the shaft in a lathe. This is only necessary if there is evidence to suspect the shaft is not true.

Check the water pump housing for corrosion on the inside and verify the impeller and base plate are in good condition. Actually, good shop practice dictates to rebuild or replace the water pump each time the lower unit is disassembled. The small cost is rewarded with "peace of mind" and satisfactory service.

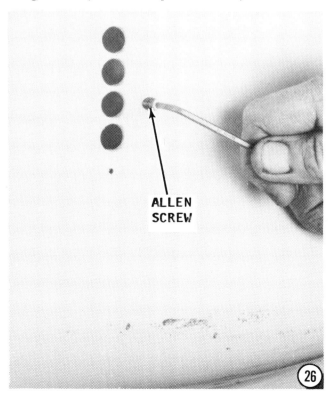

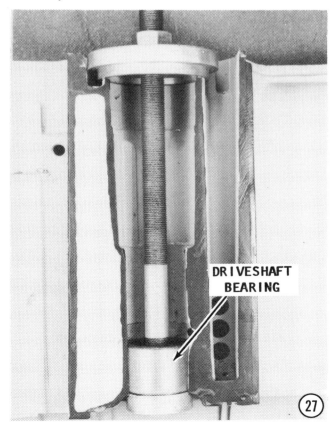

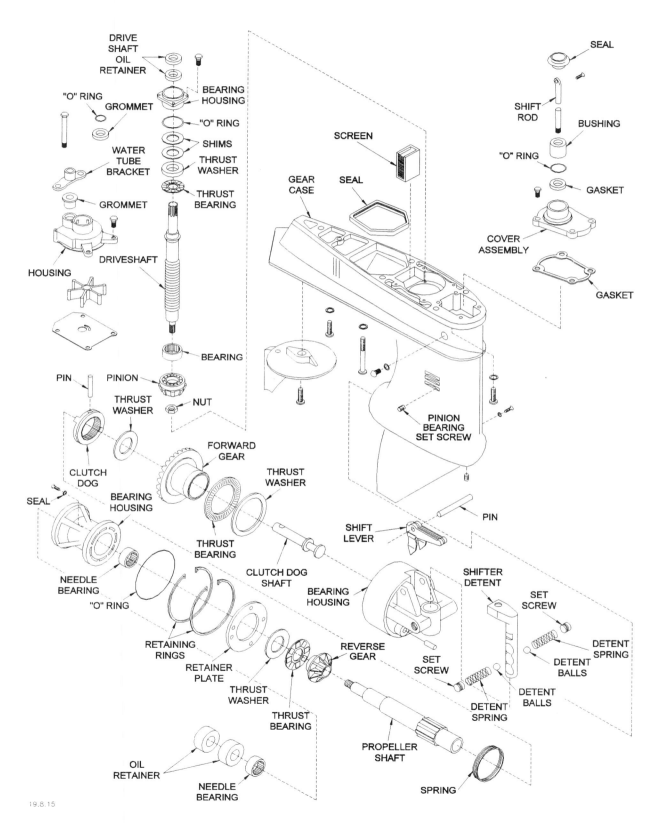

Exploded drawing of one of two lower units covered in this section. This unit has a thrust bearing and thrust washer on the driveshaft and two detent balls and springs for the shift mechanism. Enlarged details of the shift mechanism are shown in reference illustration "D" on Page 8-72. Numerous references in the text are made to these two illustrations.

8-76 LOWER UNIT

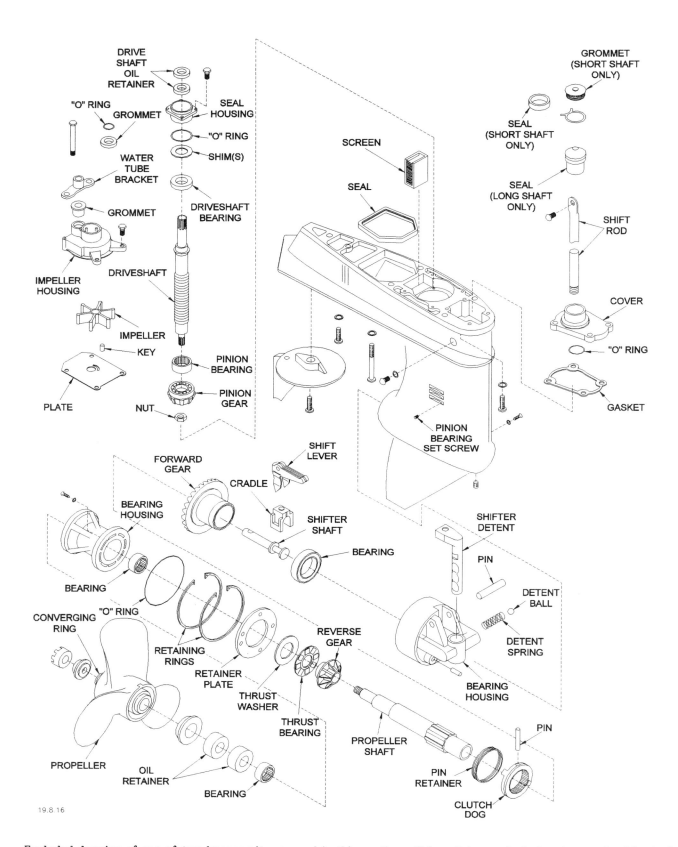

Exploded drawing of one of two lower units covered in this section. This unit has a single bearing on the driveshaft and one detent ball and spring for the shift mechanism. Enlarged details of the shift mechanism are shown in reference illustration **"E"** on Page 8-73. Numerous references in the text are made to these two illustrations.

MECHANICAL SHIFT 8-77

Inspect the lower unit housing for nicks, dents, corrosion, or other signs of damage. Nicks may be removed with No. 120 and No. 180 emery cloth. Make a special effort to ensure all old gasket material has been removed and mating surfaces are clean and smooth.

Inspect the water passages in the lower unit to be sure they are clean. The screen may be removed and cleaned.

Check the gears and clutch dog to be sure the ears are not rounded. If doubt exists as to the part performing satisfactorily, it should be replaced.

Inspect the bearings for "rough" spots, binding, and signs of corrosion or damage.

ASSEMBLING

READ AND BELIEVE

The lower unit should **NOT** be assembled in a dry condition. Coat all internal parts with OMC HI-VIS lube oil as they are assembled. All seals should be coated with OMC Gasket Seal Compound. When two seals are installed back-to-back, use Triple Guard Grease between the seal surfaces.

Propeller Shaft Assembling
SPECIAL WORDS

Two different shifter arrangements on the propeller shaft are used on the units covered in this section. A different type of lower driveshaft bearing is also used. An exploded drawing of one type is shown on Page 8-75, together with reference illustration **"F"**. An exploded drawing of the other lower unit will be found on Page 8-76 together with reference illustration **"G"**. Take special notice of how one type has a shifter shaft, bearing, one detent ball, and one spring. The other model has a clutch dog shaft, thrust washer, and two detent balls and springs. Compare the propeller shaft from the unit being serviced with the two drawings.

Now, if the unit being serviced is the one shown on Page 8-75, perform Steps 1 thru 4, then skip to Step 10.

If the unit being serviced is the one shown on Page 8-76, skip to Step 5 and perform the work thru Step 9, then continue with Step 10.

NOTE: The following four steps are to be performed if servicing a unit illustrated on Page 8-75 and reference illustration **"F"**.

1- Install the shift lever and detent. Install the clutch shaft and then the pin.

2- Insert the two detent balls into the shifter detent, then the springs. Coat the threads of the set screws with Locktite, and then secure the springs and detent balls in place with the screws. The screws should be tightened until each head is flush with the housing.

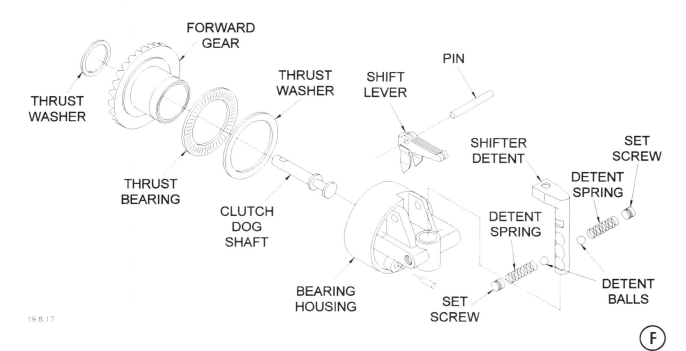

F

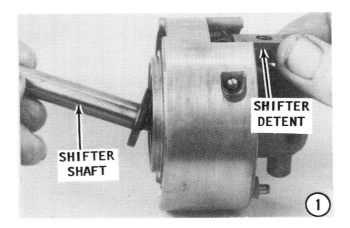

3- Install the thrust bearing and thrust washer onto the shank of the forward gear. Now, slide the assembled forward gear into the bearing housing. Slide the small thrust washer onto the propeller shaft. Insert the propeller shaft into the forward gear with the slot in the shaft aligned with the hole in the clutch dog shaft. Insert the pin, and then install the coil spring around the outside groove of the clutch dog. Check to be sure one coil of the spring does not overlap another coil. Press the shifter down to move the clutch dog into the reverse gear position. Set the unit aside for later installation.

4- Obtain bearing installer, special tool, OMC No. 391257. Assemble the washer, plate, guide sleeve, and installer portion of the tool to the screw in the order given, and with the shoulder of the tool facing down. Place the bearing onto the tool with the lettered side of the bearing **TOWARD** the shoulder. Place the lower driveshaft bearing and tool in position in the lower unit. Now, drive the bearing into the lower unit until the plate makes contact with the lower unit surface. Coat the threads of the setscrew with Loctite and then secure the bearing in place with the setscrew.

Perform the following 5 steps if servicing a unit as shown in the exploded drawing on Page 8-76 together with reference illustration "G".

5- Apply a thin coating of Needle Bearing Grease onto the detent ball and spring. Insert the detent ball and spring into the bearing housing. Observe the short arm on the shifter detent. Position the detent with

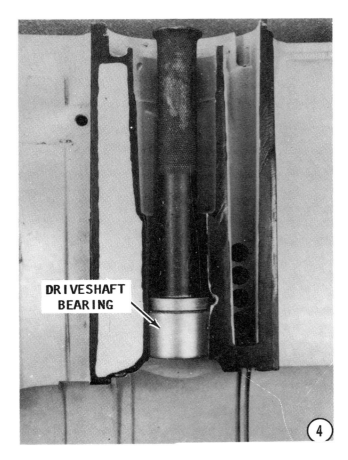

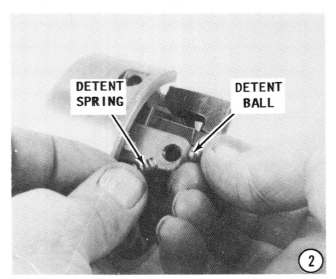

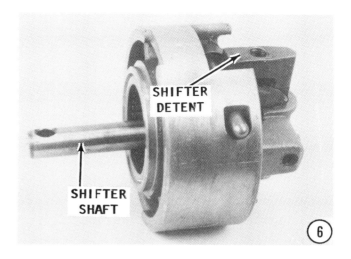

this short arm facing forward. Now, use a punch, or similar tool, and depress the ball and spring, and **AT THE SAME TIME** press the shift detent down into the bearing housing. Rotate the shifter detent 180°.

6- Place the cradle onto the shifter shaft and in position in the bearing housing. Install shifter lever with the cradle engaged with the shifter detent. Work the pin through the shifter detent and cradle.

7- Install the clutch dog onto the propeller shaft with the **PROP END** identification on the dog facing the propeller end of the shaft. Align the holes in the dog with the slot in the shaft. Slide the forward gear bearing onto the forward gear shoulder. Install the thrust washer into the bearing housing. Align the hole in the shifter shaft with the hole in the clutch dog. Now, slide the propeller shaft into the assembled forward gear and bearing housing. Insert the clutch dog retaining pin. Install a **NEW** clutch dog retaining spring. Check to be sure that one coil of the spring does not overlap another coil. Move the shifter detent down into the reverse gear position. Set the assembly aside for later installation.

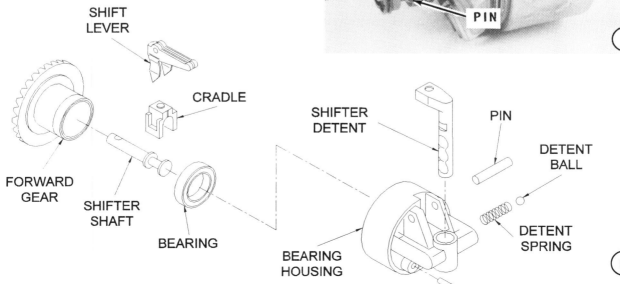

LOWER UNIT

8- Obtain bearing installer, special tool, OMC No. 391257. Assemble the parts as shown in the accompanying illustration. Place the lower driveshaft bearing on the tool with the lettered side of the bearing facing **UP** towards the driving face of the tool. Use Needle Bearing Grease to hold the bearing on the tool. Now, drive the bearing into place until the washer makes contact with the spacer. Coat the threads of the setscrew with Loctite. Secure the bearing in place with the screw.

9- Obtain special tool, OMC No. 387131 and OMC No. 320671. Use the two special tools and an arbor press to install the upper driveshaft bearing. Press with the tool against the inner race until the bearing is seated on the driveshaft shoulder.

10- Clamp the driveshaft in a vise equipped with soft jaws and in such a manner that the splines, water pump area, or other critical portions of the shaft cannot be damaged. Slide the pinion gear onto the driveshaft with the bevel of the gear teeth facing toward the lower end of the shaft. Install the pinion gear nut and tighten it to a torque value of 40 to 45 ft-lbs. No parts should be installed on the upper end of the driveshaft at this point. Slide the shims removed during disassembling onto the driveshaft and seat them against the driveshaft shoulder. If servicing a 70 hp or 75 hp V4 unit as shown in exploded drawing on Page 8-75, obtain special shimming tool, OMC No. 315767. If servicing a V6 unit, use special tool OMC No. 321520. If servicing a unit as shown in the exploded drawing on Page 8-76, obtain special shimming tool, OMC No. 320739.

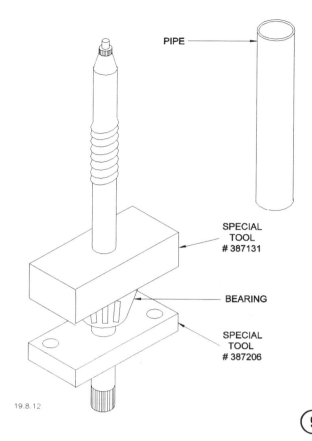

Slip the special tool down over the driveshaft and onto the upper surface of the top shim. Measure the distance between the top of the pinion gear and the bottom of the tool. The tool should just barely make

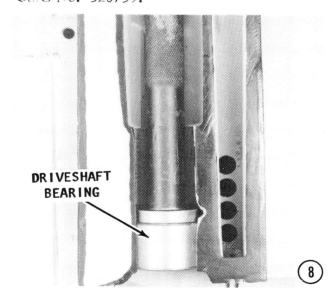

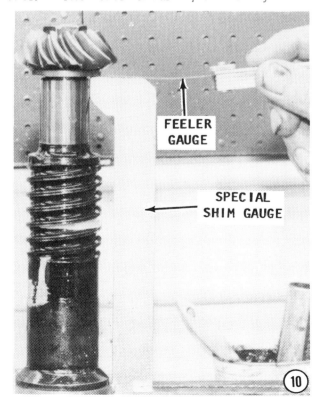

MECHANICAL SHIFT 8-81

contact with the pinion gear surface for **ZERO** clearance. Add or remove shims from the upper end of the driveshaft to obtain the required **ZERO** clearance. If servicing the model as shown on Page 8-76, remove 0.007" of shim material **AFTER** the zero clearance has been obtained. Remove the tool and set the shims aside for installation later. Back off the pinion gear nut and remove the pinion gear. Remove the driveshaft from the vise.

11- Insert the complete assembled propeller shaft into the lower unit housing with pin on the back side of the forward gear housing indexed into hole in the lower unit.

12- Thread the shift rod into the shift detent assembly. Coat both sides of the shift rod gasket with sealer. Place the gasket in position, and then install the shift rod cover onto the housing. The shift rod is adjusted in the next step. Slide the shift rod boot down over the shift rod cover.

Shift Rod Adjustment

13- Measure the distance from the top of the lower unit housing to the center of the hole in the shift rod. This distance should be as follows for a standard long shaft unit. Add exactly 5" for a unit with an extra long shaft. A purchased unit **CANNOT** be converted, either to an extra long shift shaft, nor back to a standard long shaft from extra long.

UNITS -- SINCE 1978
Model 60, 70, 75hp -- 21.72" (55.17cm)
Model 85, 90, 100, 110, 115hp -- 21.84" (55.48cm)
Model 120, 140, 200, 225hp -- 21.94" (55.73cm)
Model 150, 175, 185, 235hp -- 22.06" (56.04cm)

For 70 hp and 75 hp -- 1975 thru 1977, use special tool OMC No. 320736

For all other units, this section, use special tool OMC No. 389997

Rotate the shift rod clockwise or counterclockwise until the required dimension is obtained.

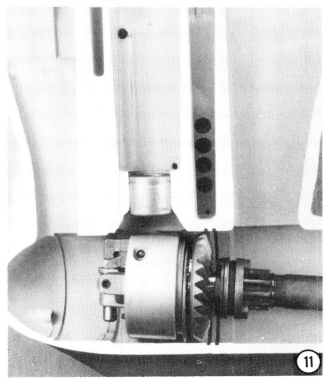

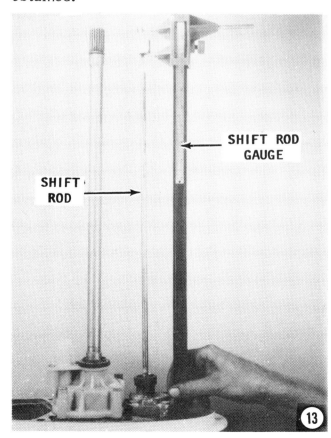

14- Insert the pinion gear into the cavity in the lower unit with the flat side of the bearing facing **UPWARD**. Hod the pinion gear in place and at the same time lower the driveshaft down into the lower unit. As the driveshaft begins to make contact with the pinion gear, rotate the shaft slightly to permit the splines on the shaft to index with the splines of the pinion gear. After the shaft has indexed with the pinion gear, thread the pinion gear nut onto the end of the shaft. Obtain OMC special tool, P/N 312752 for 75hp; P\N 311875 for all other units covered in this section. Slide the special tool over the upper end of the driveshaft with the splines of the tool indexed with the splines on the upper end of the driveshaft. Attach a torque wrench to the special tool. Now, hold the pinion gear nut with the proper size wrench and rotate the driveshaft **CLOCKWISE** with the special tool until the pinion gear nut is tightened to a torque value of 60 to 65 ft lb (81 to 88Nm). Remove the special tool.

15- If servicing a unit as shown on Page 8-75: Slide the thrust bearing, thrust washer, and the shims, set aside after the shim gauge procedure in Step 10, onto the driveshaft. If servicing a unit shown on Page 8-76: Slide the shims only onto the driveshaft. (This unit does not have the other items.)

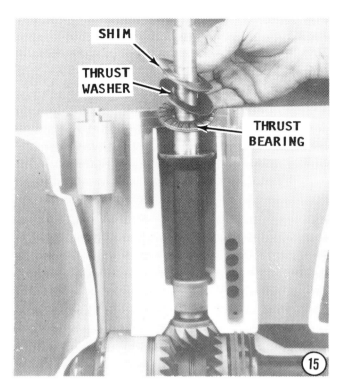

16- Install the two seals back-to-back into the opening on top of the bearing housing. Coat the outside surface of the **NEW** seals with OMC Lubricant. Press the first seal into the housing with the flat side of the seal facing **OUTWARD**. After the seal is in place, apply a coating of Triple Guard Grease to the flat side of the installed seal and the flat side of the second seal. Press the second seal into the bearing housing with the flat side of the seal facing **INWARD**. Insert a **NEW** O-ring into the bottom opening of the bearing housing.

17- Wrap friction tape around the splines of the driveshaft to protect the seals in the bearing housing as the housing is

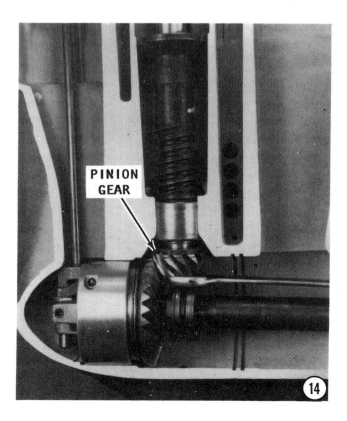

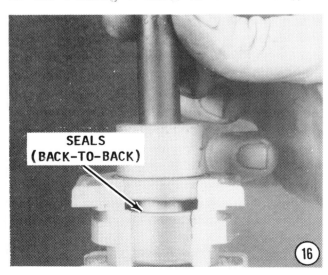

MECHANICAL SHIFT 8-83

installed. Now, slide the assembled bearing housing down the driveshaft and seat it in the lower unit housing. Secure the housing in place with the four bolts. Tighten the bolts **ALTERNATELY** and **EVENLY**.

Reverse Gear — Installation

18- Apply a light coating of grease to the inside surface of the reverse gear to hold the thrust washer in place. Insert the thrust washer into the reverse gear. Install the thrust bearing and thrust washer onto the shank on the backside of the reverse gear. Slide the reverse gear down the propeller shaft and index the teeth of the reverse gear with the teeth of the pinion gear.

19- Insert the retainer plate into the lower unit against the reverse gear.

WARNING

This next step can be dangerous. Each snap ring is placed under tremendous tension with the Truarc pliers while it is being placed into the groove. Therefore, wear **SAFETY GLASSES** and exercise care to prevent the snap ring from slipping out of the pliers. If the snap ring should slip out, it would travel with incredible speed and cause personal injury if it struck a person.

20- Install the Truarc snap rings one at-a-time following the precautions given in

the **WARNING** and the **ADVICE** given in the following paragraph.

WORDS OF ADVICE

The two snap rings index into separate grooves in the lower unit housing. As the first ring is being installed, depth perception may play a trick on your eyes. It may

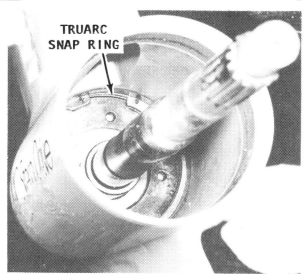

View into the lower unit showing the Truarc snap rings properly locked in the grooves.

8-84 LOWER UNIT

appear that the first ring is properly indexed all the way around in the proper groove, when in reality, a portion may be in one groove and the remainder in the other groove. Should this happen, and the Truarc pliers be released from the ring, it is extremely difficult to get the pliers back into the ring to correct the condition. If necessary use a flashlight and carefully check to be sure the first ring is properly seated all the way around **BEFORE** releasing the grip on the pliers. Installation of the second ring is not so difficult because the one groove is filled with the first ring.

Bearing Carrier Bearings & Seals Installation

21- Install the reverse gear bearing into the bearing carrier by pressing against the **LETTERED** side of the bearing with the proper size socket. Press the forward gear bearing into the bearing carrier in the same manner. Press against the **LETTERED** side of the bearing.

22- Coat the outside surfaces of the seals with a non-hardening sealant -- Adhesive "M" or "Seal 1000". Such a sealant will be a great contributor to preventing a leak at this location over a period of time. Install the first seal with the flat side facing **OUT**. Coat the flat surface of both seals with Triple Guard Grease, and then install the second seal with the flat side going in **FIRST**. The seals are then back-to-

A set of double seals showing the back side (left) and the front side (right). These seals are installed back-to-back (flat side-to-flat side) with Triple Guard Grease between the surfaces. This arrangement prevents fluid from passing in either direction.

back. The outside seal prevents water from entering the lower unit and the inside seal prevents the lubricant in the lower unit from escaping.

Bearing Carrier Installation

23- Obtain two long 1/4" rods with threads on one end. Thread the rods into the retainer plate opposite each other to act as guides for the bearing carrier. Check the bearing carrier to be sure a **NEW** O-ring has been installed. Position the carrier over the guide pins with the embossed word **UP** on the rim of the carrier facing **UP** in relation to the lower unit housing, reference illustration "H". Now, lower the bearing carrier down over the guide pins and into place in the lower unit housing.

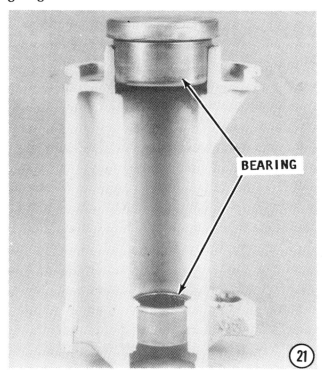

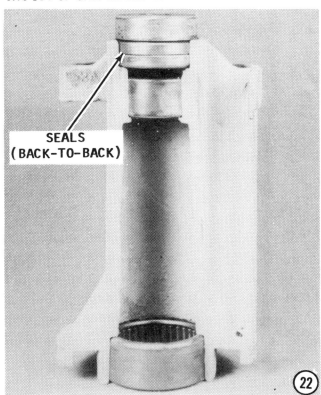

MECHANICAL SHIFT 8-85

24- Slide **NEW** little O-rings onto each bolt, and apply some OMC Sealer onto the threads. Install the bolts through the carrier and into the retaining plate. After a couple bolts are in place, remove the guide pins and install the remaining bolts. Tighten the bolts **EVENLY** and **ALTERNATELY** to the torque value given in the Appendix.

WATER PUMP INSTALLATION

FIRST, THESE WORDS

An improved water pump is available as a replacement. If the old water pump housing is unfit for further service, only the new pump housing can be purchased. It is strongly recommended to replace the water pump with the improved model while the lower unit is disassembled. The accompanying illustration shows the original equipment (left) compared with the improved pump (right), reference Illustration "J".

The new pump must be assembled before it is installed. Therefore, the following steps outline procedures for both pumps.

To assemble and install a replacement pump, perform Steps 25 thru 33, then jump to Step 36.

To install an original equipment pump, proceed directly to Step 34.

Assembling an Improved Pump Housing

25- Remove the water pump parts from the container. Insert the plate into the housing, as shown. The tang on the bottom side of the plate **MUST** index into the short slot in the pump housing.

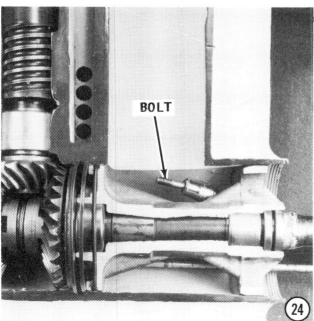

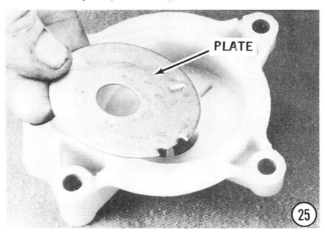

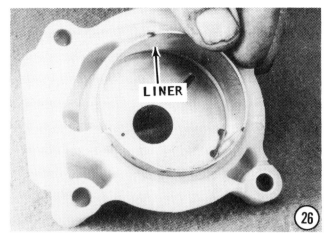

26- Slide the pump liner into the housing with the two small tabs on the bottom side indexed into the two cutouts in the plate.

27- Coat the inside diameter of the liner with light-weight oil. Work the impeller into the housing with all of the blades bent back to the right, as shown. In this position, the blades will rotate properly when the pump housing is installed. Remember, the pump and the blades will be rotating **CLOCKWISE** when the housing is turned over and installed in place on the lower unit.

28- Coat the mating surface of the lower unit with 1000 Sealer. Slide the water pump base plate down the driveshaft and into place on the lower unit. Insert the Woodruff key into the key slot in the driveshaft.

29- Lay down a very thin bead of 1000 Sealer into the irregular shaped groove in the housing. Insert the seal into the groove, and then coat the seal with the 1000 Sealer.

Water Pump Installation

30- Begin to slide the water pump down the driveshaft, and at the same time observe the position of the slot in the impeller. Continue to work the pump down the

driveshaft, with the slot in the impeller indexed over the Woodruff key. The pump must be fairly well aligned before the key is covered because the slot in the impeller is not visible as the pump begins to come close to the base plate.

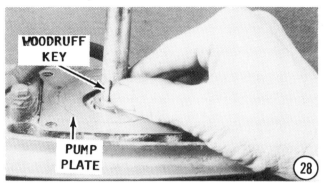

MECHANICAL SHIFT 8-87

31

31- Install the short forward bolt through the pump and into the lower unit. **DO NOT** tighten this bolt at this time. Insert the grommet into the pump housing.

32- Install the grommet retainer and water tube guide onto the pump housing. Install the remaining pump attaching bolts. Tighten the bolts **ALTERNATELY** and **EVENLY**, and at the same time rotate the driveshaft **CLOCKWISE**. If the driveshaft is not rotated while the attaching bolts are being tightened, it is possible to pinch one of the impeller blades underneath the housing.

33- Slide the large grommet down the driveshaft and seat it over the pump collar. This grommet does not require sealer. Its function is to prevent exhaust gases from entering the water pump. Proceed directly to Lower Unit Installation, Step 36.

Water Pump Installation
Original Equipment

Perform the following two steps to install an original water pump.

34- Coat the water pump plate mating surface on the lower unit with 1000 Sealer. Slide the water pump plate down the driveshaft and **BEFORE** it makes contact with the sealer check to be sure the bolt holes in the plate will align with the holes in the housing. The plate will only fit one way. If the holes will not align, remove the plate, turn it over and again slide the plate down

32

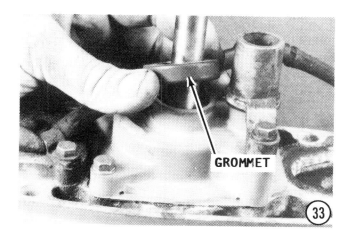

33

the driveshaft and into place on the housing. This checking will prevent accidently getting the sealer on both sides of the plate. If by chance sealer does get on the top surface it **MUST** be removed before the water pump impeller is installed.

Slide the water pump impeller down the driveshaft. Just before the impeller covers the cutout for the Woodruff key, install the key, and then work the impeller on down, with the slot in the impeller indexed over the Woodruff key. Continue working the impeller down until it is firmly in place on the surface of the pump plate.

35- Check to be sure **NEW** seals and O-rings have been installed in the water pump. Lubricate the inside surface of the water pump with light-weight oil. Lower the water pump housing down the driveshaft and over the impeller. **ALWAYS** rotate the driveshaft slowly **CLOCKWISE** as the housing is lowered over the impeller to allow the impeller blades to assume their natural and proper position inside the housing. Continue to rotate the driveshaft and work the water pump housing downward until it is seated on the plate. Coat the threads of the water pump attaching screws with sealer, and then secure the pump in place with the screws. Install the solenoid cable bracket with the

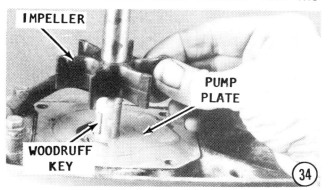

34

same bolt and in the same position from which it was removed. On some units the solenoid cable fits into a recess of the water pump and is held in place in that manner. Tighten the screws **ALTERNATELY** and **EVENLY**.

LOWER UNIT INSTALLATION

36- Check to be sure the water tubes are clean, smooth, and free of any corrosion. Coat the water pickup tubes and grommets with lubricant as an aid to installation. Check to be sure the spark plug wires are disconnected from the spark plugs. Bring the lower unit housing together with the exhaust housing, and at the same time, guide the water tube into the rubber grommet of the water pump. Continue to bring the two units together, at the same time guiding the water pickup tubes into the rubber grommet of the water pump, and, simultaneously rotating the flywheel slowly to permit the splines of the driveshaft to index with the splines of the crankshaft. This may sound like it is necessary to do four things at the same time, and so it is. Therefore, make an earnest attempt to secure the services of an assistant for this task.

37- After the surfaces of the lower unit and exhaust housing are close, dip the attaching bolts in OMC Sealer and then start them in place. Install the two bolts on each side of the two housings. Install the bolt in the recess of the trim tab. Install the bolt just forward of the trim tab that extends up through the cavitation plate. Tighten the bolts **ALTERNATELY** and **EVENLY** to the torque value given in the Appendix.

38- Install the trim tab with the mark made on the tab during disassembling aligned with the mark made on the lower unit housing.

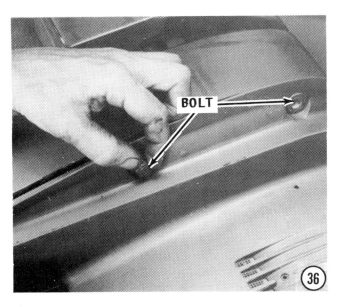

39- The shift rod is visible beneath and to the rear of the lower carburetor. Install the shift connector with a bolt or pin, depending on the model being serviced.

Filling the Lower Unit

Fill the lower unit with lubricant according to the procedures in Section 8-3.

Propeller Installation

Install the propeller, see Section 8-2.

FUNCTIONAL CHECK

Perform a functional check of the completed work by mounting the engine in a test tank, in a body of water, or with a flush attachment connected to the lower unit. If

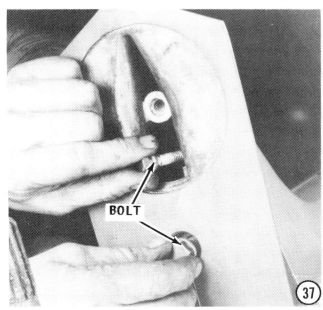

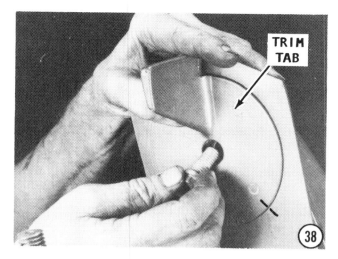

the flush attachment is used, **NEVER** operate the engine above an idle speed, because the no-load condition on the propeller would allow the engine to **RUNAWAY** resulting in serious damage or destruction of the engine.

CAUTION: Water must circulate through the lower unit to the engine any time the engine is run to prevent damage to the water pump in the lower unit. Just five seconds without water will damage the water pump.

Start the engine and observe the tattle-tale flow of water from idle relief in the exhaust housing. The water pump installation work is verified. If a "Flushette" is connected to the lower unit, **VERY LITTLE** water will be visible from the idle relief port. Shift the engine into the three gears and check for smoothness of operation and satisfactory performance. Remember, when the unit is in forward gear, it is at rest with no current flow to either solenoid.

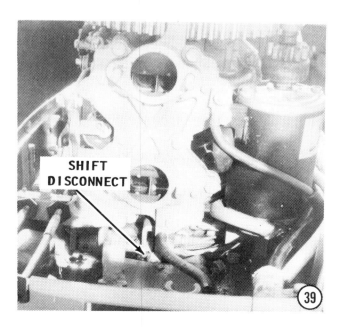

8-7 "FROZEN" PROPELLER

DESCRIPTION

If an exhaust propeller is "frozen" to the propeller shaft and the usual methods of removal fail, using a mallet to beat on the propeller blades in an effort to dislodge the propeller will have little if any affect. This is due to the cushioning affect the rubber hub has on the blow being struck.

A "frozen" propeller is caused by the inner sleeve becoming corroded and stuck to the propeller shaft. Therefore, special procedures are required to free the propeller and remove it from the shaft.

The following detailed procedures are presented as the only practical method to remove a "frozen" propeller from the shaft without damaging other more expensive parts. In almost all cases it is successful.

REMOVAL

1- Remove the cotter pin, and then the castillated nut and splined washer from the propeller shaft.

2- Heat the inside diameter of the propeller with a torch. Do not apply the heat to the outside surface of the propeller. Concentrate the heat in the area of the hub and shaft where the nut was removed and as far into the hub as possible. Continue applying heat, and at the same time have an assistant use a piece of 2" x 4" wooden block

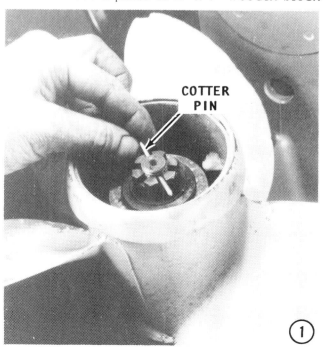

8-90 Lower Unit

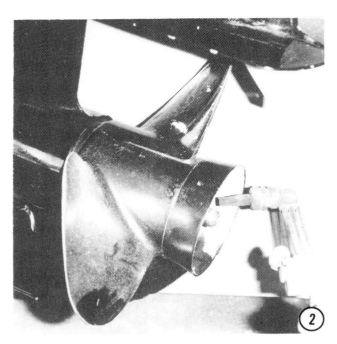

wedged between one of the blades and the lower unit housing. Use a prying force on the propeller while the heat is being applied. As the heat melts the inner rubber hub, the propeller will come free.

WARNING

As the force and heat are applied, the propeller may "pop" loose suddenly and without warning. Therefore, stand to one side while applying the heat as a precaution against personal injury.

3- After the propeller has been removed, the sleeve and the what's left of the rubber hub will still be stuck to the propeller shaft. Attach a puller to the thrust washer. Apply more heat to the sleeve and rubber hub, and at the same time take up on the puller. When the sleeve reaches the proper temperature, it will be released and come free.

4- If a puller is not available, as described in Step 3, use a sharp knife and cut the rubber hub from the sleeve. An alternate method is to use canned heat, or the equivalent, and set fire to the rubber hub. Allow the hub to burn away from the sleeve. When the fire burns out, only a small amount may be left on the sleeve. Heat the sleeve again, and then while it is still hot, use a chisel, punch, or similar tool with a hammer, and drive the sleeve free of the shaft. Allow the propeller shaft to cool, and then clean the splines thoroughly. Take time to remove any corrosion. Install the propeller with a **NEW** hub and sleeve according to the instructions outlined in Section 8-2.

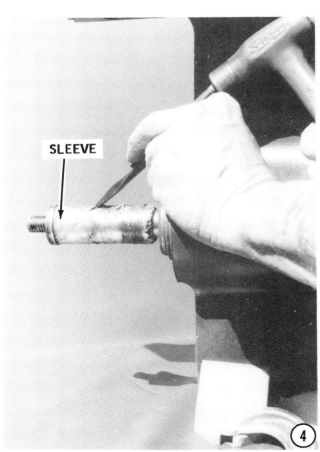

9
TRIM/TILT

9-1 SYSTEM DESCRIPTION

The power trim unit, Illustration #1, is a hydraulic/mechanical unit mounted between and attached to the stern brackets. The unit consists of two hydraulic trim cylinders, a hydraulic pump, fluid reservoir, an electric motor, a combination tilt cylinder and shock absorber, and the necessary tubing, valves, check valves and relief valves, to make it all function properly. The tilt cylinder is attached to a manifold containing most of the system parts and to the swivel bracket.

UNIQUE TRIM/TILT

A few V4 units prior to 1978 have a unique power trim/tilt system. A single trim cylinder and a single tilt cylinder is installed between the transom brackets. A combination pump, fluid reservoir and electric motor in a single unit is installed on the starboard side of the tramsom brackets, see Section 9-13 beginning on Page 9-29.

When the boat operator activates an electric switch on the control panel to the UP position, power is supplied to the electric motor which drives the hydraulic pump. The pump forces hydraulic fluid into the trim cylinders and into the tilt cylinder. The trim cylinders move the outboard motor the first 15° of movement, considered the "tilt range". At this position, the tilt cylinder takes over and moves the outboard through the final 50° of motion, the "tilt range".

The outboard may be operated up to approximately 1500 rpm while it is elevated past the 15° position for boat movement in very shallow water. However, if engine speed is increased above 1500 rpm, a relief valve will automatically open causing the outboard to lower to the fully trimmed out position. The outboard cannot be tilted while operating above 1500 rpm.

When the engine is moved in the **DOWN** direction, movement of the cylinders are all in the opposite direction to the **UP** movement. The tilt cylinder moves first until the swivel bracket is resting on the trim rods. At that point the trim cylinder continues movement in the down direction.

It is possible to raise or lower the outboard manually in the event electrical power is lost for any reason. A manual release valve on the starboard side of the unit may be activated with a screwdriver extended through an access hole in the stern bracket. With the screwdriver, the slotted valve can be opened to allow hydraulic fluid to flow from one side of the tilt cylinder to the other. This manual release valve does not affect the trim cylinder.

During movement of the boat on a trailer, an external lock, Illustration #2, is provided as a safety feature and to ensure that one of the hydraulic components is overloaded. Before movement of the trailer, the outboard motor should be tilted to the full **UP** position. The locks are then moved into position and the trim and tilt unit run in the **DOWN** position until the locks are firmly held in position on the stern brackets.

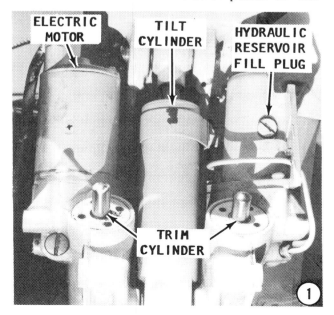

9-2 COMPONENT FEATURES

The reference numbers given for the major components are keyed to the functional diagram, Illustration #3, this page.

No. 3 Manual Release — permits manual engine tilt. When the screw is loosened, the engine will **NOT** move in the trim range.

No. 5 Trim Check Valve — works with the trim and tilt separation valve to isolate the trim cylinders from the system when the engine is not being trimmed or tilted. Allows hydraulic fluid to flow from the pump to the underside of the trim cylinders during trim **UP** operation. Also allows fluid to return to the pump and reservoir during trim and tilt **DOWN** operation.

No. 6 Tilt Check Valve — located inside the tilt piston. Provides a one way fluid passage from the hydraulic pump to the underside of the tilt piston during trimming and tilting **UP**.

No. 7 Trim UP Relief Valve — controls maximum hydraulic pressure in the system during trim and tilt **UP**. Also bleeds fluid to the reservoir when the engine is run at speeds over approximately 1500 rpm while operating in the partial tilt position. Under these conditions, this relief valve allows the

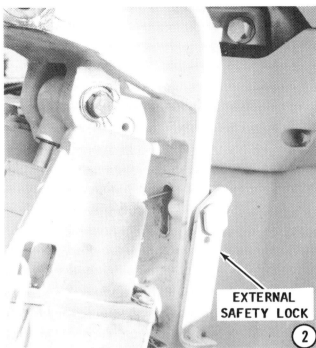

engine to return to the maximum trim **UP** position.

No. 8 Impact Letdown Valve — used **ONLY** when the engine strikes an underwater object and is forced to tilt up to pass the object. When activated, provides a fluid passage to the reservoir from the underside of the tilt piston.

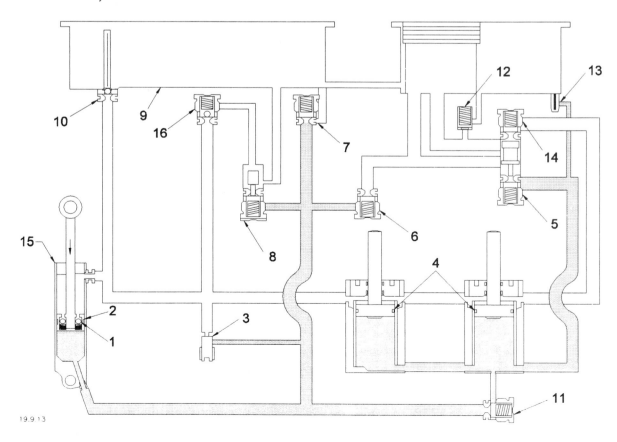

No. 10 Filter Valve — filters and allows make-up fluid to flow from the reservoir to the system during automatic letdown after tilt **UP** following an underwater strike. Valve must close during down operation of the hydraulic pump.

No. 11 Trim/Tilt Separation Valve —seals the underside of the trim cylinders from the underside of the tilt cylinder under trim load.

No. 12 Trim Down Pump Relief Valve — regulates hydraulic pressure in the system during tilt and trim **DOWN** operation.

No. 14 Reverse Lock Check Valve — permits fluid flow from the pump to top sides of the tilt and trim cylinders during trim and tilt **DOWN** operation. Also allows fluid flow to return to the pump during trim and tilt **UP** operation.

No. 15 Tilt Cylinder — moves the engine through the tilt range **ONLY**. Also absorbs the shock when the lower unit strikes an underwater object.

No. 16 Impact Sensor Valve — operates **ONLY** when the engine tilts up as a result of the lower unit striking an underwater object. Valve permits fluid from the top side of the tilt piston to flow to the letdown control piston. The piston then moves to open the impact letdown valve.

No. 17 Pump Control Piston — must be centered between the reverse lock check valve and the trim check valve when the pump is not operating.

9-3 OPERATION

Seven modes of operation are possible with the hydraulic unit:
1- Trimming or tilting **OUT**.
2- Trimming or tilting **IN**.
3- Outboard operating in **FORWARD**.
4- Outboard operating in **REVERSE**.
5- Shallow water operation or trailering.
6- Striking an underwater object.
7- Lowering outboard after striking underwater object.

Trim/Tilt OUT

See functional diagram, Illustration #4, this page.

When the hydraulic unit is operated in the trim **OUT** tilt **UP** mode, hydraulic fluid is routed from the pump as follows:

a- Hydraulic pressure is directed to the trim check valve side of the pump control piston. This action causes the piston to open the reverse lock check valve.

b- The trim check valve then opens permitting fluid to flow to the base of both

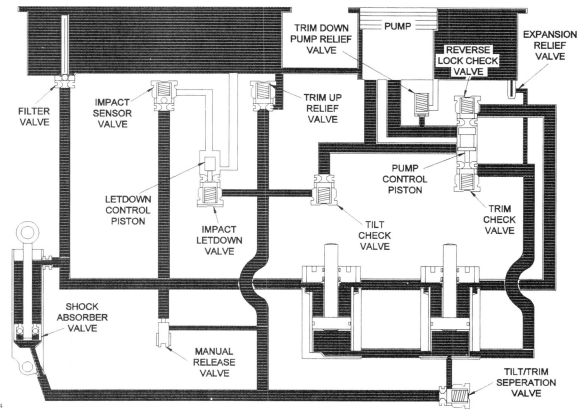

trim cylinders. At the same time, some of the fluid is directed to and opens the tilt check valve allowing fluid to move to the base of the tilt cylinder.

c- Fluid from the top of the tilt cylinder and the top of the trim cylinders passes through the reverse lock check valve. This valve is held open in this mode by the pump control piston. Hydraulic fluid thus returns to the pump. A study of the diagram shows how all fluid on the return circuit ends back at the pump. The trim and tilt rods displace fluid on the top of the cylinders, the fluid returning to the pump through the reverse lock check valve does not provide an adequate supply for the pump. Therefore, additional fluid is drawn from the reservoir to supply pump demands.

d- When hydraulic pressure exceeds 1500 psi with both trim and tilt cylinders fully extended, the trim up relief valve will open allowing fluid to pass back to the input side of the pump.

Trim/Tilt IN

See functional diagram, Illustration #5, this page.

When the hydraulic unit is operated in the trim **IN** tilt **DOWN** mode, hydraulic fluid is routed as follows:

a- Hydraulic pressure directed to the reverse lock check valve side of the pump control piston. This build-up in pressure causes the piston to open the trim check valve.

b- Hydraulic fluid then opens the reverse lock check valve and permits fluid to pass through the valve to the trim and tilt cylinders.

c- The hydraulic fluid returning from the bottom of the tilt cylinder, opens the tilt/trim separation valve. The fluid then flows through the open valve, then through the trim check valve. This valve is being held open by the pump control piston. The fluid then returns to the input side of the pump.

d- The fluid returning from the bottom of the trim cylinder flows through the trim check valve. This valve is being held open by the pump control piston. The fluid is allowed to return to the input side of the pump.

e- If the switch is held in the trim **IN** tilt **DOWN** position after the trim and tilt rods are retracted, the pressure build up will open the trim down pump relief valve and the hydraulic fluid will be bypassed back to the reservoir. The pump relief valve will release (open) at approximately 800 psi.

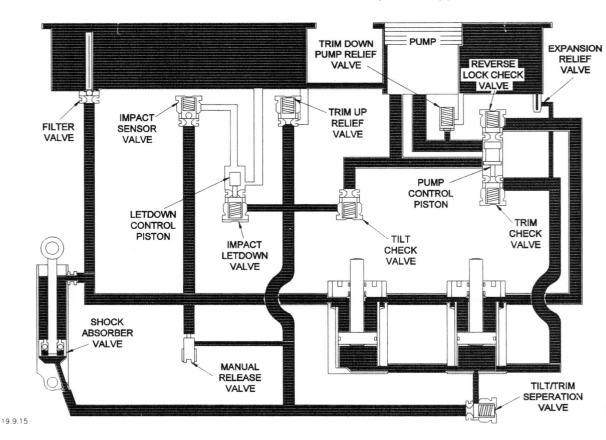

OPERATION 9-5

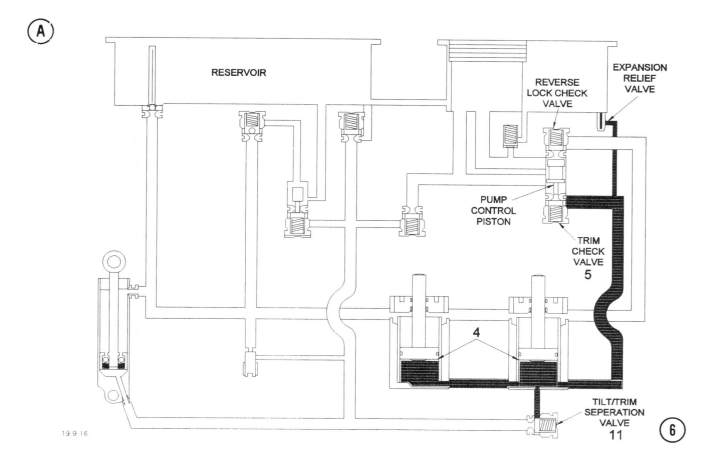

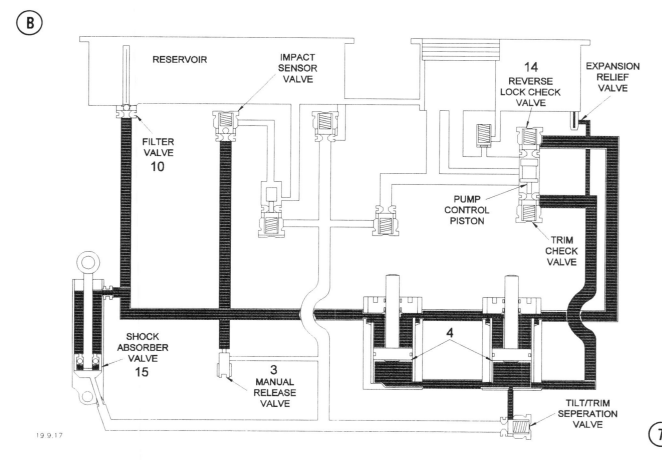

Running Forward
See functional diagram, Illustration #6, Page 9-5.

While the boat is moving in a forward direction, the trim pistons must hold the motor in the selected trim position. This is accomplished only if the seals on the trim pistons (4), the tilt trim separation valve (11), and the trim check valve (5), are **ALL** in satisfactory condition to hold hydraulic pressure.

Running in Reverse
See functional diagram, Illustration #7, Page 9-5.

When the engine is operating in **REVERSE**, the motor pulls up on the tilt cylinder and applies pressure to the top of the tilt/trim pistons. The tilt/trim separation valve prevents retraction of the trim pistons and upward movement of the tilt piston. In order for the system to hold the desired tilt/trim position while the engine is in the **REVERSE** mode, the shock absorber valves in the tilt piston (15), the manual release valve (3), the reverse lock check valve (14), and the filter valve (10), must **ALL** be in satisfactory condition to hold hydraulic pressure.

Tilting or Shallow Water Drive
See functional diagram, Illustration #8, Page 9-7.

While the engine is operating in the shallow water drive mode, or while the motor is tilted, **ALL** of the following components must be in satisfactory condition and hold hydraulic pressure: shock absorber valves in the tilt cylinder piston (1); seals on the tilt piston (2); manual release valve (3); seals on the trim pistons (4); trim check valve (5); tilt check valve (6); trim up relief valve (7); and the impact letdown valve (8).

At approximately 1500 psi (10,300 kPa), the trim up relief valve will open allowing fluid to return to the reservoir and the filter valve will open at approximately 900 to 1500 rpm, replacing fluid to the top of the tilt cylinder. The tilt cylinder will retract, lowering the motor to the fully extended trim pins. Retraction of the trim pistons under thrust is prevented by the tilt/ trim separation valve (11) and the trim check valve (5).

Return to Water
See functional diagram, Illustration #9, Page 9-7.

The weight of the motor drives the tilt rod and piston downward, forcing fluid from the bottom of the cylinder through the open impact letdown valve (1) and to the reservoir (2). The filter valve is opened and permits fluid from the reservoir to flow to the top of the cylinder (3). The letdown control piston leaks down, permitting the impact letdown valve (4) to close.

Striking an Underwater Object
See functional diagram, Illustration #10, Page 9-8.

When an underwater object is struck, the shock absorber (1) will absorb the shock and allow the motor to tilt up.

The shock absorber valves (2) in the tilt cylinder piston open and allow the fluid at the top of the piston to flow to the bottom. At the same time the tilt rod is being extended, the rest of the hydraulic unit is readied for the "return to water" function.

At 2500 psi, the high pressure fluid opens the impact sensor valve (3). Fluid passing through the impact sensor valve causes the letdown control piston (4) to move and open the impact letdown valve (5).

When the tilt cylinder piston stops moving outward, the impact sensor valve closes, keeping the letdown control piston pressurized. This action prevents the impact letdown valve (5) from closing.

At the instant the impact sensor valve closes, the letdown control piston begins to "leak down" at a definite rate.

9-4 TROUBLESHOOTING MECHANICAL COMPONENTS

Troubleshooting **MUST** be done **BEFORE** the system is opened in order to isolate the problem to one area. Always attempt to proceed with troubleshooting in a definite, orderly, and directed manner. The "shotgun" approach will only result in wasted time, incorrect diagnosis, replacement of unnecessary parts, and frustration.

The following procedures are presented in a logical sequence to check the mechanical components, with the most prevalent, easiest, and less costly items to be checked listed first. See Sections 9-5 and 9-6 for detailed procedures to troubleshoot the electrical system.

TROUBLESHOOTING 9-7

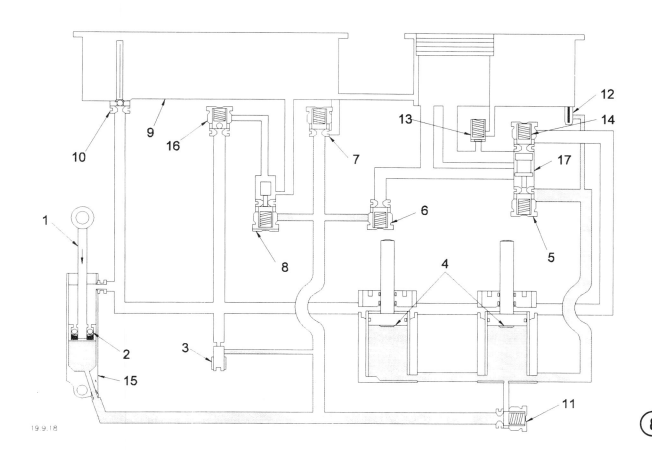

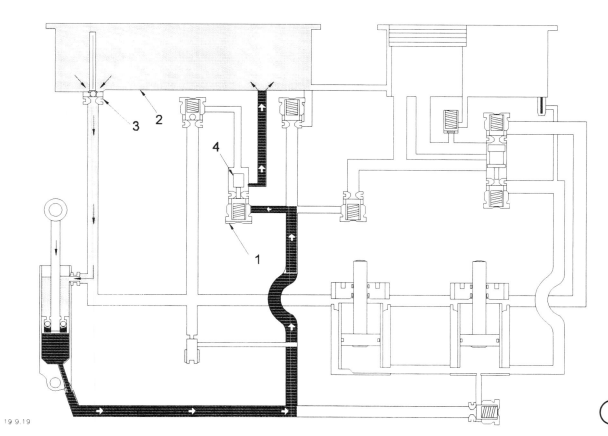

9-8 POWER TILT/TRIM

PRELIMINARY CHECKS AND INSPECTION

The following items are logical areas causing problems with the tilt/trim system. All may be checked and corrected without the use of special tools or equipment, with the exception of a torque wrench.

1- Check the battery for a full charge condition and to be sure all connections are clean and secure.

2- Check the hydraulic fluid level at the reservoir fill plug **AFTER** all air has been removed from the unit and when the tilt and trim cylinders are fully extended. The outboard unit must be in the vertical position. The fluid level should be even with the bottom of the fill hole with the motor full tilted up. Top off the reservoir to the plug level. Operate the motor and then again check the oil level. Attempt to cycle the unit several times and again check the level when the cylinders are fully extended. The system should be cycled through at least five complete movements to ensure all air has been purged from the system.

3- Add **ONLY** OMC Power Trim/Tilt Fluid or GM "Dextron" Automatic Transmission Fluid, as required. Total capacity of the system is 25 fl. oz. (740 mL).

4- Make an external (outside) inspection of the system for damage or signs of a fluid leak.

5- Seat the manual release valve by tightening the screw to a torque value of 45 to 55 in. lbs (5.1 to 6.2Nm).

6- Inspect the stern brackets for signs of binding with the swivel brackets in the thrust rod area. Check the tilt tube nut for a torque value of 24 to 26 ft lbs (32.5 to 35.2Nm), then back off (loosen) the nut 1/8 to 1/4 turn. Inspect the trim and tilt cylinders for bent rods.

7- Trailering the boat with the motor in the full tilt position and unsupported can cause a hydraulic "lock-up". To relieve such a "lock-up", loosen one trim cylinder end cap 1/4 turn with a socket type spanner wrench. Operate the unit down then up slightly. Tighten the end cap.

PROBLEMS AND CORRECTIONS

The following common problem areas are listed with several components given, any one or combination of several may cause the malfunction. Reference Illustrations **#11, #12, #13,** and **#14**.

Trim Cylinder Leak

May be caused by: trim up relief valve,

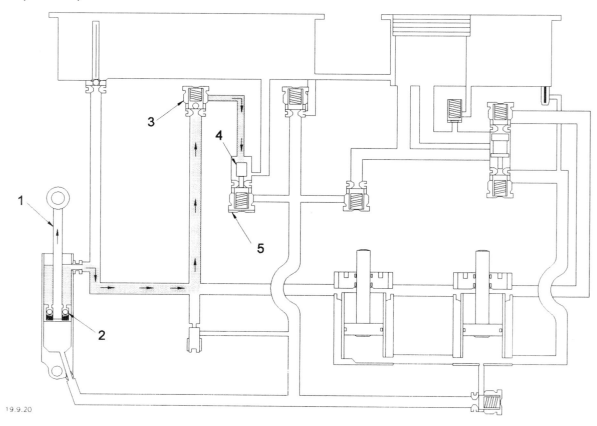

TROUBLESHOOTING 9-9

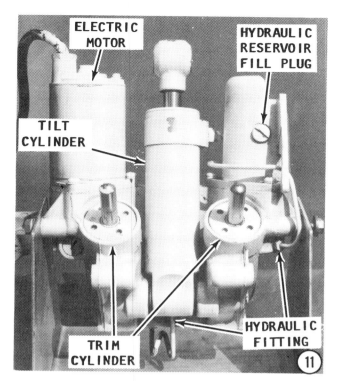

manual release, tilt cylinder valve or tilt cylinder seals, tilt check valve, impact letdown valve, or leaking hydraulic lines.

Trim and Tilt Cylinders
Both Leak

In the cylinders -- the O-rings or piston seals need replacement. The trim check valve is not holding pressure.

Reverse Lock Fails to Hold

Any one or more of these five areas: filter valve seat; impact sensor valve; manual release valve; tilt cylinder valves or seals; reverse lock check valve; or a leak in the hydraulic lines.

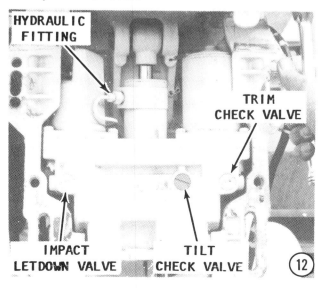

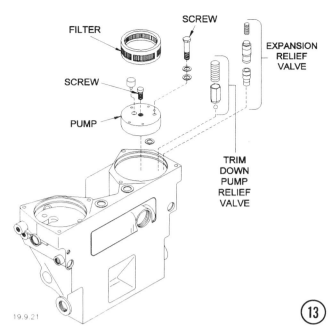

Trim Pulls IN
When Reverse Lock Fails To Hold

Trim/tilt separation valve or the trim check valve is unfit for service.

Slow Movement Up or Down

Possible mechanical bind; pump or motor malfunction; or the trim down pump relief valve may be failing (down only).

Unit Fails to Trim or
Tilt in EITHER Direction

Motor appears to run fast and there is low current draw.

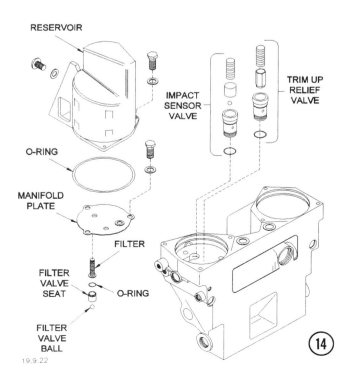

9-10 POWER TILT/TRIM

Hydraulic fluid is low -- no fluid in the pump cavity; pump coupling is broken. Motor fails to run and there is high or no current draw.

Electrical problems -- see Electrical System Troubleshooting; the motor on the pump is "frozen".

Unit Fails to Trim OUT Under Load or Fails to Tilt

In the trim cylinder -- sleeve O-rings or the piston seals need replacement; trim relief valve or the manual release no longer fit for service.

Unit Fails to Trim or Tilt Down

Trim check valve plugged; pump control piston failing to open trim check valve; or any one of the following components unfit for further service: seat in the filter valve, manual release, impact sensor valve, trim down pump relief valve.

TROUBLESHOOTING WITH EQUIPMENT

AMMETER

Hydraulic pump output pressure may be determined by measuring the amount of current in amperes required to operate the motor driving the pump. Therefore, proceed with an ammeter through the following checks.

Connect an 0-60 DC ammeter between the positive battery terminal and the heavy red cable connected to the trim/tilt solenoid.

Operate the trim switch in the UP mode and in the DOWN mode and observe the meter reading. Compare the reading with the chart on this page. The current draw and times listed represent hydraulic static requirements of the power trim/tilt. Do not take readings while the boat is under power, because such readings are not reliable.

Results of the meter reading may be interpreted as follows:

Normal Current Draw -- in both the UP and DOWN modes indicate the motor, hydraulic pump, and relief valves are functioning properly. Further troubleshooting must be directed toward the pump control piston, a malfunctioning check valve, or mechanical binding somewhere.

Low Current draw -- in both the UP and DOWN modes indicates leaking valves, weak

UNIT MOVEMENT	CURRENT DRAW (AMPS)	TIME IN SECONDS
Trim UP	23	6
Tilt UP	27	5
Full Tilt UP (Stall)	40-44	---
Trim IN to Full Tilt	---	17
Tilt DOWN	20	5
Trim DOWN	20	7
Full Trim DOWN (Stall)	28-32	---
Full Tilt To Trim IN	---	17

relief valves, pump not operating satisfactorily, a leak in an O-ring, or a leaking valve body.

High Current Draw -- in either the full UP mode or the full DOWN mode indicates the motor is malfunctioning either electrically or mechanically; or the pump may be binding. It is also possible a relief valve is failing to open at the correct pressure.

TROUBLESHOOTING WITH PRESSURE TESTER

The following tests may be performed with the engine still mounted on the boat or on the work bench. An OMC Trim/Tilt Pressure Tester No. 390010, or equivalent is required. The tester has two gauges, one measures hydraulic pressure in the UP mode, the other gauge indicates pressure in the DOWN mode.

Remove the manual release valve by first removing the retaining ring with a pair of snap ring pliers. Use the trim switch and run the unit down completely. Loosen the valve one full turn, Illustration #15.

Momentarily work the trim switch UP and DOWN, UP and DOWN a few times.

SAFETY WORD

Pressure in the system may be released when the manual release valve is removed. Therefore, pack a shop towel or rag around the valve access hole and screwdriver shank **BEFORE** removing the valve. Trapped fluid may squirt out causing eye injury. Place a drip pan in position to catch fluid released from the valve opening.

Remove the manual release valve by turning it **COUNTERCLOCKWISE** and as the valve is turned, "feel" for any residual pressure trying to push the valve out.

Make a careful check of the valve port for any metal chips or other foreign matter. Clean the opening, if necessary.

Install the **"A"** gauge into the manual release valve opening to test the **UP** operation, Illustration #16. Install the **"B"** gauge to test the **DOWN** operation.

Tighten the gauge unit **LIGHTLY**, no more than 5-10 in. lbs. (0.6-1.2 N m). After the tests are complete, remove the gauges using the same safety precautions as just desribed for removal of the manual release valve.

With the gauge properly installed, cycle the motor **UP** and **DOWN** several times and top off the fluid reservoir as outlined earlier in this section before making any tests.

Test UP Mode

With the **"A"** gauge installed, run the unit fully **UP** until the tilt piston "tops" out. Observe the gauge. The reading should not drop more than 100-200 psi (700-1400 kPa) after the motor is stopped. If the gauge reading continues to drop, one of seven areas may be leaking: trim up relief valve, impact letdown valve, tilt check valve, tilt/trim separation valve, trim check valve, shock absorber valves, or at the shock absorber piston seal.

Test DOWN Mode

With the **"B"** gauge installed in the manual release valve opening, Illustration #17, operate the unit down until the tilt piston "bottoms" out and observe the gauge. If the reading drops more then 100-200 psi (700-1400 kPa) after the motor is stopped, a leak

is indicated in one or more of six areas: impact sensor valve, filter valve, reverse lock check valve, shock absorber valves, seals on the trim rods, or the seals on the tilt piston rod.

Valve Leakage Test

A check of each valve should be made if the previous tests indicate one may be leaking. A positive check can be accomplished using an OMC Check Valve Tester, Part No. 390063 and a Stevens Gearcase Tester No. S-34, or equivalent.

Remove the valve to be tested from the system. The long and short check valves may be tested with this equipment. Lubricate both O-rings on the valve to be tested. Thread the valve into the tester. Check to be sure the outside O-ring is tight against the beveled sealing face. Tighten the valve securely to a torque value of 50-60 in.-lbs. (5.5-7.0 N m).

Thread the gearcase tester hose into the valve tested. Pump pressure to 30 psi (207 kPa). With the tester and valve pressurized, immerse the test unit in a container of clear water and check for any air bubbles, Illustration #18.

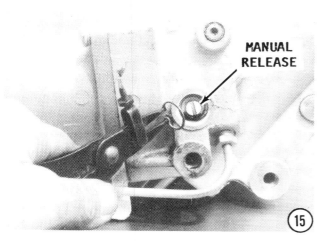

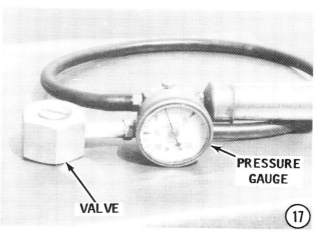

9-12 POWER TILT/TRIM

If no bubbles are visible the check valve is in satisfactory condition to hold hydraulic pressure. If any bubbles are visible, the valve is leaking. Bubbles from the edge of the valve indicates the O-rings are leaking. Bubbles from the center hole indicates the check valve is leaking and unfit for further service. The valve **MUST** be replaced.

9-5 TROUBLESHOOTING ELECTRICAL SYSTEM ALL 1978, 1979, 1980, AND CIH 1981 ONLY

This section presents detailed instructions to troubleshoot the electrical system for the model years listed in the heading. See Section 9-6 to troubleshoot the 1981 and 1982 models except the CIH. See Section 9-4 for troubleshooting procedures to check mechanical components.

SPECIAL NOTE

The accompanying functional diagram on Page 9-13, is **NECESSARY** to effectively troubleshoot the electrical system. The referenced numbers in the procedures are keyed to the numbers on the diagram.

Preliminary Checks

Check to be sure the battery is up to a full charge and the terminal connections are clean and tight. Make a visual inspection of all exposed wiring for an open circuit or other damage that might cause a problem.

Both UP and DOWN Circuits

Obtain an ohmmeter and a voltmeter. Check to be sure the battery has a full charge and the terminal connections are clean and tight.

Connect the red lead of the voltmeter to junction box terminal (1) and the black meter lead to point (14). The meter should indicate full battery voltage.

If no voltage is indicated, move the red meter lead to point (2) on the solenoid. If battery voltage is not indicated, check the 30 amp fuse in the junction box. If no voltage is indicated, check the red and black leads to the battery for an open or poor connection.

DOWN Circuit Only

Connect the red meter lead to point (3) in the junction box. Push the trim switch to the **DOWN** position. The meter should indicate battery voltage. If voltage is indicated, but the trim motor does not operate, test the motor. See procedures later in this chapter.

If no voltage is indicated, move the meter red lead to point (4) at the lever and switch assembly. Connect the black meter lead to point (13) at the trim gauge indicator. Push the trim switch to the **DOWN** position. The meter should now indicate battery voltage. If the meter indicates battery voltage, the lead between point (3) in the junction box and point (4) at the lever and switch assembly is open.

If the meter fails to indicate full battery voltage, move the red meter lead to point (5) at the lever and switch assembly. If battery voltage is not indicated, the trim switch may be defective.

If no voltage is indicated at point (5), check continuity of the ground lead from the trim gauge (13) to point (14) in the junction box. If the ground lead is satisfactory, move the red meter lead to point (1) in the junction box. Move the black meter lead to point (14) in the box. If battery voltage is now indicated, the lead is open between point (1) in the junction box and point (5) at the lever and switch assembly.

UP Circuit
1978 and 1979 Models Only

Connect the red voltmeter lead to point (6) in the junction box. Connect the black meter lead to ground at point (14) in the junction box. Push the trim switch to the **UP** position. The meter should indicate battery voltage. If the trim motor will not operate, test the motor according to procedures outlined later in this section.

If no voltage is indicated at point (6), move the red meter lead to point (7) at the solenoid. Push the trim switch to the **UP**

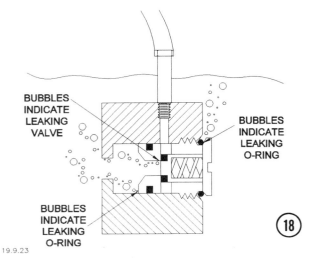

TROUBLESHOOTING 9-13

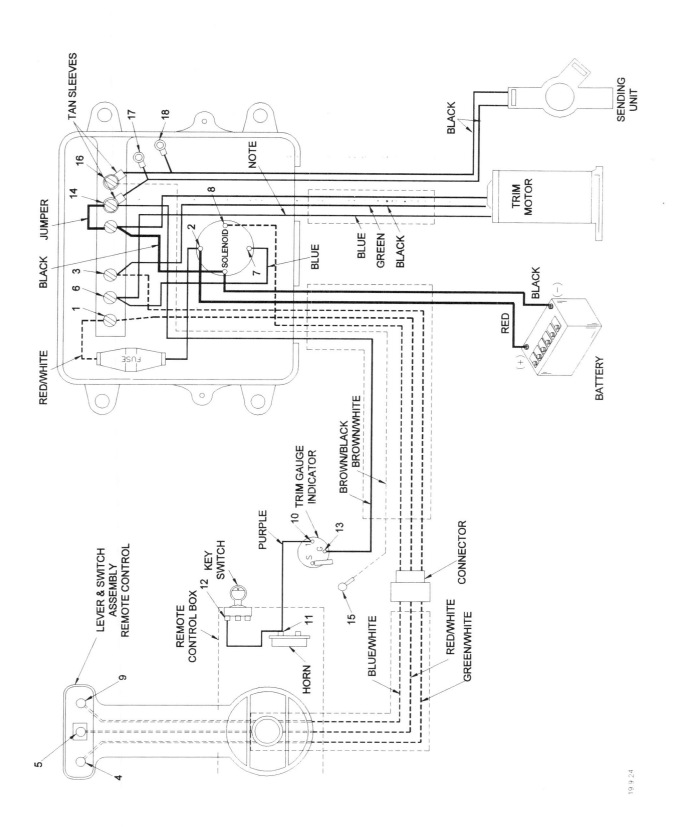

Functional electrical diagram -- Power Tilt/Trim system -- all models thru 1980 plus the CIH V4 and V6 models for 1981.

position. If no voltage is indicated the blue lead between point (6) in the junction box and point (7) at the solenoid is open. Check it out and repair the line.

UP Circuit
1980 All Models
1981 Model C1H

Connect the red voltmeter lead to point (7) in the junction box. Connect the black meter lead to ground at point (14) in the junction box. Push the trim switch to the **UP** position. The meter should indicate battery voltage. If the trim motor fails to operate, test the motor according to the procedures outlined later in this section.

If no voltage was indicated at point (7), move the red meter lead to point (8) at the solenoid. Push the trim switch to the **UP** position. The meter should indicate battery voltage. If the trim motor still fails to operate, connect the red meter lead to point (2) at the solenoid. If battery voltage is now indicated, the solenoid is defective and **MUST** be replaced.

If no voltage was indicated at point (8) at the solenoid, move the red meter lead to point (9) at the lever and switch assembly. Connect the black meter lead to ground at point 13 on the trim gauge. Push the trim switch to the **UP** position. The meter should indicate full battery voltage. If battery voltage is indicated the lead between point (8) at the solenoid and point (9) at the lever and switch assembly is open. Check it out and repair.

If no voltage was indicated at point (9), move the red meter lead to point (5) at the lever and switch assembly. If the meter now indicates battery voltage, the switch is defective and **MUST** be replaced.

If no voltage was indicated at point (5), check continuity of the ground lead between point (5) at the lever and switch assembly and point (14) in the junction box for an open circuit. Check it out and repair.

If the ground lead is good and no voltage is indicated at point (5), move the red meter lead to point (1) in the junction box. If the meter now indicates voltage, the lead is open between point (5) at the lever and switch assembly and point (1) in the junction box.

Trim Indicator Circuit

The referenced numbers in the procedures are keyed to the numbers on the electrical diagram on Page 9-13.

Connect the red voltmeter lead to point (10) on the trim gauge indicator. Connect the black meter lead to point (13) on the indicator.

If no voltage is indicated at (10), move the red meter lead to point (11) at the horn. If the meter now indicates battery voltage, the lead between the horn and the trim indicator is open. Check it out and repair.

If no voltage was indicated at point (11), move the red meter lead to point (12) at the key switch. The meter should indicate battery voltage. If the meter fails to indicate voltage, check the key switch. If the key switch is good, check the engine fuse and the lead to the key switch for an open circuit.

If battery voltage was indicated at point (13) on the trim gauge indicator, proceed with the test in this paragraph. Disconnect the tan/white lead from the "S" terminal at the trim gauge. Connect the ohmmeter to the disconnected lead at point (15) and point (13). With the motor in the full trim **IN** position, the ohmmeter should indicate 82 to 88 ohms. With the motor in the full tilt position, the ohmmeter should indicate 0 to 10 ohms. If infinite resistance is indicated, check for an open circuit between point (13) at the trim gauge indicator and point (14) in the junction box. Also check the tan/white lead with the lead disconnected from the "S" gauge terminal for an open circuit. Repair as required.

Check the circuit between point (15) at the trim gauge indicator and point (16) in the junction box. Disconnect both black sending unit leads from points (14) and (16) in the junction box. Check for a resistance of 82 to 88 ohms. If the resistance is not within this range, replace the sending unit.

9-6 TROUBLESHOOTING ELECTRICAL SYSTEM MODELS 1981 AND ON EXCEPT C1H

This section presents detailed instructions to troubleshoot the electrical system for the model years listed in the heading. See Section 9-5 to troubleshoot the 1978, 1979, 1980, and 1981 models except the C1H. See Section 9-4 for troubleshooting procedures to check mechanical components.

SPECIAL NOTE

The accompanying wiring diagram on

TROUBLESHOOTING 9-15

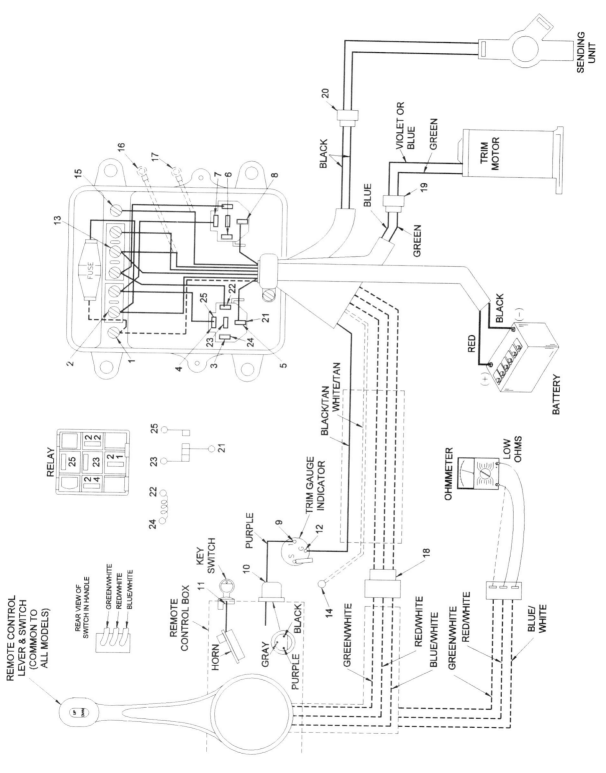

Functional electrical diagram — power tilt/trim system — 1981 and on, except CIH.

9-16 POWER TILT/TRIM

Page 9-15, is **NECESSARY** to effectively troubleshoot the electrical system. The referenced numbers in the procedures are keyed to the numbers on the diagram.

Preliminary Checks

Check to be sure the battery is up to a full charge and the terminal connections are clean and tight. Make a visual inspection of all exposed wiring for an open circuit or other damage that might cause a problem.

FUNCTIONAL CIRCUIT DESCRIPTION

Neutral Circuit

When the control switch is in the neutral position, there is no movement up or down. The contacts at point (21) and point (23) are normally closed. The contacts are normally open between point (21) and point (25).

Down Circuit

When the tilt and trim switch is moved to the **DOWN** position, current flows from the switch to the left relay, point (24), energizing the solenoid winding at point (22). The current returns to the negative side of the battery.

The energized solenoid pulls the relay points together and contact is made at points (21) and (25). It is then possible for current to flow from the positive side of the battery to point (25) through relay point (21). The current then flows from point (21) through the green wire to the motor.

From the motor, the current returns through the blue or violet wire to the right relay point (21). The current can then flow from point (21) through the relay to point (23) and return to the negative side of the battery.

UP Circuit

When the tilt and trim switch is moved to the **UP** position, current flows from the positive side of the battery through the tilt and trim switch to the right relay at point (24), through the relay winding to point (22). The current returns to the negative side of the battery.

The solenoid is energized pulling the relay points together and contact is made at points (25) and (21) and contact points between (25) and (21) are opened.

Current will flow from the positive side of the battery to point (25) of the right relay and then through the relay to point (21). From point (21) the current flows through the blue wire to the motor. From the motor, current returns through the green wire to the left relay point (21) and can continue from point (21) through the left relay to point (23). From there, the current returns to the negative side of the battery.

CIRCUIT TESTING

Relay Resistance Tests

Remove relays from connector and from the junction box. The ohmmeter to the low ohms scale. Connect one meter lead to point (22) and the other lead to point (24) on the relay. Ohmmeter should indicate approximately 85 ± 15 ohms.

Connect one meter lead to point (21) and the other meter lead at point (23) on the relay. Ohmmeter should indicate approximately zero ohms. Move the meter lead from point (23) to point (25). Meter should indicate an open circuit, no reading. If readings other than those indicated above are observed, replace the relay and proceed with the tests.

SPECIAL NOTE

The accompanying functional diagram on this page is **NECESSARY** to effectively troubleshoot the electrical system. The referenced numbers in the procedures are keyed to the numbers on the diagram.

Preliminary Checks

Check to be sure the battery is up to a full charge and the terminal connections are clean and tight. Make a visual inspection of all exposed wiring for an open circuit or other damage that might cause a problem.

Both UP and DOWN Circuits

Turn the key switch to the **ON** position. Connect the red lead of a voltmeter point (1) in the junction box and the black meter lead to a good ground. Voltmeter should indicate battery voltage.

If no voltage is indicated, move the red meter lead to point (2) on the terminal strip. If battery voltage is indicated, check the fuse in the junction box. If the meter does not indicate any voltage at point (2), check the red and black leads to the battery for poor connection or open circuit. Repair as required.

SYSTEM REMOVAL 9-17

DOWN Circuit

Connect the red meter lead to point (3) in the junction box and the black lead to a good ground. Move the trim switch to the **DOWN** position. The meter should indicate battery voltage. If voltage is indicated, move the red meter lead to point (4). If voltage at this point is indicated, move the red lead to point (5) while holding the trim switch in the **DOWN** position. If voltage is now indicated, check the connector at point (19). If the connector check is satisfactory but the motor fails to operate, test the motor as outlined later in this section.

If no voltage is indicated at point (3), check the connector at point (18) between the junction box and the remote control. Check continuity of the green/white stripe lead from the junction box to the trim switch. Check the trim switch contact with the switch in the **DOWN** position between the green/white stripe lead and the red/white stripe lead for continuity. Check the red/white stripe lead from the trim switch to the terminal strip for continuity to point (1).

UP Circuit

Connect the red meter lead to point (6) in the junction box and the black lead to a good ground. Move the trim switch to the **UP** position. The voltmeter should indicate battery voltage. If voltage is indicated, check for voltage at point (7). If voltage is indicated there, move the red meter lead to point (8) with the trim switch in the **UP** position. If voltage is indicated, check the connector point (19). If the connector checks satisfactorily, and the motor still fails to operate, test the motor according to the procedures outlined later in this section.

If no voltage is indicated at point (6), check connector point (18) between junction box and remote control. Check the blue/white stripe lead from the junction box to the trim switch for continuity. Check the trim switch contact with the switch in the **UP** position between the blue/white stripe lead and the red/white stripe lead for continuity. Check the red/white stripe lead from the trim switch to the terminal strip for continuity to point (1).

Trim Indicator Circuit

Turn the key switch to the **ON** position. Connect the red voltmeter lead to point (9) on the indicator gauge and the black meter lead to point (12) on the indicator gauge. If no voltage is indicated at point (9), move the red meter lead to point (10) at the accessory plug. If battery voltage is now indicated, the lead between the accessory plug and the trim indicator gauge is open. Check it out and repair as required.

If no voltage is indicated at point (10), move the red meter lead to point (11) at the accessory terminal on the key switch. The meter should now indicate battery voltage. If no voltage is indicated, check the key switch. If the key switch is satisfactory, check the engine fuse and the lead to the key switch for an open circuit.

If battery voltage was indicated at point (12) in the first part of this test, disconnect the white/tan lead from the "S" terminal of the trim gauge. Connect the ohmmeter to the disconnected lead and to point (12). Set the ohmmeter to the low ohm scale.

Now, with the motor in the full trim-IN position, the ohmmeter should indicate 82 to 88 ohms. with the motor in the full tilt position, the ohmmeter should indicate 0 to 10 ohms. If the meter indicates an infinite resistance, check for an open circuit between trim gauge point (12) terminal and point (13) in the junction box. Also check for an open circuit in the white/tan lead with the lead disconnected from point (14) gauge terminal. Check between point (14) and point (1) in the junction box. Disconnect both the black sending unit leads from point (15) and point (1) terminals. Check for above resistance readings between point (16) and point (17). If the readings are not within the specified limits, replace the sending unit.

9-7 SYSTEM REMOVAL

SAFETY WORD

Anytime a steel tool is struck with a hammer, there is the possibility of chips flying which could cause serious eye injury. Therefore, wear safety glasses while removing the tilt cylinder pin.

Models 1978 thru 1980

Remove the outboard motor from the boat enough to permit the tilt and trim unit to come free. Remove the starboard tilt tube nut using two wrenches, port and starboard, Illustration #19.

Mark the thrust rod location and then remove the rod. Remove the screws port and starboard, holding the power trim assembly to the stern brackets. Remove both stern brackets.

Remove the spring clip from the tilt cylinder pin. Support the weight of the power trim unit while the tilt cylinder pin is being removed. Drive the tilt cylinder pin free using a drift punch and hammer, Illustrations #20 and #21.

Models 1981 and On

Mark the angle adjusting rod location and then remove the rod. Lift the engine manually and engage the tilt trail lock. Remove the spring clip from the tilt cylinder pin.
Drive the tilt cylinder pin free using a drift punch and hammer.
Manually retract the tilt cylinder. Remove the screws port and starboard securing the unit to the stern brackets.
Remove the trim and tilt unit and at the same time pull the control cable through the stern bracket, Illustrations #20 and #21.

9-8 DISASSEMBLING

This section contains complete detailed procedures to disassemble the complete unit. However, open the system and remove **ONLY** the necessary parts to inspect, re-

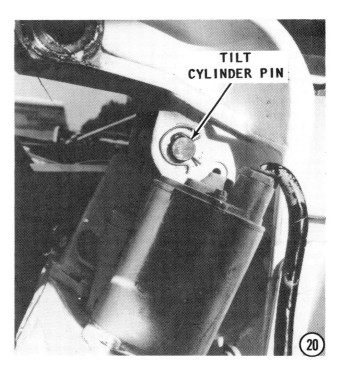

place, and restore the system to satisfactory service.

Drain the system by removing the reservoir plug and draining the hydraulic fluid into a suitable container.

Valves and Pistons

To remove the manual release valve see the Troubleshooting with Pressure Tester in Section 9-4, Page 9-10.

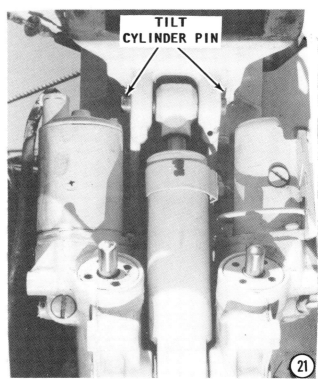

DISASSEMBLING 9-19

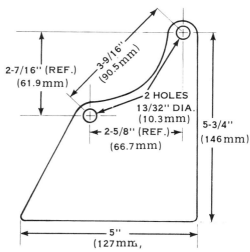

Simple drawing to assist in making a holding bracket to secure the trim and tilt assembly during service work.

Remove the external (outside) valves Illustration #22, using a drag link socket that fits the slot properly.

The letdown control piston can be reached by first removing the impact letdown valve, and then carefully removing the piston with a pair of needle nose pliers.

The pump control piston and springs are removed by first removing the reverse lock check valve and then lifting the piston and springs free with a pair of needle nose pliers. The pump control piston can only be removed from the aft end of the hole.

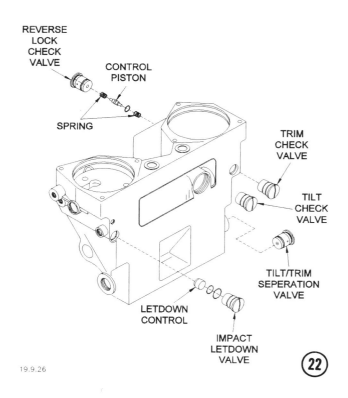

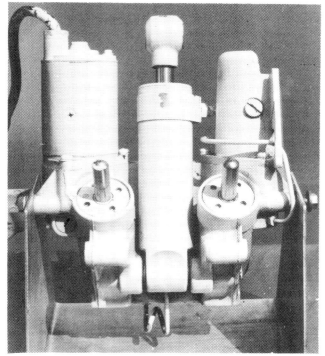

Trim and tilt assembly mounted in the bracket shown at the top of this column. Such a bracket makes service work on the unit much easier.

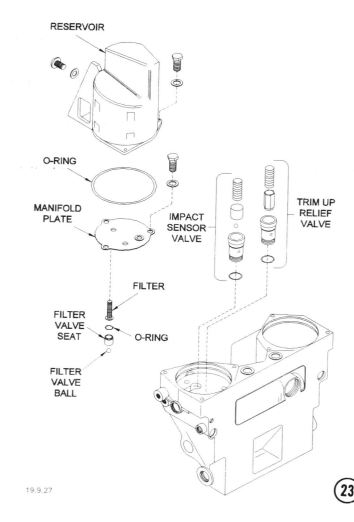

9-20 POWER TILT/TRIM

Reservoir and Valves

The reservoir, Illustration #23, is removed by first removing the upper and lower hydraulic lines and fittings. Next, remove the screws from the reservoir flange and lift the reservoir free. Remove and discard the O-ring. If further disassembly of the reservoir is desired, hold down on the reservoir manifold plate and at the same time remove the three screws securing the plate. Lift the plate free. Remove the relief valve and impact sensor valve assemblies by lifting them free of the body with a pair of needle nose pliers. The filter valve may be lifted out with a small stiff hook. Because damage usually occurs during removal, the filter valve seat and O-ring must be replaced if they are removed.

Motor and Pump

Remove the screws on the motor flange and lift the motor free. See Section 9-11 for detailed instructions for testing and to service the Prestolite and Bosch motor. Remove and discard the O-ring, Illustration #24. Remove the hydraulic pump filter. The hydraulic pump may be removed by simply removing the attaching screws and lifting the pump free. Lift the trim down pump relief valve free of the body. The expansion relief valve core can now be lifted out with a small hook shaped piece of wire.

Trim Cylinders — Disassembling

Obtain special tool Trim Cylinder End Cap Remover, OMC No. 324958 or an ad-

justable spanner wrench. Use the special tool or the spanner wrench to remove the end cap, Illustration #25.

Lift the trim piston assembly free of the cylinder. The sleeve fits snugly. Therefore, Trim Sleeve Remover, OMC No. 325065 or a

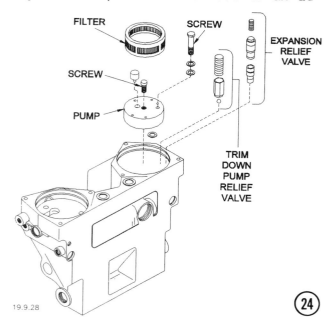

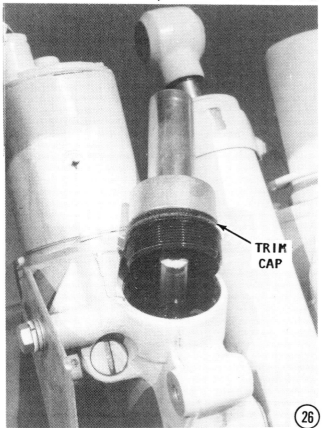

ASSEMBLING 9-21

screwdriver with the tip bent 90° must be used to remove the sleeve. Slip the tool in under the sleeve and then remove the sleeve, Illustration #26.

Tilt Cylinder — Removal

Disconnect the upper and lower hydraulic lines at the tilt cylinder and at the manifold. Push the cylinder pivot pin to one side and remove the tilt cylinder.

Tilt Cylinder — Disassembling
See Illustration #27.

SPECIAL WORDS

The Showa tilt cylinder is identified by an "S" after the part number stamped on the side of the cylinder. The Prestolite cylinder is identified with a "P" stamped on the bottom side by the part number. When ordering replacement parts, be sure to identify which cylinder is being serviced.

Clamp the cylinder in a vise at the flats of the cylinder end. Obtain and use End Cap Remover Special Tool, OMC No. 326485 with a 1/2" breaker bar to loosen the end cap assembly. With the end cap removed, the piston and rod assembly may be withdrawn free of the cylinder.

The piston contains four valves. These valves cannot be serviced separately. If the valves are worn or are not functioning properly, the piston must be replaced as an assembly.

Prestolite Piston — Disassembling

Remove and discard the O-ring from around the piston. Heat with a torch, a vise and bar through the rod end will be necessary to break the Locktite bond on the thread of the piston and piston rod. Clamp the piston securely in a vise, apply the heat, and then use the bar through the rod end to unscrew the rod from the piston. Clean all traces of old sealant from the piston rod threads.

Showa Piston — Disassembling

Clamp the piston rod end in a vise and remove the nut securing the piston to the rod. Clean all traces of sealant from the piston rod threads.

9-9 CLEANING AND INSPECTING

Discard all used O-rings and seals. Clean all parts in solvent and blow them dry with compressed air. Inspect the cylinders and sleeves for any sign of excessive wear or scoring. Inspect all parts for dirt, chips, and damage. Replace any damaged valve seats or other questionable parts.

9-10 ASSEMBLING AND INSTALLATION
See Illustration #27.

SPECIAL WORD

Lubricate all internal parts with OMC Power Trim/Tilt Fluid prior to assembling.

Tilt Cylinders — Assembling
Prestolite

Lubricate a NEW O-ring and seal with OMC Power Trim/Tilt Fluid, and then install both into the cylinder end cap. Install Seal Protector OMC No. 326005 onto the threads of the piston rod and then install the end cap onto the rod. If the seal protector is not available, wrap the threads with tape as a protection against damaging the seal when the end cap is installed to the rod. Remove the tape after the end cap is installed.

If the piston was disassembled, clean the piston rod threads with OMC Locquic Primer, OMC No. 384884, or equivalent. Clamp the rod in a vise holding the pin end. Coat the rod threads with OMC Ultra Lock, OMC No. 388517, or equivalent. Thread the piston assembly onto the rod. **TAKE CARE** not to damage the surface of the piston. Use a flywheel strap wrench to bring the piston up snug.

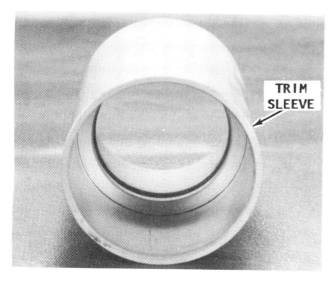

The trim cylinder sleeve must be clean and free of any scratches or other damage. The sleeve may be replaced at very modest cost.

9-22 POWER TILT/TRIM

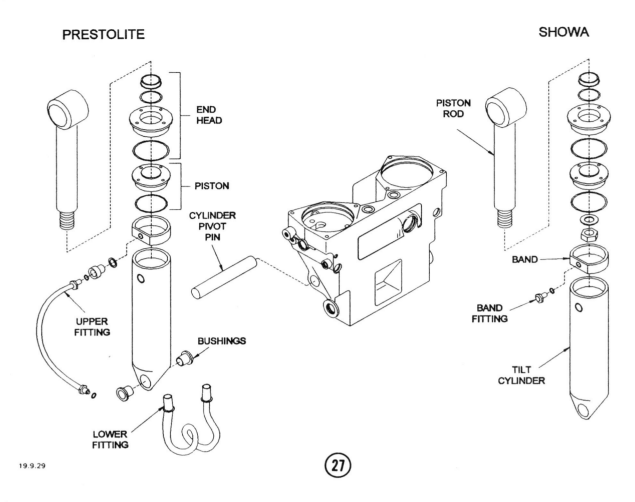

Tilt Cylinders -- Assembling
Showa

Lubricate a **NEW** O-ring and seal with OMC Power Trim/Tilt Fluid. Install the O-ring and seal into the end cap. Use Seal Protector, OMC No. 326005 and install the end cap onto the piston rod. If the seal protector is not available, wrap the threads with tape as a protection against damaging the seal when the end cap is installed. Remove the tape after the end cap is installed.

Clean the piston rod threads with OMC Locquic Primer, OMC No. 384884 and install the small O-ring, washer, and piston onto the piston rod. The small holes on the piston **MUST** face upward. Apply OMC Nut Lock, OMC No. 384849 to the piston rod threads. Secure the rod in a vise and then install and tighten the nut to a torque value of 58-87 ft-lbs (79-118 N·m).

Both Models

Lubricate the piston assembly. Install a **NEW** O-ring to the outside diameter of the piston. **CAREFULLY** insert the piston assembly into the cylinder. Tighten the end cap assembly using End Cap Remover OMC No. 326485, or an adjustable spanner wrench.

If the band has been removed, slide it into the cylinder. Use a **NEW** O-ring on the fitting and screw the fitting into the band, with the pilot on the fitting indexing into the hole in the cylinder.

Lubricate the cylinder pivot pin and pivot pin bushings with OMC Triple-Guard Grease. Install the cylinder to the manifold assembly. Attach the hydraulic line fitting on the starboard side.

Trim Cylinder -- Assembling

Insert the piston and rod assembly into the end cap. Carefully slide the piston into the trim cylinder, Illustration **#28**, until the piston and cylinder sleeve are butted against the end cap. Slide the assembled unit into the cylinder cavity and tighten the end cap to a torque value of 30-40 ft-lb (40-54 N·m) using End Cap Remover, OMC No. 324958 or an adjustable spanner wrench, Illustration **#29**.

ASSEMBLING 9-23

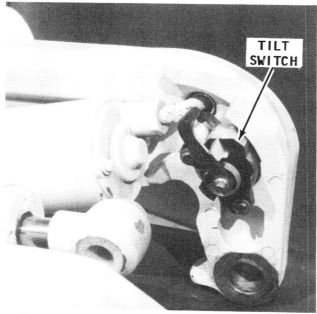

Location of the tilt/trim cutout switch. This switch has an adjusting screw for the tilt limit.

Reservoir and Valves -- Assembling
See Illustration #30.

To replace a valve, first install the valve seat. Next install the ball, core, and spring in that order. Install the filter valve ball, and then insert the filter valve seat with the O-ring end **UPWARD**. Slide the filter into the manifold plate. Shift the manifold plate until the filter is on top of the filter valve and the attaching screw holes are aligned. Thread the screws into place, and then tighten them **ALTERNATELY** and **EVENLY** to keep the valve spring positioned correctly.

Inserting the seal into the trim cylinder cap.

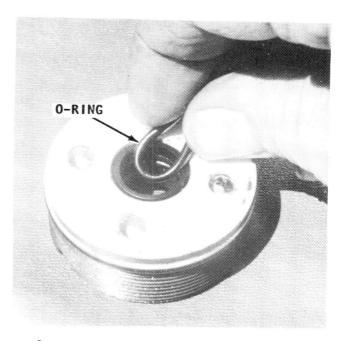

Inserting the O-ring into the trim cylinder cap.

Place a **NEW** O-ring in position. Secure the reservoir in place with the attaching screws. Tighten the screws securely.

Pistons and Valves -- Assembling
See Illustration #31.

Slide the letdown control piston into the starboard cavity with the rounded end going in first. Install the impact letdown valve. Slide the pump control piston into the port cavity from the rear with the small end going in first.

Install the reverse lock check valve and the trim check valve. Install the tilt check

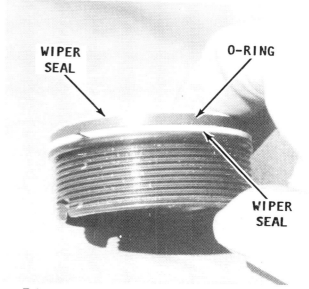

Trim cap with the O-ring and the two wiper seals properly installed.

9-24 POWER TILT/TRIM

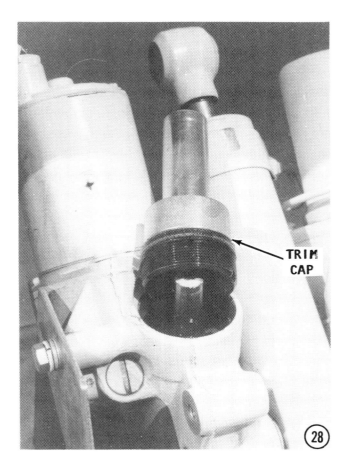

28

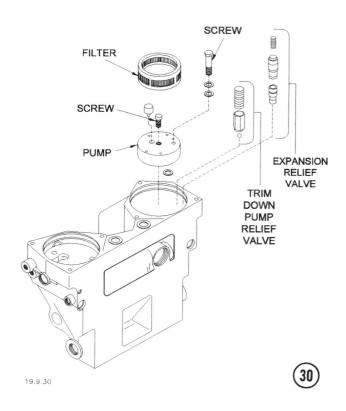

30

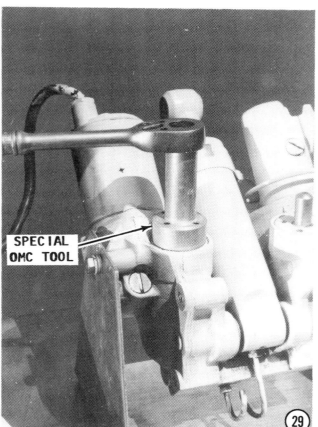

29

valve and the tilt/trim separation valve (long body valve). Tighten the valve securely.

Install the manual release valve with a **NEW** O-ring. Tighten the valve to provide operator with shock absorber protection. Install the snap ring with the flat side facing **OUT**.

Pump Assembly and Motor -- Assembling
See Illustration #32.

Install the pump relief valve and spring. Install a **NEW** O-ring. Check to be sure the pump drive tang indexes with the hole directly opposite the round locating boss.

Install the pump with the locating boss indexed into the pump cavity recess. Secure the pump in place with the three attaching screws. Tighten the screws securely. Install the pump filter and fill the filter cavity and the area over the pump with OMC Power Trim/Tilt hydraulic fluid.

CRITICAL WORDS

The pump cavity **MUST** be filled with hydraulic fluid during assembling, Illustration #33, or the unit will not operate.

Install a **NEW** O-ring onto the trim motor. Install the motor and at the same time

MOTOR TESTING AND REPAIR 9-25

rotate the motor shaft until the shaft engages with the pump shaft. Position the motor with the cable on the port side of the assembly. Install the three attaching screws. Tighten the screws securely.

Fill the reservoir with OMC Power Trim/Tilt Fluid and purge the system.

9-11 TRIM MOTOR TESTING AND REPAIR PRESTOLITE AND BOSCH

Motor Testing
Prestolite or Bosch

The condition of the motor can be tested with a current draw test on a no load test.

On a no load test, the motor should have a maximum current draw of 8 amps at a minimum of 5600 rpm at 12-volts for the Prestolite and 7 amps at 6700 rpm at 12-volts for the Bosch.

Prestolite

To make the test, connect the black wire to negative and the green wire (DOWN) to positive. The motor shaft from the drive end should turn in a COUNTERCLOCKWISE direction. Repeat the test with the blue (UP) wire to positive and the motor shaft should turn in a CLOCKWISE direction.

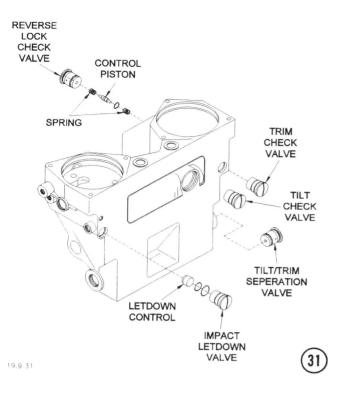
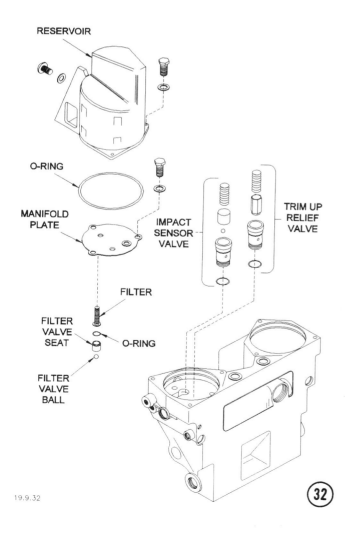
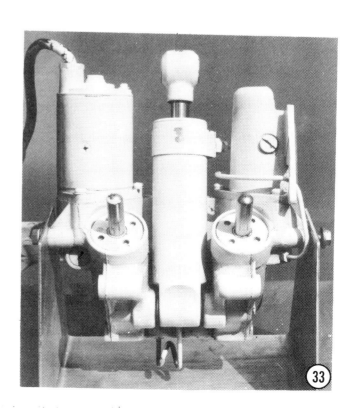

Bosch

Connect the violet or blue, **UP** wire to negative and the green, **DOWN** wire to positive. The motor shaft from the drive end should turn in a **COUNTERCLOCKWISE** direction. Repeat the test with the green (**DOWN**) wire connect to negative and the violet or blue, **UP** wire connect to positive. The motor shaft should turn in a **CLOCKWISE** direction.

If the motor fails either of the tests it must be serviced or replaced.

Motor Repair

The following procedures pick up the work after the motor has been removed.

Remove and discard the O-ring, Illustration #34. Remove the thru bolts and discard the seals on the bolts.

Remove the drive end cap from the motor. Discard the gasket. Removing the seal exercising **CARE** not to scratch the casting surfaces to ensure the new seal will seat properly. Discard the old seal.

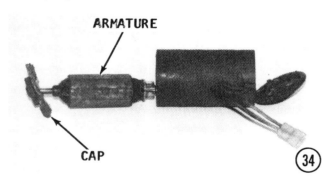

Remove the armature from the motor housing. **TAKE CARE** not to lose the fiber washer on each end of the shaft. Tip the end cap free of the motor housing. Discard the springs and the end cap gasket.

SPECIAL WORD

The end cap is serviced as a complete assembly.

CLEANING AND INSPECTION

Clean all parts with a dry cloth. **DO NOT** clean either head in solvent, because the solvent will remove the lubricating oils in the armature shaft bushings. **DO NOT** clean the armature in solvent, because the solvent will leave traces of oil residue on the commutator segments. Oil will cause arcing between the commutator and the brushes.

Brush Replacement

A new brush head may be purchased from the Local OMC dealer as a complete assembly with new brushes installed.

Armature Shorted

See Illustration #35.

The armature **CANNOT** be checked in the usual manner on a growler, because the internal connections and the low resistance of the windings. If a growler is used, all the coils will check out shorted.

HOWEVER, the armature can be tested using an AC milliammeter, five milliamperes with 100 scale divisions, and making tests between the commutator segments. Move from one segment to the next and watch closely for changes in the meter readings. The segments should all check out with almost the same reading. If a test between two segments indicates a significant lower reading, the winding is shorted.

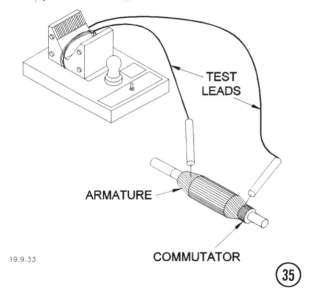

19.9.33

Armature Grounded

See Illustration #36.

Connect one lead of a continuity tester to a good ground, and then move the other lead around the entire surface of the commutator. Any indication of continuity means the armature is grounded and **MUST** be replaced. If the commutator segments are dirty or show signs of wear (roughness), clean between the bars, and then true it in a lathe. **NEVER** undercut the mica because the brushes are harder than the insulation.

After turning the armature, the insulation between the segments **MUST** be under cut to a depth of 1/32" (0.79 mm). The

MOTOR ASSEMBLING 9-27

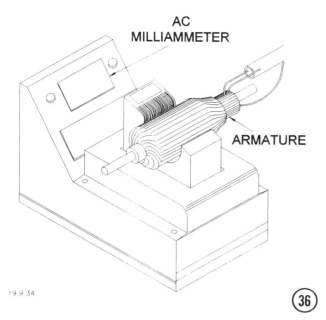

undercut **MUST** be flat at the bottom and should extend the full width of each insulated groove and beyond the brush contact in both directions. This will prevent the segment insulation from being smeared over the commutator as the segments wear.

After undercutting, the commutator should be sanded to remove the ridges left during the undercutting. Now, clean the commutator thoroughly to remove any metal chips or sanding grit.

Again perform the shorting and grounding tests on the armature.

End Head Bushings

Side play of each end head on the armature should be carefully checked. Any side play indicates bearing wear and the end head **MUST** be replaced, because the bearings are not serviced separately. If the heads with worn bearings are returned to service, the armature will rub against the pole shoes, or the armature shaft may actually bind.

To replace the commutator end head, first cut the lead connecting it to the field coil as close to the end head as possible. Next, solder the new end lead to the brush holder.

Field Coils

The field coils are series-wound and are **NOT** grounded to the frame. To test the field coils for a short, make contact with one probe of a test light to a good ground on the frame, Illustration **#37**. Make contact with the other probe to the blue or green tilt motor lead. If the test light comes on, the field coil is grounded and **MUST** be replaced. The field coils are only available as a complete field coils and frame assembly.

ASSEMBLING PRESTOLITE AND BOSCH MOTOR

Press a **NEW** seal into the end cap with the lip and seal spring facing **UP**, toward the tool. Use a flat ended bar. Continue to press the seal into place until the seal is below the chamber in the end cap.

Install a new gasket onto the end cap. Split the end of **NEW** brush lead **CAREFULLY** to 1/8" (3.18 mm) to fit over the piece of brush lead left on the motor head. Fit the end of new split brush leads over the old brush lead ends on the motor head, and then twist them slightly.

Hold the new leads with a pair of pliers and solder them using rosin core solder. The pliers will act as a heat barrier, so the solder will not spread and the rest of the brush leads will stay flexible.

Install a **NEW** end cap gasket. Slide the new gasket over the wires and the end head into position.

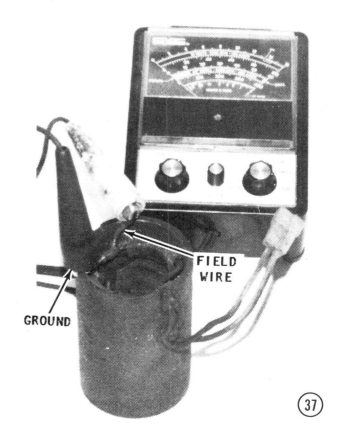

9-28 POWER TILT/TRIM

Use two paper clips bent to make brush and spring holders, as shown in the accompanying illustration. Install the new brush springs and insert the brushes into position using the paper clip as a tool to hold the brushes retracted in place.

Install the armature into the motor housing and into the motor head, with a fiber washer on the shaft at each end of the armature. Remove the paper clips to permit the brushes to ride on the armature.

Install new washer seals on each thru bolt. Tighten the thru bolts to 20 in. lbs (2.5 N·m). Apply OMC Black Neoprene Dip, or equivalent, over the bolt heads and over the two end cap gaskets to prevent any leakage.

Run the motor for a few seconds in both directions to seat the brushes.

9-12 MOUNTING TRIM/TILT ASSEMBLY MODELS 1978, 1979, AND 1980

Lubricate the tilt cylinder pin with OMC Sea Lube Anti-Corrosion Lube, or equivalent. Support the power trim unit in position and insert the tilt cylinder pin, Illustration #38. Secure the pin in place with the spring clip, Illustration #38.

Install the stern brackets, Illustration #39. Replace the tilt tube nut and tighten it to 24-26 ft.-lbs (32-36 N m) then back it off 1/8 to 1/4 turn, Illustration #40. Backing the tube nut off after it has been tightened to the proper torque value will provide the correct adjustment.

Install the thrust rod into the same location from which it was removed during the disassembling work.

MOUNTING TRIM AND TILT ASSEMBLY 1981 and On

Lubricate the tilt cylinder pin with OMC Sea Lube Anti-Corrosion Lube, or equivalent. Support the power trim unit in position and insert the tilt cylinder pin, Illustration #38. Secure the pin in place with the spring clip, Illustration #39.

Install the electrical cable through the stern brackets.

Install the bracket to manifold screws. Tighten the screws securely.

Adjusting Sending Unit

To adjust the trim sending unit, Illustration #41, to match the trim gauge indication at full UP trim position, proceed as follows.

Turn the ignition key to the ON position. Raise the engine using the tilt switch to the maximum UP trim position. Move the angle adjusting rod to the center hole. Loosen the sending unit screws until they are just snug and the unit can be pivoted. Lower the engine all the way down against the angle adjusting rod.

Observe the trim gauge. If the needle does not indicate the center position, adjust the sending unit by pivoting it up or down with a screwdriver, until the needle on the gauge does indicate the center position.

Raise the engine and tighten the sending unit screws. Lower the engine and again check the gauge indication.

Raise the engine to remove the angle adjusting rod, and then install the rod into the innermost hole.

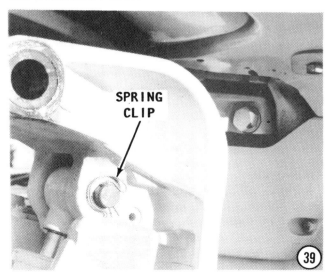

MOUNTING 9-29

9-13 UNIQUE TRIM/TILT SYSTEM SOME V4 UNITS PRIOR TO 1978

DESCRIPTION

As the heading suggests, a unique trim/tilt system was used on some V4 outboard units prior to 1978. The system consists of a single trim cylinder and a single tilt cylinder. The hydraulic pump, hydraulic reservoir, and the electric motor are all contained in a single assembly mounted outside the transom bracket on the starboard side.

Two versions of this trim/tilt system were used. One manufactured by Prestolite

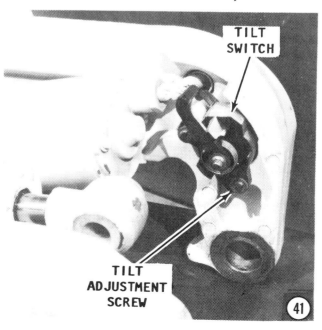

and the other by Calco, as indicated in the accompanying illustrations. The exploded drawing of the trim cylinder and the exploded drawing of the tilt cylinder are valid for both manufacturers.

A control switch similar to the other trim/tilt system controls direction of the electric motor and pump. Rotation in one direction results in the cylinders being extended while rotation in the opposite direction will retract the cylinders. As usual, a thermal overload switch on the electric motor will automatically reset in approximately one minute following an overheat condition.

A manual control valve/screw is located at the lower end of the pump. This screw may be rotated **COUNTERCLOCKWISE** to raise or lower the outboard unit in the event a malfunction in the system prevents movement using the trim/tilt system.

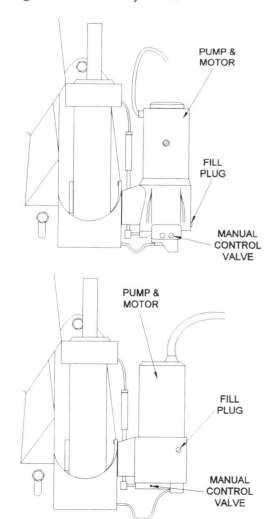

Line drawing to depict the exterior appearance of the Presotlite pump and motor (top), and the Colco pump and motor (bottom).

9-30 POWER TILT/TRIM

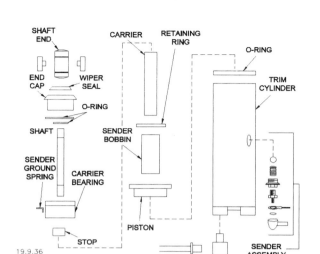

Exploded line drawing of the trim cylinder covered in this section, with major parts identified.

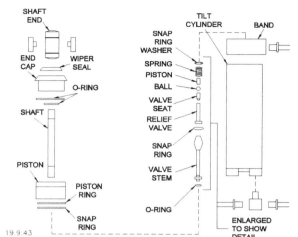

Exploded line drawing of the tilt cylinder covered in this section, with major parts identified.

FILLING SYSTEM

A fill plug is located on the outboard side of the reservoir. This plug should only be removed when the outboard unit is in the full **UP** position and held with a safe holding device to prevent accidental lowering. Such action could cause personal injury to self or others close by.

Use only approved hydraulic fluid and fill the reservoir until the fluid reaches the threads in the fill plug opening. After the correct level has been reached, install the plug snugly, remove the restraining device, and then operate the system through several complete cycles. Remove the fill plug and check the fluid level. Add fluid as required.

TROUBLESHOOTING

Troubleshooting mechanical components in the system may be performed following the procedures outlined for another system beginning on Page 9-6 and continuing on to Page 9-12.

SERVICING

Raise the outboard unit to the full **UP** position. If the trim/tilt system is not operative, rotate the manual release valve/screw **COUNTERCLOCKWISE**, and then lift the unit manually. Secure the outboard in a safe manner using a restraint or support.

Obtain a suitable container to receive the hydraulic fluid from the reservoir.

Always use flare wrenches when disconnecting or connecting hydraulic lines at the fittings to prevent "rounding" the corners, which is likely if a standard wrench is used.

Begin by disconnecting the two hydraulic lines at the trim/tilt housing base. Remove the fill plug at the reservoir and allow the hydraulic fluid to drain into the container.

After the fluid has drained, temporarily install the fill plug to prevent contaminates from entering the system.

Disconnect the trim motor electrical harness at the connector plug.

Remove the hardware securing the trim and tilt shaft ends to the outboard. Remove the bolts securing the system to the transom, and then **CAREFULLY** remove the complete unit.

The oil pump cannot be serviced. If defective, it must be replaced. The brushes in the electric motor should be replaced if they are worn to 1/4" (6mm) or less.

Follow the general procedures on Page 6-24 to service the electric motor.

NOW, THESE WORDS

In the majority of cases, service of the hydraulic items removed thus far will solve any rare problems encountered with the trim/tilt system.

The reservoir can be removed through the attaching hardware and cleaned if the system is consider to be contaminated with foreign material. The trim/tilt components comprise what is considered a "closed" system. The only route for entry of foreign material is through the fill opening.

Further disassembly and service to the system would best be left to a shop properly equipped with the proper test equipment and trained personnel with the expertise to work with high pressure hydraulic systems.

10
MAINTENANCE

10-1 INTRODUCTION

The authors estimate 75% of engine repair work can be directly or indirectly attributed to lack of proper care for the engine. This is especially true of care during the off-season period. There is no way on this green earth for a mechanical engine, particularly an outboard motor, to be left sitting idle for an extended period of time, say for six months, and then be ready for instant satisfactory service.

Imagine, if you will, leaving your automobile for six months, and then expecting to turn the key, have it roar to life, and be able to drive off in the same manner as a daily occurrence.

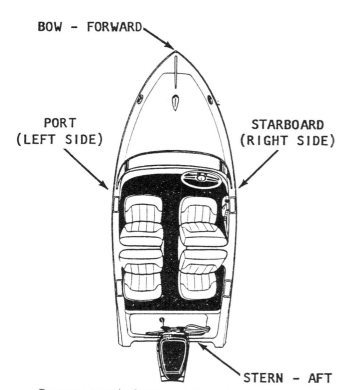

Common terminology used throughout the world for reference designation on boats. These are the terms used in this book.

It is critical for an outboard engine to be run at least once a month, preferably, in the water. At the same time, the shift mechanism should be operated through the full range several times and the steering operated from hard-over to hard-over.

Only through a regular maintenance program can the owner expect to receive long life and satisfactory performance at minimum cost.

Many times, if an outboard is not performing properly, the owner will "nurse" it through the season with good intentions of working on the unit once it is no longer being used. As with many New Year's resolutions, the good intentions are not completed and the outboard may lie for many months before the work is begun or the unit is taken to the marine shop for repair.

Suppose the cause of the problem being a blown head gasket. And let us assume water has found its way into a cylinder. This water, allowed to remain over a long period of time, will do considerably more damage than it would have if the unit had been disassembled and the repair work performed immediately. THEREFORE, if an outboard is not functioning properly, DO NOT stow it away with promises to get at it when you get time, because the work and expense will only get worse, the longer corrective action is postponed. In the example of the blown head gasket, a relatively simple and inexpensive repair job could very well develop into major overhaul and rebuild work.

Chapter Coverage

The material presented in this chapter is divided into five general areas.

1- General information every boat owner should know.

10-2 MAINTENANCE

2- Maintenance tasks that should be performed periodically to keep the boat operating at minimum cost.

3- Care necessary to maintain the appearance of the boat and to give the owner that "Pride of Ownership" look.

4- Winter storage practices to minimize damage during the off-season when the boat is not in use.

5- Preseason preparation work that should be performed to ensure satisfactory performance the first time it is put in service.

In nautical terms, the front of the boat is the **bow** amd the direction is **forward**; the rear is the **stern** and the direction is **aft**; the right side, when facing forward, is the **starboard** side; and the left side is the **port** side. All directional references in this manual use this terminology. Therefore, the direction from which an item is viewed is of no consequence, because **starboard** and **port NEVER** change no matter where the individual is located or in which direction he may be looking.

10-2 ENGINE MODEL NUMBERS

The engine model numbers are the manufacturer's key to engine changes. These numbers identify the year of manufacture, the qualified horsepower rating, and the parts book identification. If any correspondence or parts are required, the engine model number **MUST** be used or proper identification is not possible. The accompanying illustrations will be very helpful in locating the engine identification tag for the various models.

On some model engines, the serial number and model number were stamped on a plate mounted between the two swivel brackets underneath the hood.

On other model engines, the plate is mounted on the port side of the engine on the front or side of the swivel bracket. The hp and rpm range will also be found on the plate.

MODEL NUMBERS AFTER 1979

In 1980, OMC changed the model numbering system of their outboard units. The word **INTRODUCES** was used and a number assigned to each letter of the word as follows:

      I -- 1   D -- 6
      N -- 2   U -- 7
      T -- 3   C -- 8
      R -- 4   E -- 9
      O -- 5   S -- 0

The last letter of the model group on the identification plate is a code letter to indicate the plant of final assembly and is of interest **ONLY** to the manufacturer. Working backward, the next two letters (the second and third from the right end), designate the model year in which the unit was **MANUFACTURED**, not installed. Using this system, a 1980 engine would be identified with the letters "C" and "S"; in 1985, with the letters C and O, etc.

Therefore, since 1980, to establish the model year of a Johnson or Evinrude outboard unit, write the word INTRODUCES;

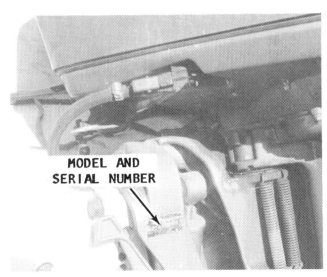

Manufacturer's identification plate mounted on the swivel housing at the front of the engine underneath the cowling.

Manufacturer's identification plate installed on the port side of the transom bracket.

assign the proper digits under each letter; then associate the letters, second and third from the right end of the group on the identification plate with the corresponding letters of "INTRODUCES" and the model year is established.

10-3 FIBERGLASS HULLS

Fiberglass reinforced hulls are tough, durable, and highly resistant to impact. However, like any other material they can be damaged. One of the advantages of this type construction is the relative ease with which it may be repaired. Because of its break characteristics, and the simple techniques used in restoration, these hulls have gained popularity throughout the world.

A fiberglass hull has almost no internal stresses. Therefore, when the hull is broken or stove-in, it retains its true form. It will not dent to take an out-of-shape set. When the hull sustains a severe blow, the impact will be either absorbed by deflection of the laminated panel or the blow will result in a definite, localized break. In addition to hull damage, bulkheads, stringers, and other stiffening structures attached to the hull, may also be affected and therefore, should be checked. Repairs are usually confined to the general area of the rupture.

10-4 ALUMINUM HULLS

Aluminum boats have become popular in recent years because they are so lightweight and may be carried with ease atop an automobile or other vehicle. These aluminum

A new fiberglass boat and trailer outfit ready for an owner and a power package.

An aluminum boat ready for an engine. Boats used with the larger horsepower engines are usually trailered to the water, instead of carrying them atop the vehicle.

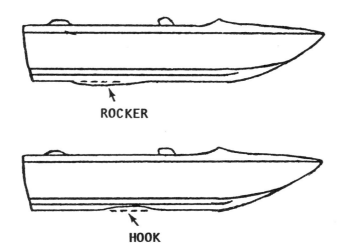

Simple drawing to illustrate two types of possible damage to the hull. Such injury to the boat will affect the boat's performance and subtract from the owner's enjoyment.

A boat and outboard used in salt water. Notice the marine growth on the lower unit and the anti-fouling bottom paint on the hull which prevented the marine growth from becoming a problem.

craft are available in sizes ranging from small 8-foot prams to twin-hulled pontoon houseboats or swimming "rafts" in excess of 30 feet. Naturally, the large units cannot be carried atop a vehicle.

One of the advantages of an aluminum hull is the easy maintenance program required, and the ability of the material to resist corrosion.

As an added protection against marine growth, the below the waterline area may be painted with an anti-fouling paint. Bottom paint sold for use on a wooden or fiberglass hull is **NOT** suitable. At the time of purchase, check to be sure the paint contains the chemical properties required for an aluminum surface. The label should clearly indicate the intended use is specifically for aluminum.

If the aluminum hull does not have anti-fouling paint but requires cleaning to remove marine growth, one method is to rub the hull with a gunny sack just as soon as the boat is removed from the water and while it is still wet. The roughness of the sack is fairly effective in cleaning the surface of marine growth, including crustaceans (barnacles for instance) that have attached themselves to the hull. As soon as the rubdown has been completed the hull should be washed with high-pressure fresh water.

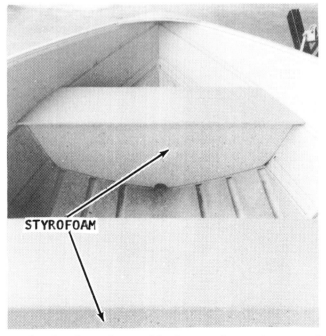

An aluminum boat with the wooden seat removed exposing the Styrofoam blocks for flotation. The seat should be removed at least once each season and the blocks thoroughly dried.

If the rubdown and wash was not accomplished immediately after the boat was removed from the water and the hull was allowed to dry, it will be necessary to cover the hull with wet blankets, gunny sacks or other suitable material and to continue soaking the covering until the growth is loosened. An easy alternate method, of course, is to return the boat to the water, if possible, and then too pull it out after it has been allowed to soak.

If an aluminum boat should strike an underwater object resulting in damage to the hull and a leak develops, the only emergency action possible is to make an attempt to reduce the amount of water being taken on by stuffing any type of available material into the opening until the boat is returned to shore. The aluminum cannot be repaired while it is wet. Repair of a damaged hull must be performed by a shop equipped for heliarc welding and other aluminum work.

Styrofoam blocks are installed under the seats of all aluminum boats. The foam blocks are designed for flotation to prevent the boat from sinking even if it should fill with water. Once each season, the wooden seat should be removed and the foam allowed to dry. Some manufacturers enclose the foam blocks in plastic bags prior to installation to protect them from moisture and loss of their flotation ability. New blocks may be purchased in a wide range of sizes. If new blocks are obtained, make an attempt to enclose the block in some form of plastic covering, then seal the package before installing it under the seat.

10-5 BELOW WATERLINE SERVICE

A foul bottom can seriously affect boat performance. This is one reason why racers, large and small, both powerboat and sail, are constantly giving attention to the condition of the hull below the waterline.

In areas where marine growth is prevalent, a coating of vinyl, anti-fouling bottom paint should be applied. If growth has developed on the bottom, it can be removed with a solution of muriatic acid applied with a brush or swab and then rinsed with clear water. **ALWAYS** use rubber gloves when working with muriatic acid and **TAKE EXTRA CARE** to keep it away from your face and hands. The **FUMES ARE TOXIC**. Therefore, work in a well-ventilated area, or if outside, keep your face on the windward side of the work.

Barnacles have a nasty habit of making their home on the bottom of boats which have not been treated with anti-fouling paint. Actually they will not harm the fiberglass hull, but can develop into a major nuisance.

If barnacles or other crustaceans have attached themselves to the hull, extra work will be required to bring the bottom back to a satisfactory condition. First, if practical, put the boat into a body of fresh water and allow it to remain for a few days. A large percentage of the growth can be removed in this manner. If this remedy is not possible, wash the bottom thoroughly with a high-pressure fresh water source and use a scraper. Small particles of hard shell may still hold fast. These can be removed with sandpaper.

10-6 SUBMERGED ENGINE SERVICE

A submerged engine is always the result of an unforeseen accident. Once the engine is recovered, special care and service procedures **MUST** be closely followed in order to return the unit to satisfactory performance.

NEVER, again we say **NEVER** allow an engine that has been submerged to stand more than a couple hours before following the procedures outlined in this section and making every effort to get it running. Such delay will result in serious internal damage. If all efforts fail and the engine cannot be started after the following procedures have been performed, the engine should be disassembled, cleaned, assembled, using new gaskets, seals, and O-rings, and then started as soon as possible.

Crankshaft from a submerged V4 engine recovered from salt water. In a very short time, the crankshaft was severely damaged by corrosion.

Rod bearing and cages badly damaged by salt water corrosion.

Submerged engine treatment is divided into three unique problem areas: submersion in salt water; submerged engine while running; and a submerged engine in fresh water, including special instructions.

The most critical of these three circumstances is the engine submerged in salt water, with submersion while running a close second.

Salt Water Submersion

NEVER attempt to start the engine after it has been recovered. This action will only result in additional parts being damaged and the cost of restoring the engine increased considerably. If the engine was submerged in salt water the complete unit

Cleaner to restore an engine recovered from fresh water after it has been submerged.

MUST be disassembled, cleaned, and assembled with new gaskets, O-rings, and seals. The corrosive effect of salt water can only be eliminated by the complete job being properly performed.

Submerged While Running
Special Instructions

If the engine was running when it was submerged, the chances of internal engine damage is greatly increased. After the engine has been recovered, remove the spark plugs and attempt to rotate the flywheel with the rewind starter. On larger horsepower engines without a rewind starter, use a socket wrench on the flywheel nut. If the attempt to rotate the flywheel fails, the chances of serious internal damage, such as, bent connecting rod, bent crankshaft, or damaged cylinder, is greatly increased. If all attempts to rotate the flywheel fail, the powerhead must be completely disassembled.

Submerged Engine — Fresh Water
SPECIAL WORD

As an aid to performing the restoration work, the following steps are numbered and should be followed in sequence. However, illustrations are not included with the procedural steps because the work involved is general in nature.

1- Recover the engine as quickly as possible.

2- Remove the hood and the spark plugs.

3- Remove the carburetor. To rebuild the carburetor, see Chapter 4.

4- Flush the outside of the engine with fresh water to remove silt, mud, sand, weeds, and other debris. **DO NOT** attempt to start the engine if sand has entered the powerhead. Such action will only result in serious damage to powerhead components. Sand in the powerhead means the unit must be disassembled.

5- Remove as much water as possible from the powerhead. Most of the water can be eliminated by first holding the engine in a horizontal position with the spark plug holes **DOWN**, and then cranking the engine with the rewind starter or with a socket wrench on the flywheel nut.

6- Alcohol will absorb water. Therefore, pour alcohol into the carburetor throat and again crank the engine.

7- Lay the engine in a horizontal position, and then roll it over until the spark plug openings are facing **UPWARD**. Pour alcohol into the spark plug openings and again crank the engine.

8- Roll the engine in the horizontal position until the spark plug openings are again facing **DOWN**. Pour engine oil into the carburetor throat and, at the same time, crank the engine to distribute oil throughout the crankcase.

9- With the engine still in a horizontal position, roll it over until the spark plug holes are again facing **UPWARD**. Pour approximately one teaspoon of engine oil into each spark plug opening. Crank the engine to distribute the oil in the cylinders.

10- Install the spark plugs and tighten them to the torque value given in the Appendix. Connect the high-tension leads to the spark plugs.

11- Install the carburetor onto the engine with a **NEW** gasket on the intake manifold.

12- Mount the engine in a test tank or body of water.

CAUTION: Water must circulate through the lower unit to the engine any time the engine is run to prevent damage to the water pump in the lower unit. Just five seconds without water will damage the water pump.

Rust preventative to be sprayed inside the engine in preparation for storage, as explained in the text.

Obtain **FRESH** fuel and attempt to start the engine. If the engine will start, allow it to run for approximately an hour to eliminate any water remaining in the engine.

13- If the engine fails to start, determine the cause, electrical or fuel, correct the problem, and again attempt to get it running. **NEVER** allow an engine to remain unstarted for more than a couple hours without following the procedures in this section and attempting to start it. If attempts to start the engine fail, the unit should be disassembled, cleaned, assembled, using new gaskets, seals, and O-rings, just as soon as possible.

10-7 WINTER STORAGE

Taking extra time to store the boat properly at the end of each season, will increase the chances of satisfactory service for the next season. **REMEMBER**, idleness is the greatest enemy of an outboard motor. The unit should be run on a monthly basis. The boat steering and shifting mechanism should also be worked through complete cycles several times each month. The owner who spends a small amount of time involved in such maintenance will be rewarded by satisfactory performance, and greatly reduced maintenance expense for parts and labor.

V4 engine mounted on the boat in a test tank. The engine can be safely operated in preparation for winter storage, as explained in the text.

ALWAYS remove the drain plug and position the boat with the bow higher than the stern. This will allow any rain water and melted snow to drain from the boat and prevent "trailer sinking". This term is used to describe a boat that has filled with rain water and ruined the interior, because the plug was not removed or the bow was not high enough to allow the water to drain properly.

Proper storage for the engine involves adequate protection of the unit from physical damage, rust, corrosion, and dirt.

The following steps provide an adequate maintenance program for storing the unit at the end of a season.

1- Remove the hood. Start the engine and allow it to warm to operating temperature.

CAUTION: Water must circulate through the lower unit to the engine any time the engine is run to prevent damage to the water pump in the lower unit. Just five seconds without water will damage the water pump.

Disconnect the fuel line from the engine and allow the unit to run at **LOW** rpm and, at the same time, inject about 4 ounces of rust preventative spray through each carburetor throat. Allow the engine to run until it shuts down from lack of fuel, indicating the carburetor/s are dry of fuel.

2- Drain the fuel tank and the fuel lines. Pour approximately one qt. of benzol (benzene) into the fuel tank, and then rinse the tank and pickup filter with the benzol.

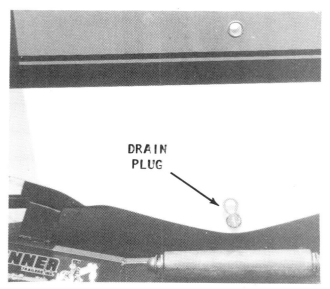

Drain plug removed from the transom to allow rain and melted snow to drain from the boat. Failure to remove this plug during long periods of storage can cause "boku" problems.

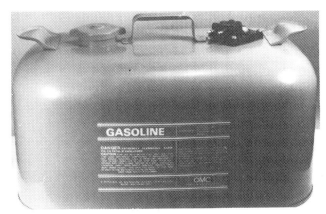

Standard OMC fuel tank. During periods of storage, the tank should be empty and the cap "cracked" open to allow the tank to "breathe".

Drain the tank. Store the fuel tank in a cool dry area with the vent **OPEN** to allow air to circulate through the tank. **DO NOT** store the fuel tank on bare concrete. Place the tank to allow air to circulate around it. If the fuel tank containing fuel is to be stored for more than a month, a commercial additive such as Sta-Bil should be added to the fuel. This type of additive will maintain the fuel in a "fresh" condition for up to a full year.

OMC fuel conditioner added to the fuel will keep it fresh for up to one full year.

Rust preventative to be used when preparing the engine for long periods of non-use and storage.

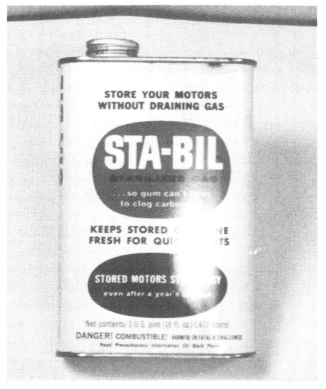

Chemical additives, such as Sta-Bil and the OMC conditioner shown at the top of the page will prevent fuel from "souring" for up to twelve months.

LOWER UNIT 10-9

3- Clean the carburetor fuel filter/s with benzol, see Chapter 4, Carburetor Repair Section.

4- Drain, and then fill the lower unit with OMC Lower Unit Gear Lubricant, as outlined in Section 10-8.

5- Lubricate the throttle and shift linkage. Lubricate the swivel pin and the tilt tube with Multipurpose Lubricant, or equivalent.

Clean the engine thoroughly. Coat the powerhead with Corrosion and Rust Preventative spray. Install the hood and then apply a thin film of fresh engine oil to all painted surfaces.

Remove the propeller. Apply Perfect Seal or a waterproof sealer to the propeller shaft, and then install the propeller back in position.

FINAL WORDS: Be sure all drain holes in the gear housing are open and free of obstruction. Check to be sure the **FLUSH** plug has been removed to allow all water to drain. Trapped water could freeze, expand, and cause expensive castings to crack.

ALWAYS store the engine off the boat with the lower unit below the powerhead to prevent any water from being trapped inside. The ideal storage position for an outboard is to hang it with the lower unit down. If hanging is not practical, lay the outboard on its back. This will place the lower unit below the powerhead.

10-8 LOWER UNIT SERVICE PROPELLERS

With Exhaust

Propellers with the exhaust passing through the hub **MUST** be removed more frequently than the standard propeller. Removal after each weekend use or outing is not considered excessive. These propellers do not have a shear pin. The shaft and propeller have splines which **MUST** be coated with an anit-corrosion lubricant prior to installation as a aid to removal the next time the propeller is pulled. Even with the lubricant applied to the shaft splines, the propeller may be difficult to remove.

The propeller with the exhaust hub is more expensive than the standard propeller and therefore, the cost of rebuilding the unit, if the hub is damaged, is justified.

A replaceable diffuser ring on the backside of the propeller dispurses the exhaust away from the propeller blades. If the ring becomes broken or damaged "ventillation" would be created pulling the exhaust gases

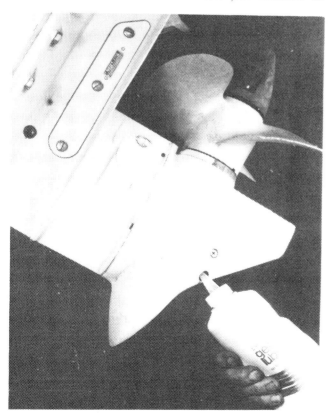

Adding lubricant to the lower unit. The lubricant must always be added through the drain plug after the upper vent plug has been removed.

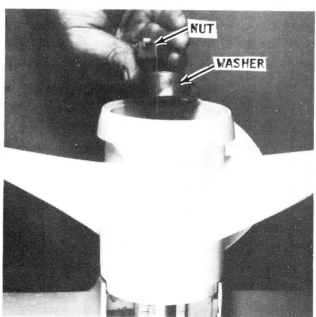

Installing a "prop exhaust" propeller. The propeller is installed onto the shaft, followed by the thrust washer, propeller nut, and finally the shear pin through the nut.

back into the negative pressure area behind the propeller. This condition would create considerable air bubbles and reduce the effectiveness of the the propeller.

Propeller With Exhaust Removal

First, disconnect the high tension leads to the spark plugs to prevent accidental engine start. Next, pull the cotter pin from the propeller nut. Wedge a piece of wood between one of the propeller blades and the cavitation plate to prevent the propeller from rotating. Back off the castillated propeller nut. Pull the propeller straight off the shaft. It may be necessary to carefully tap on the front side of the propeller with a soft headed mallet to jar it loose. If the propeller appears to be "frozen" to the shaft, see Chapter 8 for special removal instructions. The thrust washer does not have to be removed unless it appears damaged.

Propeller Exhaust Installation

First, check to be sure the high tension leads have been disconnected from the spark plugs to prevent accidental engine start. Install the thrust washer onto the propeller shaft, if it was removed. Coat the splines of the driveshaft with anti-corrosion lubricant. The lubricant **MUST** be applied to the shaft **EACH** time the propeller is installed to prevent it from "freezing" to the shaft. The propeller may "freeze" to the shaft in a short time in fresh water and much sooner in salt water.

Slide the propeller onto the shaft with the splines on the shaft indexed with the splines in the propller hub. Force the propeller onto the shaft until it is tight against the thrust washer. If the propeller cannot be moved tight against the thrust washer, the splines in the hub or on the shaft are dirty must be cleaned.

After the propeller is in place, wedge a piece of wood between one of the blades and the cavitation plate to prevent the propeller from rotating. Thread the castillated nut onto the shaft and bring it up tight against the propeller. Insert the cotter pin through the nut and propeller shaft. If the hole in the shaft does not align with one of the holes in the nut, **TIGHTEN** the nut until one of the holes is aligned. **NEVER** loosen the nut to align the holes. Install the cotter pin, remove the piece of wood, and connect the high tension leads to the spark plugs.

Draining Lower Unit

Remove the **VENT** plug just above the anti-cavitation plate **FIRST**, and then the

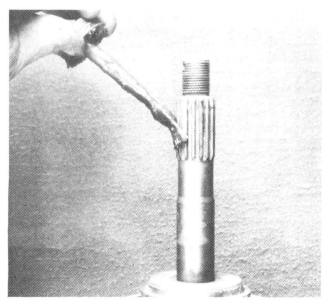

Applying gasket sealer to the shaft splines prior to installing a "prop exhaust" propeller.

OMC anti-corrosion lubricant that should be applied to the propeller shaft before the propeller is installed. Such lubricant on the shaft will not only fight corrosion, but assist in propeller removal.

FILL plug from the gear housing. **NEVER** remove the vent or filler plug when the drive unit is hot. Expanded lubricant would be released through the plug hole.

Allow the gear lubricant to drain into the container. As the lubricant drains, catch some with your fingers, from time-to-time, and rub it between your thumb and finger to determine if any metal particles are present. If metal is detected in the lubricant, the unit must be completely disassembled, inspected, and the damaged parts replaced.

If the lubricant appears milky brown, or if large amounts of lubricant must be added to bring the lubricant up to the full mark, a thorough check should be made to determine the cause of the loss.

Filling Lower Unit

Add only OMC lower unit lubricant. Lubricant, for engines covered in this manual, is as follows: Use **ONLY** Type C, now known as Premium Blend Gearcase Lube, in all electric shift models. Use either Premium Blend Lube or the OMC Hi-Vis Gearcase Lube for all other engines. **NEVER** use regular automotive-type grease in the lower unit because it expands and foams too much. Lower units do not have provisions to accommodate such expansion.

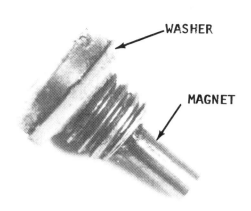

Late model lower unit drain plugs have a magnet to attract metal particles in the lubricant before they cause damage. If an unusual amount of particles are discovered, disassembly and search should be made without delay to discover the source.

The gearcase lubricant should be changed twice each year or season. If the lubricant is purchased in a large container, say the one-gallon size, a considerable savings can be realized. What is not used this season will be used in the next or the one

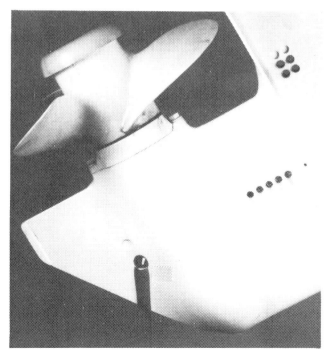

10-11
Draining lubricant from the lower unit. Check the lubricant for evidence of metal particles. A milky color indicates water in the lubricant.

The two types of lubricant used for Johnson/Evinrude engines. The text clearly identifies the lubricant to be used on the different model engines covered in this manual.

after. A small inexpensive pump can be be purchased to move the lubricant from the large container to the lower unit.

Position the drive unit approximately vertical and without a list to either port or starboard. Insert the lubricant tube into the **FILL/DRAIN** hole at the bottom plug hole, and inject lubricant until the excess begins to come out the **VENT** hole. Install the **VENT** and **FILL** plugs with **NEW** gaskets.

After the lower plug has been installed, remove the vent plug again and using a squirt-type oil can, add lubricant through this vent hole. A squirt-type oil can must be used to allow the trapped air in the lower unit to escape at the same time the final lubricant is added. Once the unit is completely full, install and tighten the vent plug.

Check to be sure the vent and drain plug gaskets are properly positioned to prevent water from entering the housing.

See the Appendix for lower unit capacities.

10-9 BATTERY STORAGE

Remove the batteries from the boat and keep them charged during the storage period. Clean the batteries thoroughly of any dirt or corrosion, and then charge them to full specific gravity reading. After they are fully charged, store them in a clean cool dry place where they will not be damaged or knocked over.

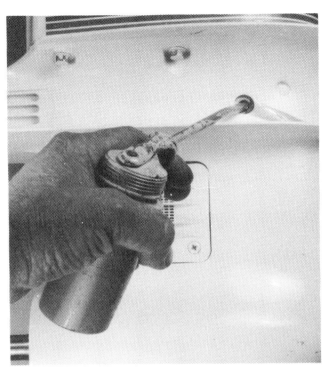

Using a squirt can to "top off" the lubricant in the lower unit, as explained in the text.

A check of the electrolyte in the battery should be on the maintenance schedule for any boat. A hydrometer reading of 1.300 or in the green band, indicates the battery is in satisfactory condition. If the reading is 1.150 or in the red band, the battery needs to be charged. Observe the six safety points given in the text, when using a hydrometer.

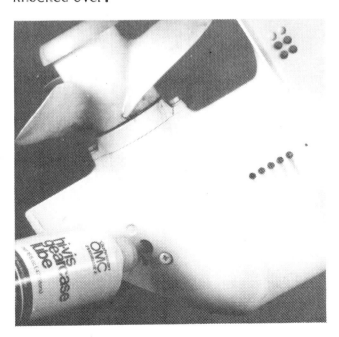

Filling a "prop exhaust" lower unit with lubricant.

Neoprene sealer used to waterproof electrical connections. The sealer is almost a "MUST" on screw-type connections at a terminal board.

NEVER store the battery with anything on top of it or cover the battery in such a manner as to prevent air from circulating around the filler caps. All batteries, both new and old, will discharge during periods of storage, more so if they are hot than if they remain cool. Therefore, the electrolyte level and the specific gravity should be checked at regular intervals. A drop in the specific gravity reading is cause to charge them back to a full reading.

In cold climates, **EXERCISE CARE** in selecting the battery storage area. A fully-charged battery will freeze at about 60 degrees below zero. A discharged battery, almost dead, will have ice forming at about 19 degrees above zero.

10-10 PRESEASON PREPARATION

Satisfactory performance and maximum enjoyment can be realized if a little time is spent in preparing the engine for service at

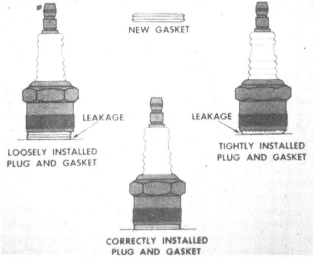

Correct and incorrect spark plug installation. The plugs **MUST** be installed properly and tightened to the proper torque value for satisfactory performance.

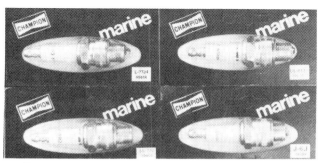

Today, numerous type spark plugs are available for service. **ALWAYS** check with the local OMC dealer to be sure you are purchasing the proper plugs for the engine being serviced.

the beginning of the season. Assuming the unit has been properly stored, as outlined in Section 10-7, a minimum amount of work is required to prepare the engine for use.

The following steps outline an adequate and logical sequence of tasks to be performed before using the engine the first time in a new season.

1- Lubricate the engine according to the manufacturer's recommendations. Remove, clean, inspect, adjust, and install the spark plugs with new gaskets if they require gaskets. Make a thorough check of the ignition system. This check should include: the points, coil, condenser, condition of the wiring, and the battery electrolyte level and charge.

2- If a built-in fuel tank is installed, take time to check the tank and all of the fuel lines, fittings, couplings, valves, and the flexible tank fill and vent. Turn on the fuel supply valve at the tank. If the fuel was not drained at the end of the previous season, make a careful inspection for gum formation. If a six-gallon fuel tank is used, take the same action. When gasoline is allowed to stand for long periods of time, particularly in the presence of copper, gum-

Typical fuel hose with squeeze bulb. The hose and bulb must remain flexible. The O-rings **MUST** prevent fuel leakage.

10-14 MAINTENANCE

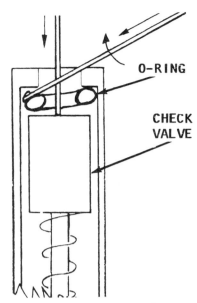

Method of removing an O-ring from a connector. The connector is also replaceable.

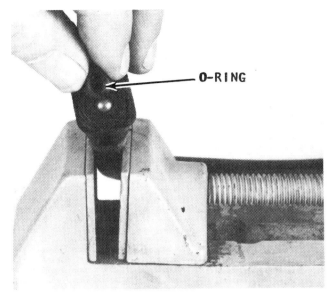

Using a punch to depress the check ball while installing an oiled O-ring.

my deposits form. This gum can clog the filters, lines, and passageways in the carburetor. See Chapter 4, Fuel System Service.

3- Check the oil level in the lower unit by first removing the vent screw on the port side just above the anti-cavitation plate. Insert a short piece of wire into the hole and check the level. Fill the lower unit according to procedures outlined in Section 10-8.

4- Close all water drains. Check and replace any defective water hoses. Check to be sure the connections do not leak. Replace any spring-type hose clamps, if they have lost their tension, or if they have distorted the water hose, with band-type clamps.

5- The engine can be run with the lower unit in water to flush it. If this is not practical, a flush attachment may be used. This unit is attached to the water pick-up in the lower unit. Attach a garden hose, turn on the water, allow the water to flow into the engine for awhile, and then run the engine.

OMC lubricants used on all Johnson/Evinrude outboard engines.

Adding OMC oil to the fuel. Only a high grade oil should be added to the fuel to ensure proper lubrication.

PRESEASON PREPARATION 10-15

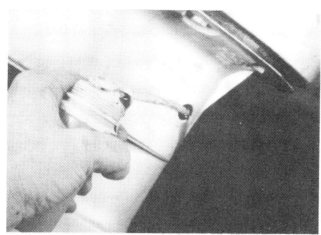
Using a squirt can to "top off" the lubricant in the lower unit, as explained in the text.

CAUTION: Water must circulate through the lower unit to the engine any time the engine is run to prevent damage to the water pump in the lower unit. Just five seconds without water will damage the water pump.

Check the idle exhaust port for water discharge. At idle speed, only a fine water mist will be visible. Check for leaks. Check operation of the thermostat. After the engine has reached operating temperature, tighten the cylinder head bolts to the torque value given in the Specifications in the Appendix.

The battery should be located near the engine and well secured to prevent even the slightest amount of movement. The battery, including the terminals, MUST be kept clean for maximum performance.

Testing a 100 hp engine in a test tank. Such testing after adjustments, ensures the owner of maximum performance and enjoyment from his unit.

An OMC degreaser widely used for cleaning the boat and engine.

6- Check the electrolyte level in the batteries and the voltage for a full charge. Clean and inspect the battery terminals and cable connections. **TAKE TIME** to check the polarity, if a new battery is being installed. Cover the cable connections with grease or special protective compound as a prevention to corrosion formation. Check all electrical wiring and grounding circuits.

7- Check all electrical parts on the engine and electrical fixture or connections in the lower portions of the hull inside the boat to be sure they are not of a type that could cause ignition of an explosive atmosphere. Rubber caps help keep spark insulators clean and reduce the possibility of arcing. Starters, generators, distributors, alternators, electric fuel pumps, voltage regulators, and high-tension wiring harnesses should be of a marine type that cannot cause an explosive mixture to ignite.

ONE FINAL WORD

Before putting the boat in the water, **TAKE TIME** to **VERIFY** the drain plugs are installed. Countless number of boating excursions have had a very sad beginning because the boat was eased into the water only to have water begin filling the inside.

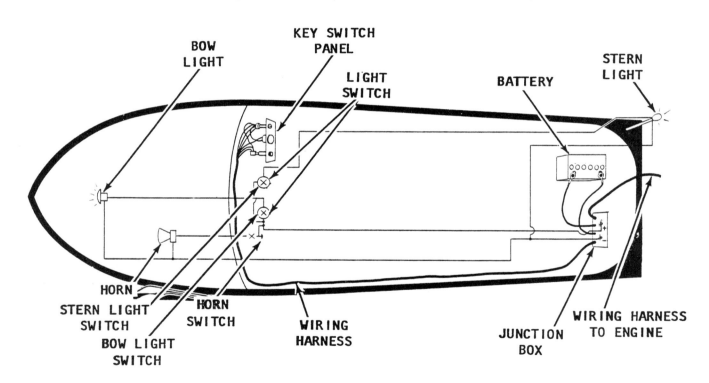

Principle electrical points requiring a careful check at the start of each season.

11
OUTBOARD JET DRIVE

11-1 INTRODUCTION

Outboard Jet Drive units are manufactured by Specialty Manufacturing Co., San Leandro, California. The units have been designed to permit boating in areas prohibited to a boat equipped with a conventional propeller outboard drive system. The housing of the jet drive barely extends below the hull of the boat allowing passage in ankle deep water, white water rapids, and over sand bars or in shoal water which would foul a propeller drive.

Description and Operation

The Outboard Jet Drive provides reliable propulsion with a minimum of moving parts. The units operate under the same laws and principles employed for the other jet drive units covered in this manual. Simply stated, water is drawn into the unit through an intake grille by an impeller driven by a driveshaft off the crankshaft of the powerhead. The water is immediately expelled under pressure through an outlet nozzle directed away from the stern of the boat.

As the speed of the boat increases and reaches planing speed, the jet drive discharges water freely into the air, and only the intake grille makes contact with the water.

The jet drive is provided with a gate arrangement and linkage to permit the boat to be operated in reverse. When the gate is moved downward over the exhaust nozzle, the pressure stream is reversed by the gate and the boat moves sternward.

Conventional controls are used for powerhead speed, movement of the boat, shifting and power trim and tilt.

General Information

The manufacturer has replaced the stock lower unit with a jet pump. Custom adaptor plates are manufactured to mate the jet pump with existing holes in the outboard intermediate housing.

The accompanying table on the last page of this chapter -- Page 11-18, lists outboard units by horsepower and model years and the jet drive unit to be used with each model.

Model Identification and Serial Numbers

A model letter identification indicating "size" is stamped on the rear, port side of the jet drive housing. A serial number for the unit is stamped on the starboard side of the jet drive housing, as indicated in the accompanying illustration.

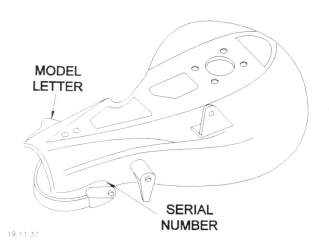

The model letter designation and the serial numbers are embossed on the jet drive housing.

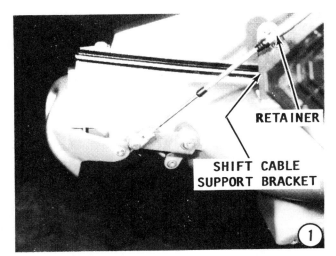

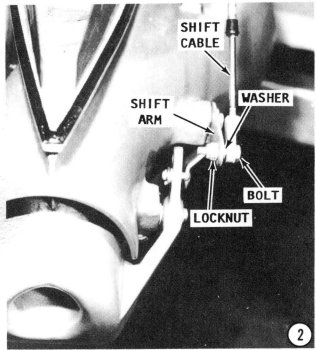

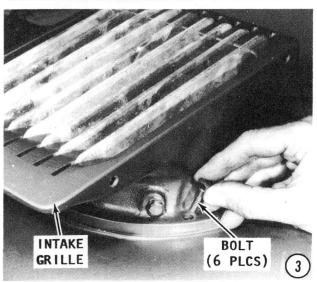

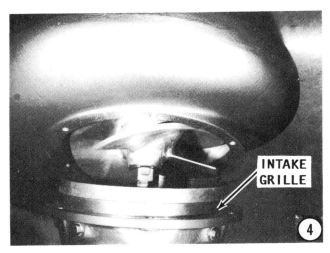

The following procedures are valid for all four size jet drive units, because they are identical in design, function and operation. Differences lie in size and securing hardware.

11-2 REMOVAL & DISASSEMBLING

1- Remove the two bolts and retainer securing the shift cable to the shift cable support bracket.

2- Remove the locknut, bolt, and washer securing the shift cable to the shift arm. Try not to disturb the length of the cable.

3- Remove the six bolts securing the intake grille to the jet casing.

4- Ease the intake grille from the jet drive housing.

5- Pry the tab or tabs of tabbed washer away from the nut to allow the nut to be removed.

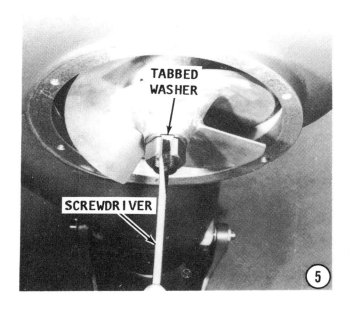

DISASSEMBLING 11-3

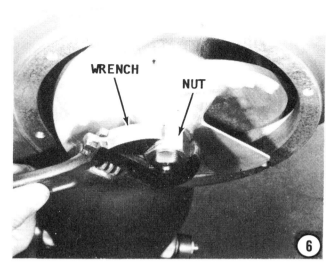

6- Loosen, and then remove the nut.

7- Remove the tabbed washer and spacers. Make a careful count of the spacers behind the washer. If the unit is relatively new, there could be as many as eight or nine spacers behind the washer. If less than eight or nine spacers, depending on the model being serviced, are removed from behind the washer, the others will be found behind the jet impeller, which is removed in the following step. A **TOTAL** of either **EIGHT** or **NINE** spacers will be found.

8- Remove the jet impeller from the shaft. If the impeller is "frozen" to the shaft, obtain a block of wood and a hammer.

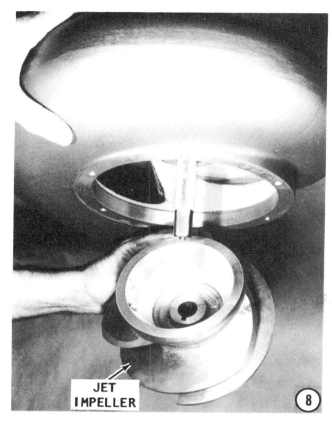

Tap the impeller in a **CLOCKWISE** direction to release the shear key.

9- Slide the nylon sleeve and shear key free of the driveshaft and any spacers found behind the impeller. Make a note of the number of spacers at both locations --behind the impeller **AND** on top of the impeller, under the nut and tabbed washer.

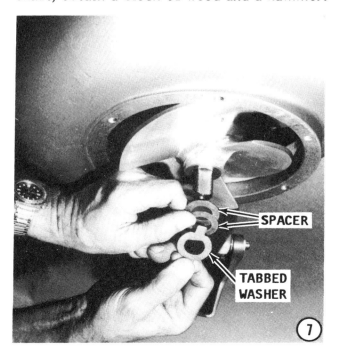

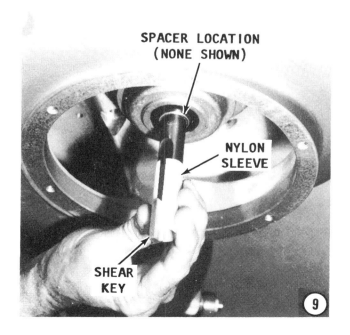

11-4 OUTBOARD JET

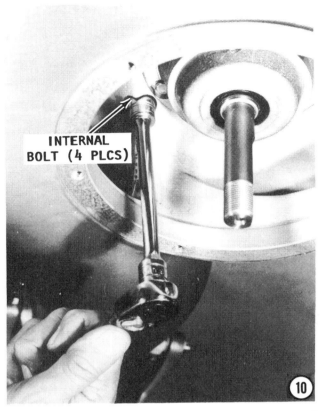

10- One external bolt and four internal bolts are used to secure the jet drive to the intermediate housing. The external bolt is located at the aft end of the anti-cavitation plate. The four internal bolts are located inside the jet drive housing, as indicated in the accompanying illustration. Remove the five attaching bolts.

Location of the one exterior bolt securing the jet drive to the outboard intermediate housing.

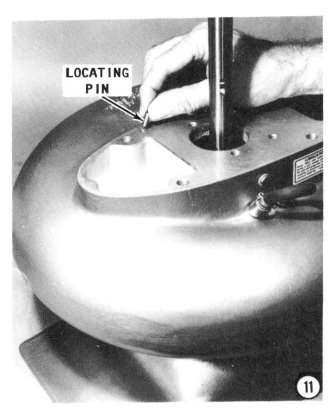

11- Lower the jet drive from the intermediate housing. Remove the locating pin from the forward starboard side (or center forward, depending on the model being serviced) of the upper jet housing.

SPECIAL WORDS

There will be a total of **SIX** locating pins to be removed in the following steps. Make careful note of the size and location of each

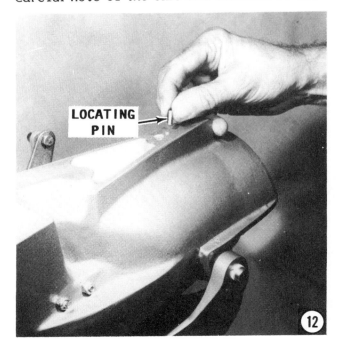

DISASSEMBLING 11-5

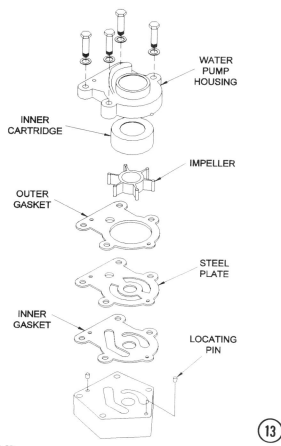

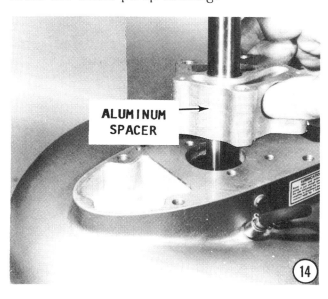

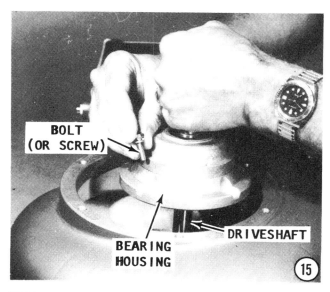

when they are removed, as an assist during assembling.

12- Remove the locating pin from the aft end of the housing. This pin and the one removed in the previous step should be of identical size.

13- Remove the four bolts and washers from the water pump housing.

SPECIAL WORDS

Two different length bolts are used at this location on some models, and one of these models uses special "D" shaped washers.

Pull the water pump housing, the inner cartridge and the water pump impeller, up and free of the driveshaft. Remove the Woodruff key from its recess in the driveshaft. Next, remove the outer gasket, the steel plate and the inner gasket.

14- Remove the two small locating pins and lift the aluminum spacer up and free of the driveshaft.

15- Remove the driveshaft and bearing assembly from the housing. One model has two #10-24x5/8" screws and lockwashers securing the driveshaft assembly to the housing. All other models have four ¼-20x7/8" bolts and lockwashers securing the driveshaft assembly to the housing.

16- Remove the large thick adaptor plate from the intermediate housing. This plate is secured with five bolts and lockwashers. The aft bolt is smaller than the

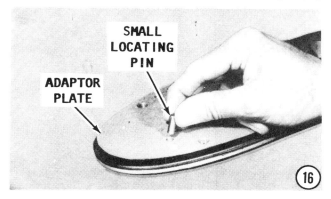

other four. Lower the adaptor plate from the intermediate housing and remove the two small locating pins, one on the forward port side and another from the last aft hole in the adaptor plate. Both pins are identical size.

11-3 CLEANING AND INSPECTING

Wash all parts, except the driveshaft assembly, in solvent and blow them dry with compressed air. Rotate the bearing assembly on the driveshaft to inspect the bearings for "rough" spots, binding, and signs of corrosion or damage.

Saturate a shop towel with solvent and wipe both extensions of the driveshaft.

Bearing Assembly

Lightly wipe the exterior of the bearing assembly with the same shop towel. Do not allow solvent to enter the three lubricant passages of the bearing assembly. The best way to clean these passages is **NOT** with solvent - because any solvent remaining in the assembly after installation will continue to dissolve good useful lubricant and leave bearings and seals dry. This condition will cause bearings to fail through friction and seals to dry up and shrink - losing their sealing qualities.

The only way to clean and lubricate the bearing assembly is after installation to the jet drive - via the exterior lubrication fitting. This procedure is described in Section 11-6 at the end of this chapter and explains how the old lubricant may be completely replaced with new.

If the old lubricant emerging from the hose coupling is a dark, dirty, grey color, the seals have already broken down and water is attacking the bearings. If such is the case, it is recommended the entire driveshaft bearing assembly be taken to the dealer for service of the bearings and seals.

Dismantling Bearing Assembly

A complicated procedure must be followed to dismantle the bearing assembly including "torching" off the bearing housing. Naturally, excessive heat might ruin the seals and bearings. Therefore, the best recommendation is to leave this part of the service work to the experts at your local dealership.

Driveshaft and Associated Parts

Inspect the threads and splines on the driveshaft for wear, rounded edges, corrosion and damage.

Carefully check the driveshaft to verify the shaft is straight and true without any sign of damage.

Inspect the jet drive housing for nicks, dents, corrosion, or other signs of damage. Nicks may be removed with No. 120 and No. 180 emery cloth.

Reverse Gate

Inspect the gate and its pivot points. Check the swinging action to be sure it moves freely the entire distance of travel without binding.

Inspect the slats of the water intake grille for straightness. Straighten any bent slats, if possible. Use the utmost care when prying on any slat, as they tend to break if

Take extra precautions to PREVENT cleaning solution from entering the three lubrication passages.

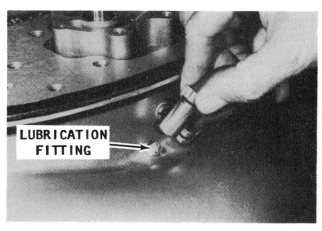

Cleaning and lubricating the bearing assembly is best accomplished by completely replacing the old lubricant with new, as described in the text.

CLEANING & INSPECTING 11-7

excessive force is applied. Replace the intake grille if a slat is lost, broken, or bent, and cannot be repaired. The slats are spaced evenly and the distance between them is critical, to prevent large objects from passing through and becoming lodged between the jet impeller and the inside wall of the housing.

Jet Impeller

The jet impeller is a precisely machined and dynamically balanced aluminum spiral. Observe the drilled recesses at exact locations to achieve this delicate balancing. Some of these drilled recesses are clearly shown in the accompanying illustration.

Excessive vibration of the jet drive may be attributed to an out-of-balance condition caused by the jet impeller being struck excessively by rocks, gravel or cavitation "burn".

The term cavitation "burn" is a common expression used throughout the world among people working with pumps, impeller blades, and forceful water movement.

"Burns" on the jet impeller blades are caused by cavitation air bubbles exploding with considerable force against the impeller blades. The edges of the blades may develop small "dime size" areas resembling a porous sponge, as the aluminum is actually "eaten" by the condition just described.

Excessive rounding of the jet impeller edges will reduce efficiency and performance. Therefore, the impeller should be inspected at regular intervals.

If rounding is detected, the impeller should be placed on a work bench and the edges restored to as sharp a condition as possible, using a file. Draw the file in only one direction. A back-and-forth motion will not produce a smooth edge. **TAKE CARE** not to nick the smooth surface of the jet impeller. Excessive nicking or pitting will create water turbulence and slow the flow of water through the pump.

Inspect the shear key. A slightly distorted key may be reused although some difficulty may be encountered in assembling the jet drive. A cracked shear key should be discarded and replaced with a new key.

Water Pump

Clean all water pump parts with solvent, and then blow them dry with compressed air. Inspect the water pump housing for cracks and distortion, possibly caused from overheating. Inspect the steel plate, the thick aluminum spacer and the water pump cartridge for grooves and/or rough spots. If possible **ALWAYS** install a new water pump impeller while the jet drive is disassembled. A new water pump impeller will ensure extended satisfactory service and give "peace of mind" to the owner. If the old water pump impeller must be returned to service, **NEVER** install it in reverse of the original direction of rotation. Installation in reverse will cause premature impeller failure.

If installation of a new water pump impeller is not possible, check the sealing surfaces and be satisfied they are in good condition. Check the upper, lower, and ends of the impeller vanes for grooves, cracking, and wear. Check to be sure the indexing notch of the impeller hub is intact and will not allow the impeller to slip.

The slats of the grille must be carefully inspected and any bent slats straightened, for maximum performance of the jet drive.

The edges of the jet impeller should be kept as sharp as possible for maximum jet drive efficiency.

11-8 OUTBOARD JET

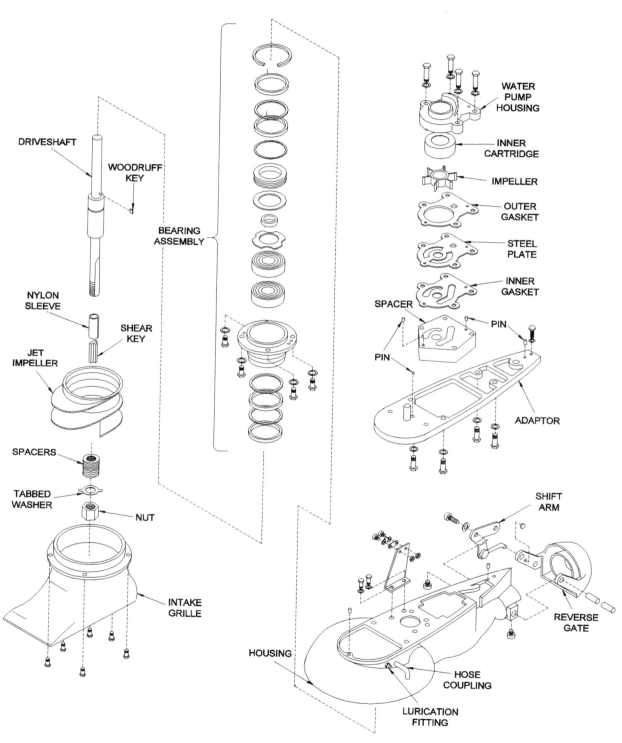

Exploded drawing of a jet drive lower unit, with major parts identified.

After 60 seconds at 1500 rpm.

After 90 seconds at 1500 rpm.

After 30 seconds at 2000 rpm.

After 45 seconds at 2000 rpm.

After 60 seconds at 2000 rpm

Cautions throughout this manual point out the danger of operating the engine without water passing through the water pump. The above photographs are self evident.

11-4 ASSEMBLING

1- Identify the two small locating pins used to index the large thick adaptor plate to the intermediate housing. Insert one pin into the last hole aft on the topside of the plate. Insert the other pin into the hole forward toward the port side, as shown.

Lift the plate into place against the intermediate housing with the locating pins indexing with the holes in the intermediate housing.

On some smaller models: Secure the plate with the five bolts. One of the five bolts is shorter than the other four. Install the short bolt in the most aft location.

Tighten the long bolts to a torque value of 22 ft lbs (30Nm). Tighten the short bolt to a torque value of 11 ft lbs (15Nm).

On some larger models: Secure the plate with seven bolts and tighten them to a torque value of 22 ft lb (30Nm).

2- Place the driveshaft bearing assembly into the jet drive housing. Rotate the bearing assembly until all bolt holes align. There is only **ONE** correct position.

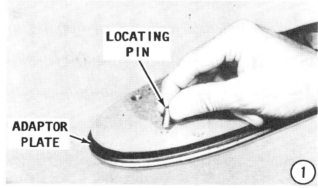

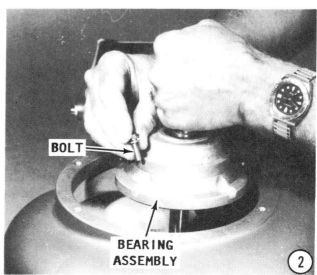

One model has two securing screws and lockwasher. Tighten these screws just good and snug by hand. All other models have four securing bolts and lockwasher. Tighten these four bolts to a torque value of 5 ft lbs (7Nm).

SPECIAL WORDS

If installing a new jet impeller, place all the spacers (eight or nine, depending on the model being serviced) at the lower or "nut" end of the impeller, and skip the following step.

Shimming Jet Impeller

The clearance between the outer edge of the jet drive impeller and the water intake housing cone wall should be maintained at approximately 1/32" (0.8mm). This distance can be visually checked by shining a flashlight up through the intake grille and estimating the distance between the impeller and the casing cone, as indicated in the accompanying illustrations. It is not humanly possible to accurately measure this clearance, but by observing closely and estimating the clearance, the results should be fairly accurate.

After continued use, the clearance will increase. The spacers removed in Steps 7 & 8 are used to position the impeller along the driveshaft with a desired clearance of 1/32" (0.8mm) between the jet impeller and the housing wall.

3- A total of either **EIGHT** or **NINE** spacers are used depending on the model being serviced. When new, all spacers are located at the tapered (or "nut") end of the impeller. As the clearance increases, the spacers are transferred from the tapered ("nut") end and placed at the wide ("intermediate housing") end of the jet impeller.

This procedure is best accomplished while the jet drive is removed from the intermediate housing.

Secure the driveshaft with the attaching hardware. Installation of the shear key and nylon sleeve is not vital to this procedure. Place the unit on a convenient work bench. Shine a flashlight through the intake grille into the housing cone and "eyeball" the clearance between the jet impeller and the cone wall, as indicated in the accompanying line drawing. Move spacers one-at-a-time from the tapered end to the wide end to obtain a satisfactory clearance. Dismantle the driveshaft and note the exact count of spacers at both ends of the bearing assembly. This count will be recalled in Steps 9 & 11 of Assembly to properly install the jet impeller.

Water Pump Assembling

4- Place the aluminum spacer over the driveshaft with the two holes for the indexing pins facing **UPWARD**. Fit the two locating pins into the holes of the spacer.

GOOD WORDS

The manufacturer recommends **NO** sealant be used on either side of the water pump gaskets.

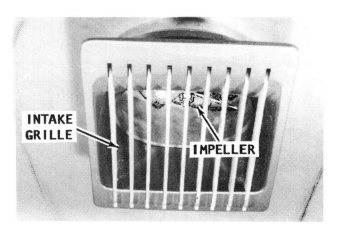

The clearance between the jet impeller and the casing cone can be fairly well estimated by shining a flashlight up through the grille and visually checking the distance between the impeller and the cone.

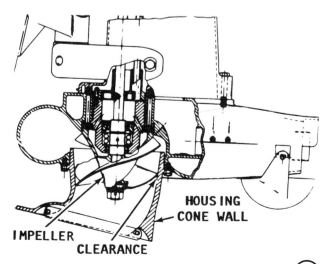

ASSEMBLING 11-11

5- Slide the inner water pump gasket (the gasket with two curved openings) over the driveshaft. Position the gasket over the two locating pins. Slide the steel plate down over the driveshaft with the tangs on the plate facing **DOWNWARD** and with the holes in the plate indexed over the two locating pins.

Check to be sure the tangs on the plate fit into the two curved openings of the gasket beneath the plate. Now, slide the outer gasket (the gasket with the large center hole) over the driveshaft. Position the gasket over the two locating pins.

Fit the Woodruff key into the driveshaft. Just a dab of grease on the key will help to hold the key in place. Slide the water pump impeller over the driveshaft with the rubber membrane on the top side and the keyway in the impeller indexed over the Woodruff key. **TAKE CARE** not to damage the membrane. Coat the impeller blades with water resistant lubricant.

Install the insert cartridge, the inner plate, and finally the water pump housing over the driveshaft. Rotate the insert cartridge **COUNTERCLOCKWISE** over the impeller to tuck in the impeller vanes. Seat all parts over the two locating pins.

On some smaller models two different length bolts are used at this location, and one of these models uses special **"D"** shaped washers. All other models use plain washers.

All models: Tighten the four bolts to a torque value of 11 ft lbs (15Nm).

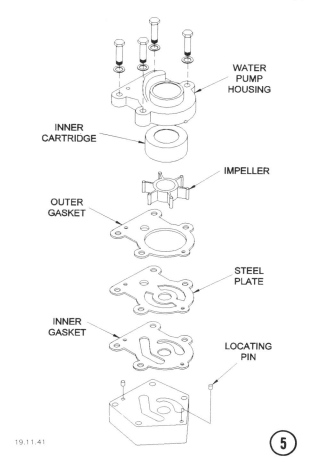

6- Install one of the small locating pins into the aft end of the jet drive housing.

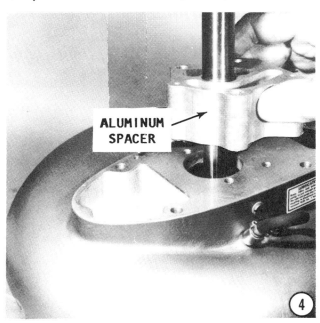

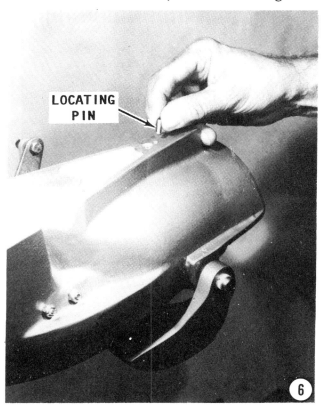

11-12 OUTBOARD JET

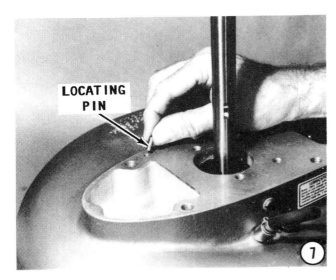

Jet Drive Installation

7- Install the other small locating pin into the forward starboard side (or center forward end, depending on the model being serviced).

8- Raise the jet drive unit up and align it with the intermediate housing, with the small pins indexed into matching holes in the adaptor plate. Install the four internal bolts and the one external bolt. Location of the external bolt is at the aft end of the anticavitation plate. Tighten the five bolts to a torque value of 11 ft lbs (15Nm).

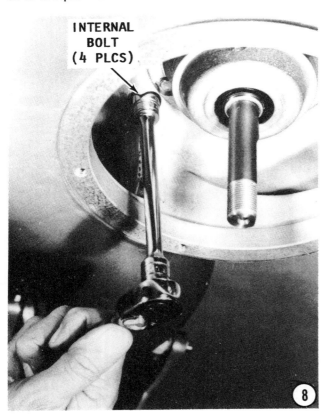

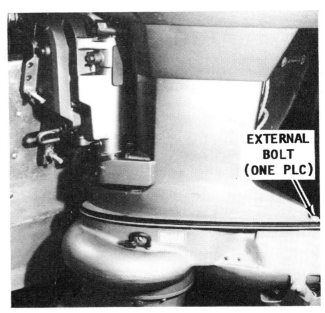

Location of the one exterior bolt securing the jet drive to the outboard intermediate housing.

9- Place the required number of spacers (if any) as determined in Step 3, **OR** in the paragraphs following "Shimming the Jet Impeller", up against the bearing housing. Slide the nylon sleeve over the driveshaft and insert the shear key into the slot of the nylon sleeve with the key resting against the flattened portion of the driveshaft.

10- Slide the jet impeller up onto the driveshaft, with the groove in the impeller collar indexing over the shear key.

11- Place the remaining spacers over the driveshaft. The number of spacers will

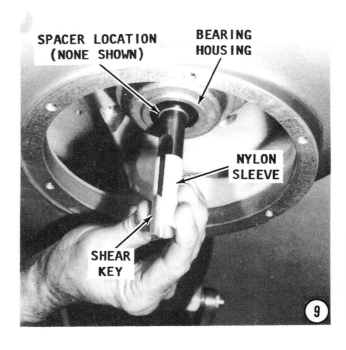

ASSEMBLING 11-13

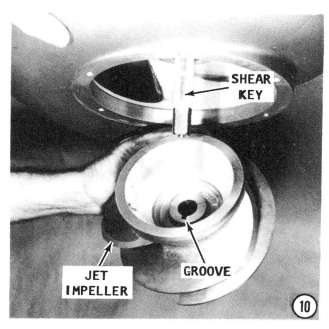

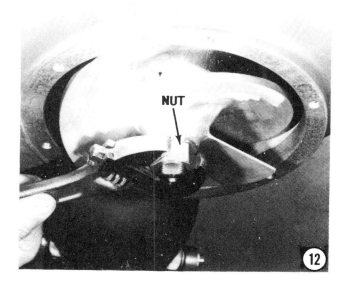

be eight or nine less the number used in Step 9, depending on the model being serviced.

12- Tighten the nut to a torque value of 17 ft lbs (23Nm). If neither of the two tabs on the tabbed washer aligns with the sides of the nut, remove the nut and washer. Invert the tabbed washer. Turning the washer over will change the tabs by approximately 15°. Install and tighten the nut to the required torque value. The tabbed washer is designed to align with the nut in one of the two positions described.

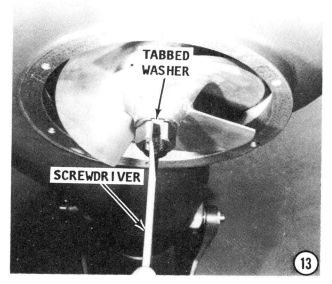

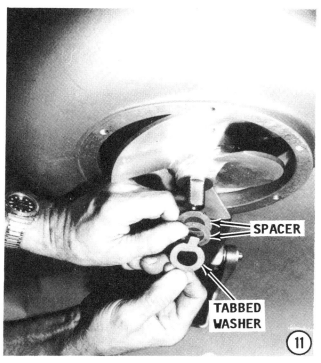

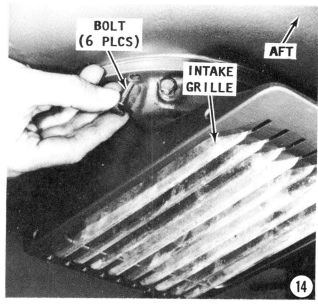

11-14 OUTBOARD JET

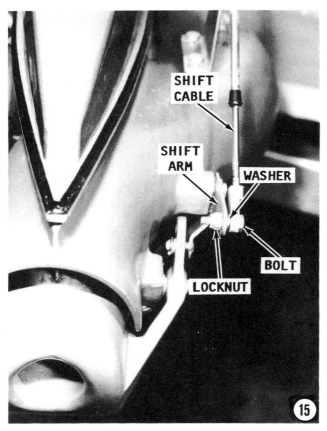

13- Bend the tabs up against the nut to prevent the nut from backing off and becoming loose.

14- Install the intake grille onto the jet drive housing with the slots facing aft. Install and tighten the six securing bolts. Some models use a few 1/4" bolts. Tighten the 1/4" bolts to a torque value of 5 ft lbs (7Nm). Tighten the remaining 5/16" bolts used on all models to 11 ft lbs (15Nm).

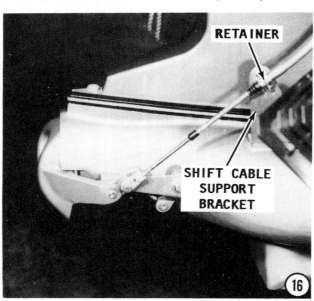

15- Slide the bolt through the end of the shift cable, washer, and into the shift arm. Install the locknut onto the bolt and tighten the bolt securely.

16- Install the shift cable against the shift cable support bracket and secure it in place with the two bolts.

11-5 JET DRIVE ADJUSTMENTS

Cable Alignment and Free Play Adjustment

1- Move the shift lever downward into the **FORWARD** position. The leaf spring should snap over on top of the lever to lock it in position.

2- Remove the locknut, washer, and bolt from the threaded end of the shift cable. Push the reverse gate firmly against the rubber pad on the underside of the jet drive housing.

Check to be sure the link between the reverse gate and the shift arm is hooked into the **LOWER** hole on the gate.

Hold the shift arm up until the link rod and shift arm axis form an imaginary straight line, as indicated in the accompanying illustration. Adjust the length of the

ADJUSTMENTS 11-15

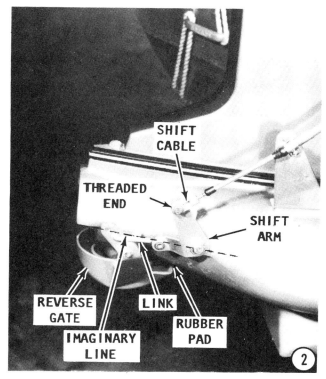

shift cable by rotating the threaded end, until the cable can be installed back onto the shift arm **WITHOUT** disturbing the imaginary line. Pass the nut through the cable end, washer, and shift arm. Install and tighten the locknut.

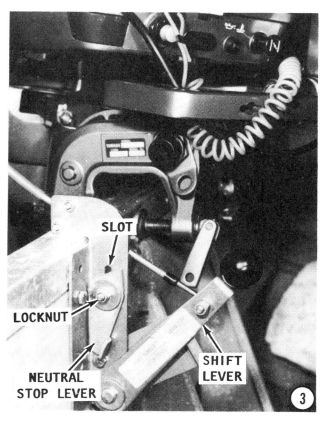

Neutral Stop Adjustment
FIRST, THESE WORDS

In the **FORWARD** position, the reverse gate is neatly tucked underneath and clear of the exhaust jet stream.

In the **REVERSE** position, the gate swings up and blocks the jet stream deflecting the water in a forward direction under the jet housing to move the boat sternward.

In the neutral position, the gate assumes a "happy medium" -- a balance between forward and reverse when the powerhead is operating at **IDLE** speed. Actually, the gate is deflecting some water to prevent the boat from moving forward, but not enough volume to move the boat sternward.

WARNING
THE GATE MUST BE PROPERLY ADJUSTED FOR SAFETY OF BOAT AND PASSENGERS. IMPROPER ADJUSTMENT COULD CAUSE THE GATE TO SWING UP TO THE REVERSE POSITION WHILE THE BOAT IS MOVING FORWARD CAUSING SERIOUS INJURY TO BOAT OR PASSENGERS.

3- Loosen, but do **NOT** remove the locknut on the neutral stop lever. Check to be sure the lever will slide up and down along the slot in the shift lever bracket.

CRITICAL WORDS
The following procedure **MUST** be performed with the boat and jet drive in a body

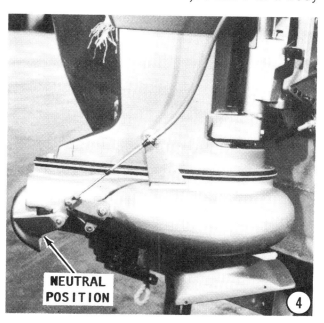

of water. Only with the boat in the water can a proper jet stream be applied against the gate for adjustment purposes.

CAUTION

Water must circulate through the lower unit to the powerhead anytime the powerhead is operating to prevent damage to the water pump in the lower unit. Just five seconds without water will damage the water pump impeller.

Start the powerhead and allow it to operate **ONLY** at **IDLE** speed. With the neutral stop lever in the down position, move the shift lever until the jet stream forces on the gate are "balanced". "Balanced" means the water discharged is divided in both directions and the boat moves neither forward nor sternward. The gate is then in the neutral position with the powerhead at idle speed.

4- Move the neutral stop lever up against the shift lever until the stop lever barely makes contact with the shift lever. Tighten the locknut to maintain this new adjusted position. Shut down the powerhead.

GOOD WORDS

The reverse gate may not swing to the full **UP** position in reverse gear after Steps 1

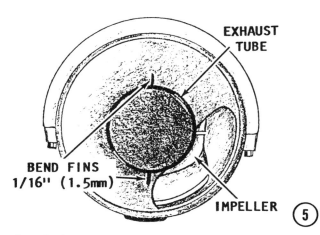

thru 4 have been performed. Do **NOT** be concerned. This condition is acceptable, because water pressure in reverse will close the gate fully under normal operation.

Trim Adjustment

5- During operation, if the boat tends to "pull" to port or starboard, the flow fins may be adjusted to correct the condition. These fins are located at the top and bottom of the exhaust tube.

If the boat tends to "pull" to starboard, bend the trailing edge of each fin approximately 1/16" (1.5mm) toward the starboard side of the jet drive. Naturally, if the boat tends to "pull" to port, bend the fins toward the port side.

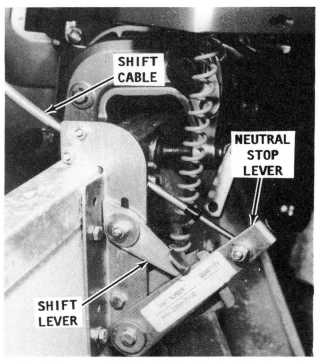

Typical arrangement with the neutral stop lever in place against the shift lever.

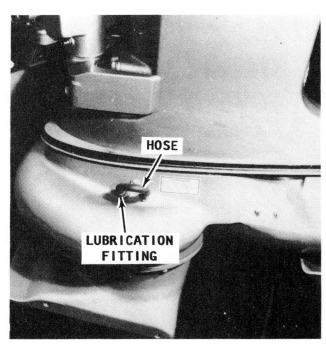

Removing the hose is accomplished by deflecting the hose to one side to snap it free of the fitting. Do NOT attempt to "pull" the hose off the fitting.

11-6 JET DRIVE LUBRICATION

The bearing and seal housing installed on the jet unit drive shaft is lubricated through an externally mounted zerk fitting. This fitting is located at the top of the jet drive casing on the port side.

To lubricate through the fitting, first push the coupling aside to release the coupling and the hose from the fitting. **DO NOT** attempt to pull on the coupling or the hose. Pushing the coupling aside is the answer.

Pump multi-purpose water resistant marine lubricant into the fitting until the old grease emerges from the hose coupler. Internal passageways are provided through the outer jet drive housing and the bearing and the seal housing to route the lubricant. When the old grease emerges from the hose coupling, there is no doubt the system has been properly lubricated. Wipe off the excess grease, and then snap the coupler back into place over the fitting.

The fitting should be lubricated every ten hours of jet drive operation.

Every 50 hours of jet drive operation, pump enough new grease into the fitting to replace the old grease. A distinct change in color between the old grease and the new grease will be noted, indicating the unit is filled with the new lubricant.

When the unit is new, a slight discoloration may be expected, as the new seals are "broken in".

If the old grease contains tiny beads of water, the seals are beginning to break down. If the old grease emerging from the hose coupling is a dark dirty grey color, the seals have already broken down and water is attacking the bearings.

Lubricate the gate control linkage pivot points at regular intervals. These points are indicated in the accompanying illustration.

Jet Drive and Flushing Device

Regular flushing of the jet drive will prolong the life of the powerhead, by clearing the cooling system of possible obstructions.

Some jet drives manufactured after January, 1987, are equipped with a plug on the port side just above the lubrication hose.

Remove the plug and gasket, install the flush adaptor, connect the garden hose and turn on the water supply.

Start and operate the powerhead at a fast idle for about 15 minutes. Disconnect the flushing adaptor and replace the plug.

SPECIAL WORDS

The procedure just described will only flush the powerhead cooling system, not the jet drive. To flush the jet drive unit, direct a stream of high pressure water through the intake grille.

Models Without A Flushing Plug

Units manufactured prior to January, 1987 must have a small hole drilled and tapped in the port side of the jet drive unit to accept a flush adaptor. Proceed as follows:

First, the jet drive must be "dropped" from the intermediate housing and disassembled according to the procedures outlined in this chapter.

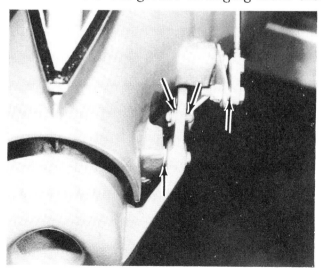

The arrows on this illustration indicate pivot points to be serviced on a regular basis with a good grade of water resistant multi-purpose lubricant.

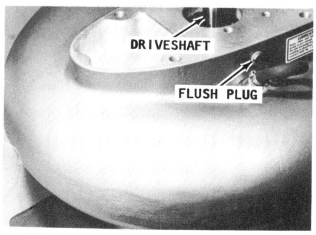

All jet drive units, 1987 and on, are provided with a flush plug on the port side of the housing.

11-18 OUTBOARD JET

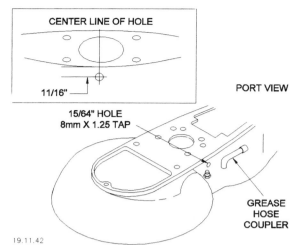

On jet drive units prior to 1987, a hole should be drilled and tapped to receive a flush plug. The dimensions indicate exactly where the hole is to be drilled.

Deflect the lubrication hose coupler sideways to remove it from the grease fitting and secure the hose out of the way.

Drill a 15/64" hole above the grease fitting, as shown in the accompanying illustration. Using a 8mm x 1.25 tap thread the hole. Blow away all metal chips from the interior and exterior of the housing, using compressed air.

Install the plug and gasket (available at the local dealer). Assemble and install the jet drive to the intermediate housing according to the procedures outlined in this chapter.

The powerhead may now be flushed as described in the previous paragraphs for units with the flush plug.

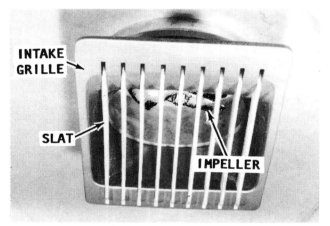

The clearance between the impeller and the casing core should be checked on a regular basis and should never exceed 1/32" (0.8mm).

The edges of the impeller should be kept sharp for maximum efficiency and performance.

OUTBOARD JET CHART

H.P	YEAR	NO. CYL.	DISPL. CU. IN.	JET MODEL SHAFT LENGTH 15"	20"
55-75	1968-88	3	49.7-56.1	L60	L60L
55-75	1968-88	3	49.7	L60C	L60CL
60-70	1989 & On	3	56.1	---	L60L-89
60-70	1989 & On	3	56.1	---	L60CL-89
85-100	1969-72	V4	92.6	---	M100
115-125	1969-72	V4	96.1	---	M125
85	1973-77	V4	92.6	---	M85-73
115-135	1973-77	V4	99.6	---	M135-73
85-100	1978 & On	V4	92.6-99.6	---	R85
110-140	1977 & On	V4	99.6-122.0	---	R140
150-235	1977 & On	V6	149.4-165.0	---	R6

APPENDIX

METRIC CONVERSION CHART

LINEAR
inches	X 25.4	= millimetres (mm)
feet	X 0.3048	= metres (m)
yards	X 0.9144	= metres (m)
miles	X 1.6093	= kilometres (km)
inches	X 2.54	= centimetres (cm)

AREA
inches^2	X 645.16	= millimetres^2 (mm^2)
inches^2	X 6.452	= centimetres^2 (cm^2)
feet^2	X 0.0929	= metres^2 (m^2)
yards^2	X 0.8361	= metres^2 (m^2)
acres	X 0.4047	= hectares (10^4 m^2) (ha)
miles^2	X 2.590	= kilometres^2 (km^2)

VOLUME
inches^3	X 16387	= millimetres^3 (mm^3)
inches^3	X 16.387	= centimetres^3 (cm^3)
inches^3	X 0.01639	= litres (l)
quarts	X 0.94635	= litres (l)
gallons	X 3.7854	= litres (l)
feet^3	X 28.317	= litres (l)
feet^3	X 0.02832	= metres^3 (m^3)
fluid oz	X 23.57	= millilitres (ml)
yards^3	X 0.7646	= metres^3 (m^3)

MASS
ounces (av)	X 28.35	= grams (g)
pounds (av)	X 0.4536	= kilograms (kg)
tons (2000 lb)	X 907.18	= kilograms (kg)
tons (2000 lb)	X 0.90718	= metric tons (t)

FORCE
ounces - f (av)	X 0.278	= newtons (N)
pounds - f (av)	X 4.448	= newtons (N)
kilograms - f	X 9.807	= newtons (N)

ACCELERATION
feet/sec^2	X 0.3048	= metres/sec^2 (m/S^2)
inches/sec^2	X 0.0254	= metres/sec^2 (m/s^2)

ENERGY OR WORK (watt-second - joule - newton-metre)
foot-pounds	X 1.3558	= joules (j)
calories	X 4.187	= joules (j)
Btu	X 1055	= joules (j)
watt-hours	X 3500	= joules (j)
kilowatt - hrs	X 3.600	= megajoules (MJ)

FUEL ECONOMY AND FUEL CONSUMPTION
miles/gal	X 0.42514	= kilometres/litre (km/l)

Note:
235.2/(mi/gal) = litres/100km
235.2/(litres/100 km) = mi/gal

LIGHT
footcandles	X 10.76	= lumens/metre^2 (lm/m^2)

PRESSURE OR STRESS (newton/sq metre - pascal)
inches HG (60°F)	X 3.377	= kilopascals (kPa)
pounds/sq in	X 6.895	= kilopascals (kPa)
inches H2O (60°F)	X 0.2488	= kilopascals (kPa)
bars	X 100	= kilopascals (kPa)
pounds/sq ft	X 47.88	= pascals (Pa)

POWER
horsepower	X 0.746	= kilowatts (kW)
ft-lbf/min	X 0.0226	= watts (W)

TORQUE
pound-inches	X 0.11299	= newton-metres (N·m)
pound-feet	X 1.3558	= newton-metres (N·m)

VELOCITY
miles/hour	X 1.6093	= kilometres/hour (km/h)
feet/sec	X 0.3048	= metres/sec (m/s)
kilometres/hr	X 0.27778	= metres/sec (m/s)
miles/hour	X 0.4470	= metres/sec (m/s)

TEMPERATURE

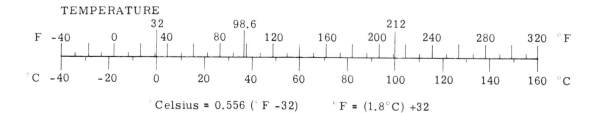

Celsius = 0.556 (°F -32)   °F = (1.8°C) +32

POWERHEAD SPECIFICATIONS

HP	YEAR	STROKE	BORE STD	BORE OVERSIZE	BORE OVERSIZE	PISTON TO CYL CLEARANCE MAX	MIN
60	1986 & On	2.344	3.00	.020	.030	.0055	.0035
65	1972-73	2.344	3.00	.020	.030	.0055	.0035
70	1974-78	2.344	3.00	.020	.030	.0055	.0035
70	1979-81	2.344	3.00	.020	.030	.0065	.0045
70	1982-85	2.344	3.00	.020	.030	.0055	.0045
70	1986 & On	2.344	3.187	.020	.030	.0055	.0045
75	1975-78	2.344	3.00	.020	.030	.0055	.0035
75	1979-81	2.344	3.00	.020	.030	.0065	.0045
75	1982-85	2.344	3.00	.020	.030	.0065	.0045
75	1987	2.344	3.00	.020	.030	.0065	.0045
75	1989 & On	2.344	3.00	.020	.030	.0065	.0045
85	1972-73	2.588	3.37	.020	.030	.0045	.0025
85	1974-77	2.588	3.37	---	.030	.0045	.0025
85	1978	2.588	3.37	---	.030	.0075	.0055
85	1979-80	2.588	3.50	.020	.030	.0075	.0045
90	1981-86	2.588	3.50	.020	.030	.0075	.0045
90	1987 & On	2.858	3.50	.020	.030	.0075	.0045
100	1979-80	2.588	3.50	.020	.030	.0075	.0045
110	1985-89	2.858	3.50	.020	.030	.0075	.0045
115	1973-77	2.588	3.50	.020	.030	.0055	.0035
115	1978	2.588	3.50	.020	.030	.0075	.0055
115	1979	2.588	3.50	.020	.030	.0075	.0045
115	1980-84	2.588	3.50	.020	.030	.0075	.0045
115	1990 & On	2.858	3.50	.020	.030	.0075	.0045
120	1985 & On	2.858	3.50	.020	.030	.0075	.0045
135	1973-76	2.588	3.50	.020	.030	.0055	.0035
140	1977	2.588	3.50	.020	.030	---	---
140	1978	2.588	3.50	.020	.030	.0075	.0055
140	1979-84	2.588	3.50	.020	.030	.0075	.0045
140	1985	2.860	3.50	.020	.030	.0075	.0045
140	1986 & On	2.858	3.50	.020	.030	.0075	.0045
150	1978	2.588	3.50	.020	.030	.0075	.0055
150	1979-85	2.588	3.50	.020	.030	.0075	.0045
150	1986-88	2.588	3.625	.020	.030	.0075	.0045
150	1989 & On	2.588	3.50	.020	.030	.0075	.0045
175	1977-78	2.588	3.50	.020	.030	.0075	.0055
175	1979-83	2.588	3.50	.020	.030	.0075	.0045
185	1985	2.588	3.50	.020	.030	.0075	.0045
200	1976-78	2.588	3.50	.020	.030	.0075	.0055
200	1979-83	2.588	3.50	.020	.030	.0075	.0045
200	1986 & On	2.858	3.50	.020	.030	.0075	.0045
225	1989	2.858	3.50	.020	.030	.0075	.0045
235	1978	2.588	3.50	.020	.030	.0075	.0055
235	1979	2.588	3.50	.020	.030	.0075	.0045
235	1980-85	2.588	3.625	---	.030	.0075	.0045

TORQUE SPECIFICATIONS

TORQUE VALUES

Torque values given in both Inch/pounds and Foot/pounds.
Divide by 12 for Foot/pounds.
Multiply by 12 for Inch/pounds.

HORSEPOWER & MODEL YEAR	FLYWHEEL NUT Ft./Lbs	CONNECTING ROD BOLT Ft./Lbs	HEAD BOLT In./Lbs	MAIN BEARING BOLT In./Lbs	CRANKCASE HEAD BOLT In./Lbs
60 hp 1985 & On	100	30	228	228	70
65 hp 1972 & 1973	100	30	228	228	70
70 hp 1974 and On	100	30	228	228	70
75 hp 1975 and On	100	30	228	228	70
85 hp 1972 thru 1980	100	30	228	228	120
90 hp 1981 and On	100	30	228	228	120
100 hp 1979 thru 1980	100	30	228	228	120
110 hp 1985 and On	100	30	228	228	120
115 hp 1973 thru 1984	100	30	228	228	120
115 hp 1990 and On	100	30	228	228	120
120 hp 1985 and On	100	42	228	336	85
135 hp 1973 thru 1976	100	30	228	336	85
140 hp 1977 and On	100	42	228	336	85
150 hp 1978 and On	145	30	228	228	85
175 hp 1977 thru 1983	145	30	228	228	85
175 hp 1986 and On	145	30	228	228	85
185 hp 1984 & 1985	145	30	228	228	85
200 hp 1976 and On	145	42	228	336	85
225 hp 1986 and On	145	42	228	336	85
235 hp 1978 and On	145	42	228	336	85

All spark plugs, all models -- 17.5 to 22.0 Ft. - lbs.

STANDARD BOLTS AND NUTS

Torque value for special bolts may vary.

Bolt Size	In./Lbs	Ft/Lbs	Newton Meters
No. 6	7-10	--	0.8-1.2
No. 8	15-22	--	1.6-2.4
No. 10	25-35	2-3	2.8-4.0
No. 12	35-40	3-4	4.0-4.6
1/4"	60-80	5-7	7-9
5/16"	120-140	10-12	14-16
3/8"	220-240	18-20	24-27
7/16"	340-360	28-30	38-40

ENGINE SPECIFICATIONS

SEE GENERAL AND SPECIAL NOTES, APPENDIX PAGE A-12

JOHNSON MODEL	EVINRUDE MODEL	NO. CYL.	HP	CU IN DISPL.
1972	**1972**			
65ES; 65ESL-72	65272-73	3	65	49.7
1973	**1973**			
65ES; 65ESL; 65SLR-73	65372-73	3	65	49.7
85ESL73	85393	V4	85	92.6
115ESL-73	115393	V4	115	99.6
135ESL73	135383	V4	135	99.6
1974	**1974**			
70ES; 70ESL; 70ELR-74	70442-43; 70472-73	3	70	49.7
85ESL74	85493	V4	85	92.6
115ESL; ETL-74	115493	V4	115	99.6
135ESL; ETL-74	135443; 135483-89	V4	135	99.6
1975	**1975**			
70ES; 70ESL-75	70572-73	3	70	49.7
75ESR; 75ESLR-75	75542-43	3	75	49.7
85ESLR-75	85593	V4	85	92.6
115ESL; 115ETL-75	115593	V4	115	99.6
135ESL; 135ETL-75	135583	V4	135	99.6
1976	**1976**			
70EL-76	70673	3	70	49.7
75ER; 75ELR-76	75642-43	3	75	49.7
85EL; 85ETLR-76	85693	V4	85	92.6
115EL; 115ETL-76	115693	V4	115	99.6
135EL; 135ETL-76	135643; 135683	V4	135	99.6
200TL; 200TXL-76	200640; 200649	V6	200	149.4
1977	**1977**			
70EL-77	70773	3	70	49.7
75ER; 75ELR-77	75742-43	3	75	49.7
85EL; 85ETLR; 85TXLR-77	85790; 93; 99	V4	85	92.6
115EL; 115ETL; 115TXL-77	115790; 93; 99	V4	115	99.6
140ML; 140TL; 140TXL-77	140740; 43; 83	V4	140	99.6
175TL; 175TXL-77	176740; 49	V6	175	149.4
200TL; 200TXL-77	200740; 49	V6	200	149.4
1978	**1978**			
70EL-78	70873	3	70	49.7
75ER; 75EL-78	75842, 43	3	75	49.7
85ML; 85ETLR; 85TXLR-78	85890, 95, 99	V4	85	92.6
115ML; 115ETL; 115TXL-78	115890, 93, 99	V4	115	99.6
140ML; 140TL; 140TXL-78	140840, 43, 83	V4	140	99.6
150TL-78	150849	V6	150	149.4
175TL; 175TXL-78	175840, 49	V6	175	149.4
200TL; 200TXL-78	200840, 49	V6	200	149.4
235TL; 235TXL-78	235840, 49	V6	235	149.4

AND TUNE-UP ADJUSTMENTS

SEE GENERAL AND SPECIAL NOTES, APPENDIX PAGE A-12

IDLE RPM	W O T RPM	BORE INCH	SPARK PLUG TYPE	SPARK PLUG GAP	THROTTLE PICKUP	MAXIMUM ADVANCE BTDC	CARB. TYPE
					1972		
600-700	5000	3.00	L77J4	0.030"	8°BTDC	22°	I
					1973		
600-700	5000	3.00	L77J4	0.030"	8°BTDC	26°	I
600-700	5000	3.37	L77J4	0.030"	5°BTDC	28°	II
600-700	5000	3.50	L77J4	0.030"	5°BTDC	26°	II
600-700	5000	3.50	UL77V	--	5°BTDC	22°	II
					1974		
700-750	5000	3.00	L77J4	0.030"	TDC	20°	I
600-700	5000	3.37	L77J4	0.030"	5°BTDC	28°	II
600-700	5000	3.50	L77J4	0.030"	5°BTDC	21°	II
600-700	5000	3.50	UL77V	--	5°BTDC	20°	II
					1975		
700-750	5000	3.00	L77J4	0.030"	TDC	17°	I
700-750	5500	3.00	L77J4	0.030"	TDC	16°	I
600-700	5000	3.37	L77J4	0.030"	5°BTDC	26°	II
600-700	5000	3.50	L77J4	0.030"	5°BTDC	24°	II
600-700	5000	3.50	UL77V	--	5°BTDC	20°	II
					1976		
700-750	5000	3.00	L77J4	0.030"	TDC	17°	I
700-750	5500	3.00	L77J4	0.030"	TDC	16°	I
600-700	5000	3.37	L77J4	0.030"	5°BTDC	26°	II
600-700	5000	3.50	L77J4	0.030"	5°BTDC	24°	II
600-700	5000	3.50	UL77V	--	5°BTDC	20°	II
600-700	5250	3.50	UL77V	--	5°BTDC	28°	II
					1977		
700-750	5000	3.00	L77J4	0.030"	TDC	17°	I
700-750	5500	3.00	L77J4	0.030"	TDC	16°	I
600-700	5000	3.37	L77J4	0.030"	3-5°BTDC	26°	II
600-700	5000	3.50	L77J4	0.030"	0-3°BTDC	28°	II
600-700	5000	3.50	UL77V	--	0-3°BTDC	28°	II
600-700	5000	3.50	UL77V	--	5°BTDC	28°	II
600-700	5250	3.50	UL77V	--	5°BTDC	28°	II
					1978		
700-750	5000	3.00	L77J4	0.030"	TDC	17°	I
700-750	5500	3.00	L77J4	0.030"	TDC	16°	I
600-700	5000	3.37	L77J4	0.030"	5°BTDC	28°	II
600-700	5000	3.50	L77J4	0.030"	5°BTDC	28°	II
600-700	5000	3.50	UL77V	--	5°BTDC	28°	II
600-700	5000	3.50	UL77V	--	5°BTDC	28°	II
600-700	5000	3.50	UL77V	--	5°BTDC	28°	II
600-700	5250	3.50	UL77V	--	5°BTDC	28°	II
600-700	5250	3.50	UL77V	--	5°BTDC	28°	II

ENGINE SPECIFICATIONS

SEE GENERAL AND SPECIAL NOTES, APPENDIX PAGE A-12

JOHNSON MODEL	EVINRUDE MODEL	NO. CYL.	HP	CU IN DISPL.
1979	**1979**			
70EL-79	70973	3	70	49.7
75ER; 75ELR-79	75942, 43	3	75	49.7
85ML; 85TL; 85TXL-79	85990, 95, 99	V4	85	99.6
100ML; 100TLR; 100TXLR-79	100990, 93, 99	V4	100	99.6
115ML; 115TL; 115TXL-79	115990, 93, 99	V4	115	99.6
140ML; 140TL; 140TXL-79	140940, 43, 83	V4	140	99.6
150TL; 150TXL-79	150940, 49	V6	150	149.4
175TL; 175TXL-79	175940, 49	V6	175	149.4
200TL; 200TXL-79	200940, 49	V6	200	149.4
235TL; 235TXL-79	235940, 49	V6	235	149.4
1980	**1980**			
70ELCS	E70ELCS	3	70	49.7
75ERCS; 75ERLCS; 75TRLCS	E75ERCS; E75ERLCS	3	75	49.7
85MLCS; 85TLCS	E85MLCS; E85TLCS; E85TXCS	V4	85	99.6
100MLCS; 100TRLCS; 100TRXCS	E100MLCS, TLCS, TXCS	V4	100	99.6
115LMCS; TLCS; TXCS	E115MLCS, TLCS, TXCS	V4	115	99.6
140MLCS; TLCS; TXCS	E140MLCS, TRLCS, TRXCS	V4	140	99.6
150TLCS; TXCS	E150TRLCS, TRXCS	V6	150	149.4
175TLCS; TXCS	E175TRLCS, TRXCS	V6	175	149.4
200TLCS; TXCS	E200TRLCS, TRXCS	V6	200	149.4
235TRLCS; TXCS	E235TRLCS, TRXCS	V6	235	160.3
1981	**1981**			
J70ELCIH, IM; J70TLCIM	E70ELCIH, IM; E70TLCIM	3	70	49.7
J75ERCIH, CIM, LCIH, LCIM J75TRLCIH, CIM	E75ERCIH, CIM, LCIH, E75ERLCIM, TRLCIH, TRLCIM	3	75	49.7
J90MLCIH, IM; J90TLCIH; J90TLCIM; J90TXCIH, IM	E90MLCIH, IM; E90TLCIH; E90TLCIM; E90TXCIH, IM	V4	90	99.6
J115TLCIH, IM; J115MLCIH; J115MLCIM; J115TXCIH, IM	E115TLCIH, IM; E115MLCIH, IM; E115TXCIH, IM	V4	115	99.6
J140TLCIH, IM; J140MLCIH: J140MLCIM; J140TXCIH, IM	E140TRLCIH, IM; E140MLCIH, IM; E140TRXCIH, IM	V4	140	99.6
J150TLIH, IA; J150TXCIH, IA	E150TRLCIH, IA; E150TRXCIH, IA	V6	150	149.4
J175TLCIH, IM; J175TXCIH, IM	E175TRLCIH, IM; E175TRXCIH, IM	V6	175	149.4
J200TLCIH, IB; J200TXCIH, IB	E200TRLCIH, IB; E200TRXCIH, IB	V6	200	149.4
J235TLCIH, IB; J235TXCIH, IB	E235TRLCIH, IB; E235TRXCIH, IB	V6	235	160.3
1982	**1982**			
J70ELCNB; J70TLCNB	E70ELCNB; E70TLCNB	3	70	49.7
J75ELCNB, ECNB, TLCNB	E75ERLCNB, ERCNB, TRLCNB	3	75	49.7
J90MLCNB, TLCNB, TXCNB	E90MLCNB, TLCNB, TXCNB	V4	90	99.6
J115MLCNB, TLCNB, TXCNB	E115MLCNB, TLCNB, TXCNB	V4	115	99.6
J140MLCNB, TLCNB, TXCNB	E140MLCNB, TRLCNB, TRXCNB	V4	140	99.6
J150TLCNM; J150TXCNM	E150TRLCNM; E150TRXCNM	V6	150	149.4
J175TLCNB; J175TXCNB	E175TRLCNB; E175TRXCNB	V6	175	149.4
J200TLCNE; J200TXCNE	E200TRLCNE; E200TRXCNE	V6	200	149.4
J235TLCNE; J235TXCNE	E235TRLCNE; E235TRXCNE	V6	235	160.3

ENGINE SPECIFICATIONS A-7
AND TUNE-UP ADJUSTMENTS
SEE GENERAL AND SPECIAL NOTES, APPENDIX PAGE A-12

IDLE RPM	WOT RPM	BORE INCH	SPARK PLUG TYPE	SPARK PLUG GAP	THROTTLE PICKUP	MAXIMUM ADVANCE	CARB. TYPE
					1979		
700-750	5000	3.00	L77J4	0.030"	TDC	19°	I
700-750	5500	3.00	L77J4	0.030"	TDC	19°	I
600-700	5000	3.50	L77J4	0.030"	4-6°BTDC	28°	II
600-700	5000	3.50	L77J4	0.030"	4-6°BTDC	28°	II
600-700	5000	3.50	L77J4	0.030"	4-6°BTDC	28°	II
600-700	5000	3.50	UL77V	--	4-6°BTDC	28°	II
600-700	5000	3.50	UL77V	--	5°BTDC	26°	III
600-700	5000	3.50	UL77V	--	5°BTDC	28°	III
600-700	5250	3.50	UL77V	--	5°BTDC	28°	III
600-700	5250	3.50	UL77V	--	5°BTDC	28°	III
					1980		
700-750	5500	3.00	L77J4	0.030"	TDC	19°	I
700-750	5500	3.00	L77J4	0.030"	TDC	19°	I
600-700	5000	3.50	L77J4	0.030"	4-6°BTDC	28°	III
600-700	5000	3.50	L77J4	0.030"	4-6°BTDC	28°	III
600-700	5000	3.50	L77J4	0.030"	4-6°BTDC	28°	III
600-700	5000	3.50	UL77V	--	4-6°BTDC	28°	III
600-700	5000	3.50	UL77V	--	6-8°BTDC	28°	III
600-700	5000	3.50	UL77V	--	6-8°BTDC	28°	III
600-700	5250	3.50	UL77V	--	6-8°BTDC	26°	III
600-700	5250	3.62	UL77V	--	6-8°BTDC	24°	III
					1981		
700-750	5500	3.00	L77J4	0.030"	TDC	19°	I
700-750	5500	3.00	L77J4	0.030"	TDC	19°	I
600-700	5500	3.50	L77J4	0.030"	4-6°BTDC	28°	III
600-700	5500	3.50	L77J4	0.030"	4-6°BTDC	28°	III
600-700	5500	3.50	UL77V	--	4-6°BTDC	28°	III
600-700	5500	3.50	UL77V	--	6-8°BTDC	28°	III
600-700	5500	3.50	UL77V	--	6-8°BTDC	28°	III
600-700	5750	3.50	UL77V	--	6-8°BTDC	28°	III
600-700	5750	3.625	UL77V	--	6-8°BTDC	28°	III
					1982		
700-750	5500	3.00	L77J4	0.040"	TDC	19°	I
700-750	5800	3.00	L77J4	0.040"	TDC	19°	I
600-700	5500	3.50	L77J4	0.040"	4-6°BTDC	28°	III
600-700	5500	3.50	L77J4	0.040"	4-6°BTDC	28°	III
600-700	5500	3.50	UL77V	--	4-6°BTDC	28°	III
600-700	5500	3.50	UL77V	--	6-8°BTDC	28°	III
600-700	5500	3.50	UL77V	--	6-8°BTDC	28°	III
600-700	5750	3.50	UL77V	--	6-8°BTDC	28°	III
600-700	5750	3.63	UL77V	--	6-8°BTDC	30°	III

ENGINE SPECIFICATIONS

SEE GENERAL AND SPECIAL NOTES, APPENDIX PAGE A-12

JOHNSON MODEL	EVINRUDE MODEL	NO. CYL.	HP	CU IN DISPL.
1983	**1983**			
J70ELCT, TLCT	E70ELCT, TLCT	3	70	49.7
J75ELTC, ECT, TLCT	E75ELCT, ECT, TLCT	3	75	49.7
J90MLCT, TLCT, TXCT	E90MLCT, TLCT, TXCT	V4	90	99.6
J115MLCT, TLCT, TXCT	E115MLCT, TLCT, TXCT	V4	115	99.6
J140MLCT, TLCT, TXCT	E140MLCT, TLCT, TXCT	V4	140	99.6
J150TLCT, TXCT	E150TLCT, TXCT	V6	150	149.4
J175TLCT, TXCT	E175TLCT, TXCT	V6	175	149.4
J200TLCT, TXCT	E200TLCT, TXCT	V6	200	149.4
J235TLCT, TXCT	E235TLCT, TXCT	V6	235	160.3
1984	**1984**			
J70ELCR, TLCR	E70ELCR, TLCR	3	70	49.7
J75ELCR, ECR, TLCR	E75ELCR, ECR, TLCR	3	75	49.7
J90MLCR, TLCR, TXCR	E90MLCR, TLCR, TXCR	V4	90	99.6
J115MLCR, TLCR, TXCR	E115MLCR, TLCR, TXCR	V4	115	99.6
J140MLCR, TLCR, TXCR	E140MLCR, TLCR, TXCR	V4	140	99.6
J150TLCR, TXCR, STLCR	E150TLCR, TXCR, STLCR	V6	150	149.4
J185TLCR, TXCR	E185TLCR, TXCR	V6	185	149.4
J235TLCR, TXCR, STLCR	E235TLCR, TXCR, STLCR	V6	235	160.3
1985	**1985**			
J70ELCO, TLCO	E70ELCO, TLCO	3	70	49.7
J75ELCO, ECO, TLCO	E75ELCO, ECO, TLCO	3	75	49.7
J90MLCO, TLCO, TXCO	E90MLCO, TLCO, TXCO	V4	90	99.6
J120TLCO, TXCO	E120TLCO, TXCO	V4	120	110
J140TLCO, TXCO	E140TLCO, TXCO	V4	140	110
J150TLCO, TXCO, STLCO	E150TLCO, TXCO, STLCO	V6	150	149.4
J185TLCO, TXCO	E185TLCO, TXCO	V6	185	149.4
J235TLCO, TXCO, STLCO	E235TLCO, TXCO, STLCO	V6	235	160.3
1986	**1986**			
J60ELCD, TLCD	E60ELCD, TLCD	3	60	49.7
J70ELCD, TLCD	E70ELCD, TLCD	3	70	56.1
J90MLCD, TLCD, TXCD	E90MLCD, TLCD, TXCD	V4	90	99.6
J110MLCD, TLCD	E110MLCD, TLCD	V4	110	99.6
J120TLCD, TXCD	E120TLCD, TXCD	V4	120	110
J140TLCD, TXCD	E140TLCD, TXCD	V4	140	110
J150TLCD, TXCD	E150TLCD, TXCD	V6	150	149.4
J175TLCD, TXCD	E175TLCD, TXCD	V6	175	160.3
J200TXCD	E200TXCD	V6	200	165
J225TLCD, TXCD	E225TLCD, TXCD	V6	225	165
PTLCD, PTXCD	PTLCD, PTXCD			

ENGINE SPECIFICATIONS AND TUNE-UP ADJUSTMENTS

SEE GENERAL AND SPECIAL NOTES, APPENDIX PAGE A-12

IDLE RPM	WOT RPM X100	BORE INCH	SPARK PLUG TYPE	SPARK PLUG GAP	THROTTLE PICKUP	MAXIMUM ADVANCE BTDC	CARB. TYPE
1983							
700-750	45-55	3.000	L77J4	0.040	TDC	NOTE 2	III
700-750	52-58	3.000	L77J4	0.040	TDC	NOTE 2	III
600-700	45-55	3.500	L77J4	0.040	3°-5°BTDC	NOTE 2	III
600-700	45-55	3.500	L77J4	0.040	3°-5°BTDC	NOTE 2	III
600-700	45-55	3.500	UL77V	--	3°-5°BTDC	NOTE 2	III
600-700	45-55	3.500	UL77V	--	6°-8°BTDC	NOTE 2	III
600-700	45-55	3.500	UL77V	--	6°-8°BTDC	NOTE 2	III
600-700	47-57	3.500	UL77V	--	6°-8°BTDC	NOTE 2	III
600-700	47-57	3.625	UL77V	--	6°-8°BTDC	NOTE 2	III
1984							
700-750	45-55	3.000	L77J4	0.040	TDC	NOTE 2	III
700-750	52-58	3.000	L77J4	0.040	TDC	NOTE 2	III
600-700	45-55	3.500	L77J4	0.040	3°-5°BTDC	NOTE 2	III
600-700	45-55	3.500	L77J4	0.040	3°-5°BTDC	NOTE 2	III
600-700	45-55	3.500	UL77V	--	3°-5°BTDC	NOTE 2	III
600-700	45-55	3.500	UL77V	--	6°-8°BTDC	NOTE 2	III
600-700	47-57	3.500	UL77V	--	6°-8°BTDC	NOTE 2	III
600-700	47-57	3.6250	UL77V	--	6°-8°BTDC	NOTE 2	III
1985							
700-750	45-55	3.000	L77J4	0.040	TDC	NOTE 2	III
700-750	52-58	3.000	L77J4	0.040	TDC	NOTE 2	III
600-700	45-55	3.500	L77J4	0.040	3°-5°BTDC	NOTE 2	III
600-700	50-60	3.500	L77J4	0.040	3°-5°BTDC	NOTE 2	IV
600-700	50-60	3.500	L77J4	0.040	3°-5°BTDC	NOTE 2	IV
600-700	45-55	3.500	UL77V	--	6°-8°BTDC	NOTE 2	III
600-700	47-57	3.500	UL77V	--	6°-8°BTDC	NOTE 2	III
600-700	47-57	3.625	UL77V	--	6°-8°BTDC	NOTE 2	III
1986							
4500-5500	70-75	3.000	QL77JC4	0.040	TDC	18-20°BTDC	III
5000-6000	70-75	3.187	QL77JC4	0.040	3-5°ATDC	18-20°BTDC	III
4500-5500	60-70	3.500	QL77JC4	0.040	3-5°BTDC	27-29°BTDC	III
4500-5500	60-70	3.500	QL77JC4	0.040	3-5°BTDC	27-29°BTDC	III
5000-6000	60-70	3.500	QL77JC4	0.040	3-5°BTDC	17-19°BTDC	IV
5000-6000	60-70	3.500	QL77JC4	0.040	3-5°BTDC	17-19°BTDC	IV
4500-5500	60-70	3.500	QUL77V	--	6-8°BTDC	31-33°BTDC	III
4750-5750	60-70	3.625	QUL77V	--	6-8°BTDC	27-29°BTDC	III
5000-6000	60-70	3.500	QL77JC4	0.040	6-8°BTDC	17-19°BTDC	IV
5000-6000	60-70	3.500	QL77JC4	0.040	6-8°BTDC	17-19°BTDC	IV

ENGINE SPECIFICATIONS

SEE GENERAL AND SPECIAL NOTES, APPENDIX PAGE A-12

ALL DIMENSIONS GIVEN IN INCHES

JOHNSON MODEL	EVENRUDE MODEL	NO. CYL.	H.P.	CU. IN. DISPL.
1987	**1987**			
J60ELCU, TLCU	E60ELCU, TLCU	3	60	49.7
J70ELCU, TLCU	E70ELCU, TLCU	3	70	56.1
J90MLCU, TLCU, TXCU	E90MLCU, TLCU, TXCU	V4	90	99.6
J110MLCU, TLCU	E110MLCU, TLCU	V4	110	99.6
J120TLCU, TXCU	E120TLCU, TXCU	V4	120	110
J140TLCU, TXCU	E140TLCU, TXCU	V4	140	110
J150TLCU, TXCU	E150TLCU, TXCU	V6	150	149.4
J175TLCU, TXCU	E175TLCU, TXCU	V6	175	160.3
J200TXCU	E200TXCU	V6	200	165
J225TLCU, TXCU	E225TLCU, TXCU	V6	225	165
PTXCU, TLCU, TXCU	PTXCU, TLCU, TXCU			
1988	**1988**			
J60ELCC, TLCC	E60ELCC, TLCC	3	60	49.7
J70ELCC, TLCC	E70ELCC, TLCC	3	70	56.1
J90TLCC	E90TLCC	V4	90	99.6
J110MLCC, TLCC, TXCC	E110MLCC, TLCC, TXCC	V4	110	99.6
J120TLCC, TXCC	E120TLCC, TXCC	V4	120	122
J140TLCC, TXCC, CXCC	E140TLCC, TXCC, CXCC	V4	140	122
J150TLCC, TXCC, CXCC	E150TLCC, TXCC, CXCC	V6	150	149.4
J175TLCC, TXCC	E175TLCC, TXCC	V6	175	160.3
J200TXCC, SCLCC, CXCC	E200TXCC, SCLCC, CXCC	V6	200	183
J225TLCC, TXCC, CLCC	E225TLCC, TXCC, CLCC	V6	225	183
PLCC, PXCC, CXCC	PLCC, PXCC, CXCC			
1989	**1989**			
J60ELCE, TLCE, TTLCE	E60ELCE, TLCE, TTLCE	3	60	56.1
J70ELCE, TLCE	E70ELCE, TLCE	3	70	56.1
J90TLCE, TXCE	E90TLCE, TXCE	V4	90	99.6
J110TLCE, TXCE	E110TLCE, TXCE	V4	110	99.6
J120TLCE, TXCE	E120TLCE, TXCE	V4	120	122
J140TLCE, TXCE	E140TLCE, TXCE	V4	140	122
J150TLCE, TXCE, CXCE	E150TLCE, TXCE, CXCE	V6	150	149.4
J175TLCE, TXCE, STLCE	E175TLCE, TXCE, STLCE	V6	175	160.3
J200STLCE, TXCE, CXCE	E200STLCE, TXCE, CXCE	V6	200	183
J225TLCE, TXCE, CXCE, PLCE, PXCE, SCXCE, STLCE, SPLCE, SPXCE	E225TLCE, TXCE, CXCE PLCE, PXCE, SCXCE	V6	225	183

ENGINE SPECIFICATONS A-11
AND TUNE-UP ADJUSTMENTS

SEE GENERAL AND SPECIAL NOTES, APPENDIX PAGE A-12

W.O.T. RPM	IDLE RPM	SPARK PLUG CHAMP.	SPARK PLUG GAP	THROTTLE PICKUP	MAXIMUM ADVANCE BTDC	CARB. TYPE
1987						
4500-5500	700-750	QL77JC4	0.040	TDC	18-20°	III
5000-6000	700-750	QL77JC4	0.040	3-5°ATDC	18-20°	III
4500-5500	600-700	QL77JC4	0.040	3-5°BTDC	27-29°	III
4500-5500	600-700	QL77JC4	0.040	3-5°BTDC	27-29°	III
5000-6000	600-700	QL77JC4	0.040	3-5°BTDC	17-19°	IV
5000-6000	600-700	QL77JC4	0.040	3-5°BTDC	17-19°	IV
4500-5500	600-700	QUL77V	--	6-8°BTDC	31-33°	III
4750-5750	600-700	QUL77V	--	6-8°BTDC	27-29°	III
5000-6000	600-700	QL77JC4	0.040	6-8°BTDC	17-19°	IV
5000-6000	600-700	QL77JC4	0.040	6-8°BTDC	17-19°	IV
1988						
4500-5500	700-750	QL77JC4	0.040	TDC	18-20°	III
5000-6000	700-750	QL77J3C4	0.040	3-5°ATDC	18-20°	III
4500-5500	600-700	QL77JC4	0.040	3-5°BTDC	27-29°	III
4500-5500	600-700	QL77JC4	0.040	3-5°BTDC	27-29°	III
5000-6000	600-700	QL77JC4	0.040	3-5°BTDC	17-19°	IV
5000-6000	600-700	QL77JC4	0.040	3-5°BTDC	17-19°	IV
4500-5500	600-700	QUL77V	--	6-8°BTDC	31-33°	III
4750-5750	600-700	QUL77V	--	6-8°BTDC	27-29°	III
5000-6000	600-700	QL77JC4	0.040	6-8°BTDC	17-19°	IV
5000-6000	600-700	QL77JC4	0.040	6-8°BTDC	17-19°	IV
1989						
5000-6000	700-750	QL77JC4	0.030	TDC	18-20°	III
5000-6000	700-750	QL77JC4	0.030	TDC-2°BTDC	18-20°	III
4500-5500	600-700	QL77JC4	0.030	6-8°ATDC	27-29°	III
4500-5500	600-700	QL77JC4	0.030	6-8°ATDC	27-29°	III
5000-6000	600-700	QL77JC4	0.030	3-5°BTDC	17-19°	IV
5000-6000	600-700	Ql77JC4	0.030	3-5°BTDC	17-19°	IV
4500-5500	600-700	QL77JC4	0.030	4-6°ATDC	17-19°	III
4750-5750	600-700	QL77JC4	0.030	4-6°ATDC	17-19°	III
5000-6000	575-700	QL77JC4	0.030	4-6°ATDC	17-19°	IV
5000-6000	575-700	QL77JC4	0.030	4-6°ATDC	17-19°	IV

APPENDIX A-12

ENGINE SPECIFICATIONS

JOHNSON MODEL	EVENRUDE MODEL	NO. CYL.	H.P.	CU. IN. DISPL.
1990 & 1991	**1990 & 1991**			
J60ELES, TLES, TTLES	E60ELES, TLES, TTLES	3	60	56.1
J70ELES, TLES	E70ELES, TLES	3	70	56.1
J90TLES, TXES	E90TLES, TXES	V4	90	99.6
J115TLES, TXES	E115TLES, TXES	V4	115	99.6
J120TLES, TXES	E120TLES, TXES	V4	120	122
J140TLES, TXES	E140TLES, TXES	V4	140	122
J150TLES, TXES, CXES	E150TLES, TXES, CXES	V6	150	149.4
J175TLES, TXES, STLES	E175TLES, TXES, STLES	V6	175	160.3
J200STLES, TXES, CXES	E200STLES, TXES, CXES	V6	200	183
J225TLES, TXES, CXES, PLES, PXES, SCXES, STLES, SPLES, SPXES	E225TLES, TXES, CXES, PLES, PXES, SCXES	V6	225	183

GENERAL NOTES

In-line powerheads -- top cylinder is No. 1. **V4 powerheads** -- starboard bank -- No. 1 top, No. 3 lower. Port bank -- No. 2 top, No. 4 lower. **V6 powerheads** -- starboard bank -- No. 1 top, No. 3 center, No. 5 lower. Port bank -- No. 2 top, No. 4 center, No. 6 lower.

Spark Plug UL77V -- Gap adjustment not necessary.

TDC -- Top Dead Center. **BTDC** -- Before Top Dead Center. **ATDC** -- After Top Dead Center.

SPECIAL NOTES

1- Carburetor Type
   Type I — single barrel with low speed and high speed fixed orifices. Three carburetors installed per powerhead.
   Type II — dual barrel with low speed and high speed fixed orifices. Two carburetors on V4 powerheads; three carburetors on V6 powerheads.
   Type III — dual barrel with low speed, intermediate speed, and high speed fixed orifices. Two carburetors on V4 powerheads; three carburetors on V6 powerheads.
   Type IV — single barrel with low speed, intermediate speed, and high speed orifices. Four carburetors on V4 powerheads; six carburetor on V6 powerheads.
2- See decal affixed on front of powerhead.
3- Ignition Type
   Single magneto flash capacitor discharge (CD) installed on all 65 hp, 70 hp, and 75 hp units covered in this manual.
   Dual magneto flash capacitor discharge (CD) installed on all units covered in this manual except those listed in above note.

Ignition Timing

	Model 150 & 175hp,	Model 200 hp	Model 235 hp
CIH Models, timer base with black sleeves.	28°	26°	30°
CIH Models, timer base with red sleeves	32°	30°	---
CIA, CIB, and CIM Models, timer base with black sleeves	28°	28°	30°
	28°	28°	30°

AND TUNE-UP ADJUSTMENTS

W.O.T. RPM	IDLE RPM	SPARK PLUG CHAMP.	SPARK PLUG GAP	THROTTLE PICKUP	MAXIMUM ADVANCE BTDC	CARB. TYPE
1990 & 1991				*1990 & 1991*		
5000-6000	700-750	QL77JC4	0.030	TDC	18-20°	III
5000-6000	700-750	QL77JC4	0.030	TDC-2°BTDC	18-20°	III
4500-5500	600-700	QL77JC4	0.030	6-8°ATDC	27-29°	III
4500-5500	600-700	QL77JC4	0.030	6-8°ATDC	27-29°	III
5000-6000	600-700	QL77JC4	0.030	3-5°BTDC	17-19°	IV
5000-6000	600-700	Ql77JC4	0.030	3-5°BTDC	17-19°	IV
4500-5500	600-700	QL77JC4	0.030	4-6°ATDC	17-19°	III
4750-5750	600-700	QL77JC4	0.030	4-6°ATDC	17-19°	III
5000-6000	575-700	QL77JC4	0.030	4-6°ATDC	17-19°	IV
5000-6000	575-700	QL77JC4	0.030	4-6°ATDC	17-19°	IV

GEAR OIL CAPACITIES

ENGINE SIZE	MODEL	CAPACITY OUNCES	ENGINE SIZE	MODEL	CAPACITY OUNCES
60	1985 & On	20-22	120	1985 & On	23-26
65	1972	25.3	135	1973-74	27.9
65	1973	21.3	135	1975-76	29.0
70	1974 & On	25.4	140	1977	34.5
75 SS	1975 & On	21.3	140	1978	37.8
75 LS	1975 & On	25.4	140	1979 & On	26.6
85	1972-76	27.9	150	1978	42.3
85 LS	1977	29.0	150	1979-85	28.3
85 ELS	1977	34.5	175	1977	44.0
85	1978	37.8	175	1978	42.3
85	1979-80	26.6	175	1979-83	28.3
90	1981 & On	26.6	175	1985 & On	30-33
100	1979-80	26.6	185	1984-85	28.3
110	1985-89	23-26	200	1976-77	44.0
115	1973-76	27.9	200	1978	42.3
115 LS	1977	29.0	200	1979-83	28.3
115 ELS	1977	34.5	200	1986 & On	30-33
115	1978	37.8	225	1986 & On	30-33
115	1979-84	26.6	235	1978	42.3
115	1990 & On	26.6	235	1979-85	28.3

OIL/FUEL MIXTURE

1- For all 1971-1984 engines in this manual -- use 50:1 oil mixture (1 pint oil per 6 gallons of fuel).

2- Since 1985 -- units without VRO system: during break-in and after -- use 50:1 oil mixture.

3- Units with VRO, for first 10 hours of operation, use 50:1 oil mixture in gasoline in **ADDITION** to VRO system.

4- **ALWAYS** use OMC or BIA certified TC-W oil lubricant.

5- **NEVER** use: automotive oils, premixed fuel of unknown oil quantity, or premixed fuel richer than 50:1.

APPENDIX A-14

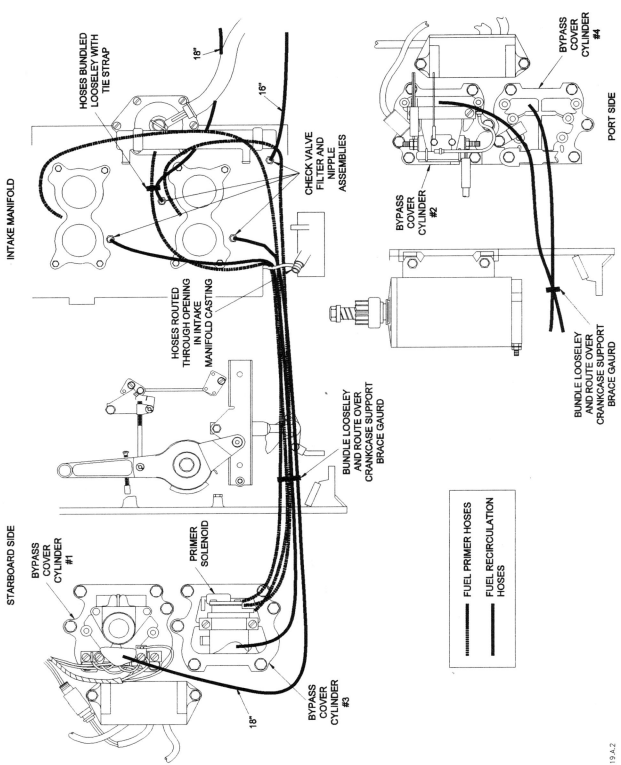

Hose Routing — V4 Engines with Primer Choke System

HOSE ROUTING A-15

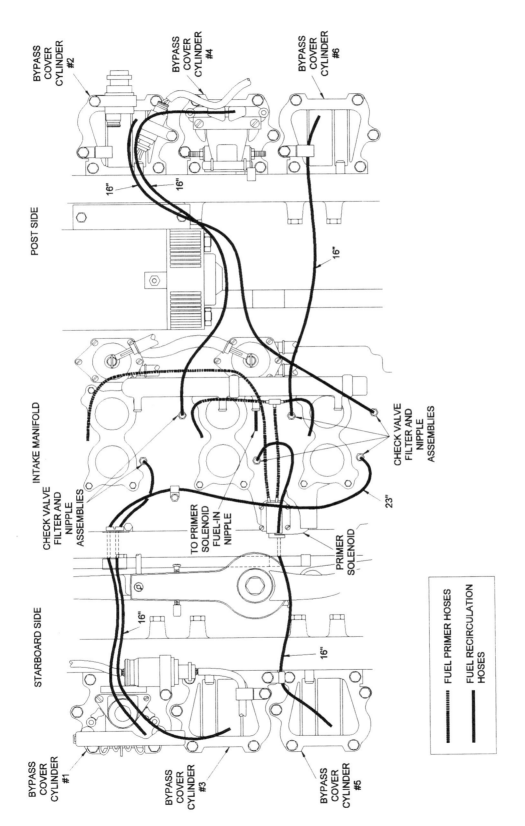

Hose Routing — V6 Engines with Primer Choke System

APPENDIX A-16

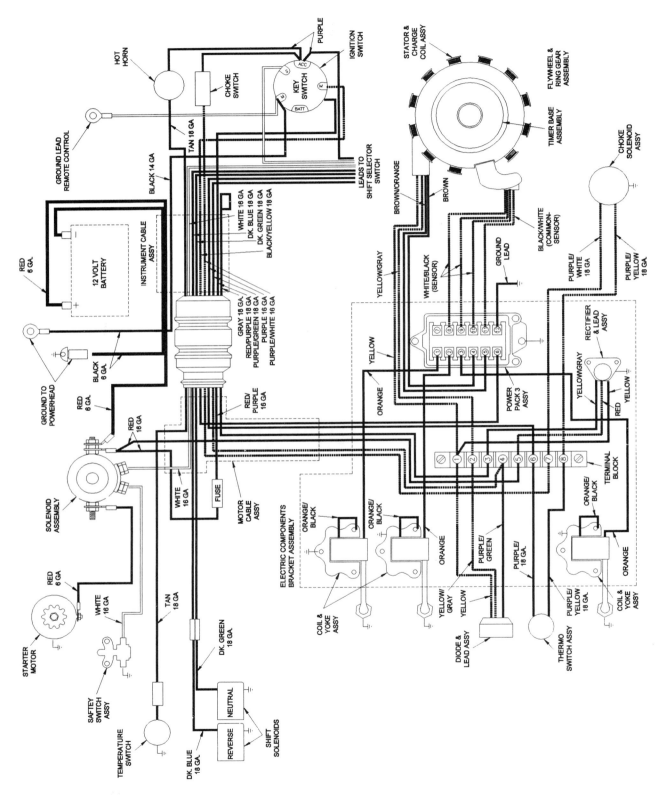

Wire Identification — 65 hp 3-Cylinder — with alternator — 1972

WIRE IDENTIFICATION A-17

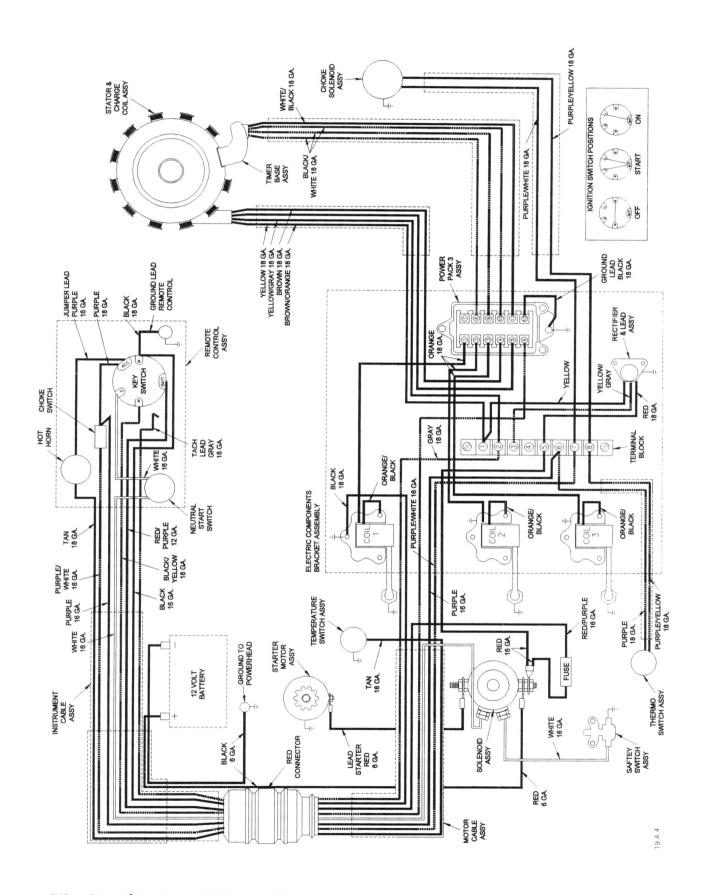

Wire Identification — 70 hp — 1974-1975 without Safety Switch and 75 hp — 1975

APPENDIX A-18

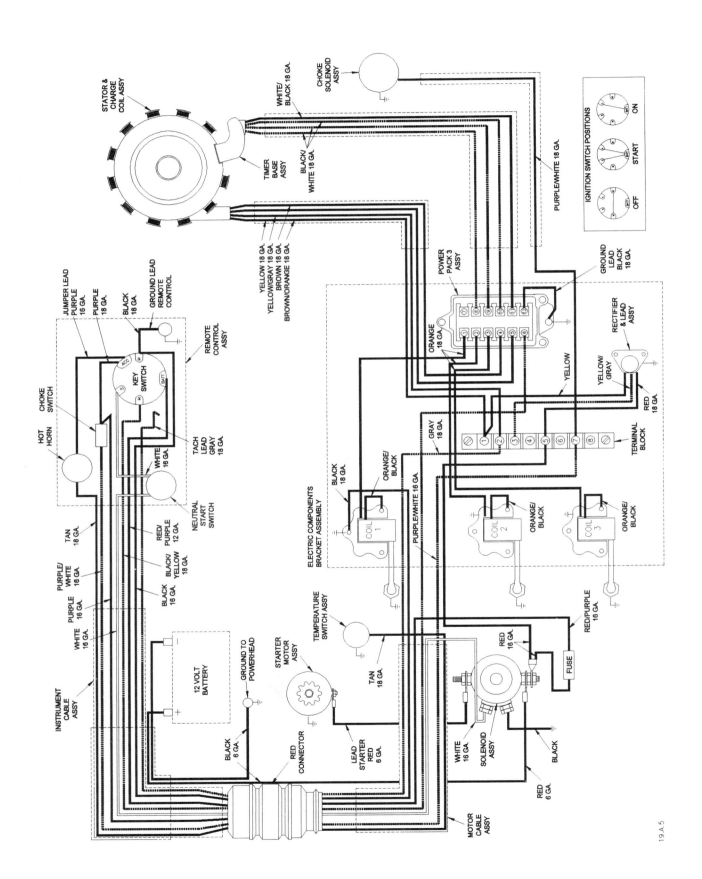

Wire Identification — 70 hp and 75 hp — 1976 and 1977

WIRE IDENTIFICATION A-19

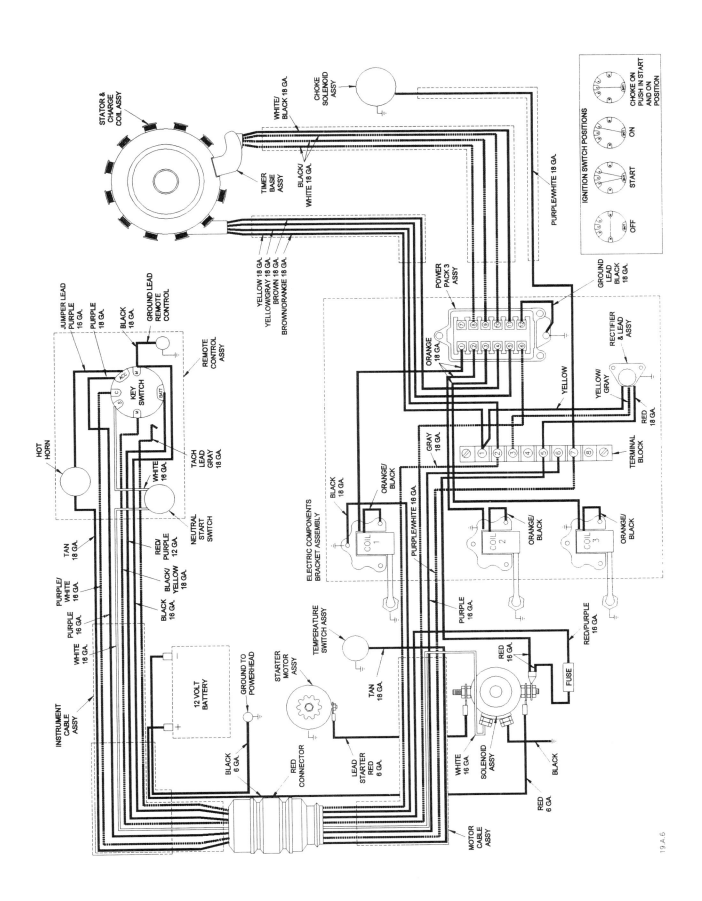

Wire Identification — 70 hp and 75 hp — 1978

APPENDIX A-20

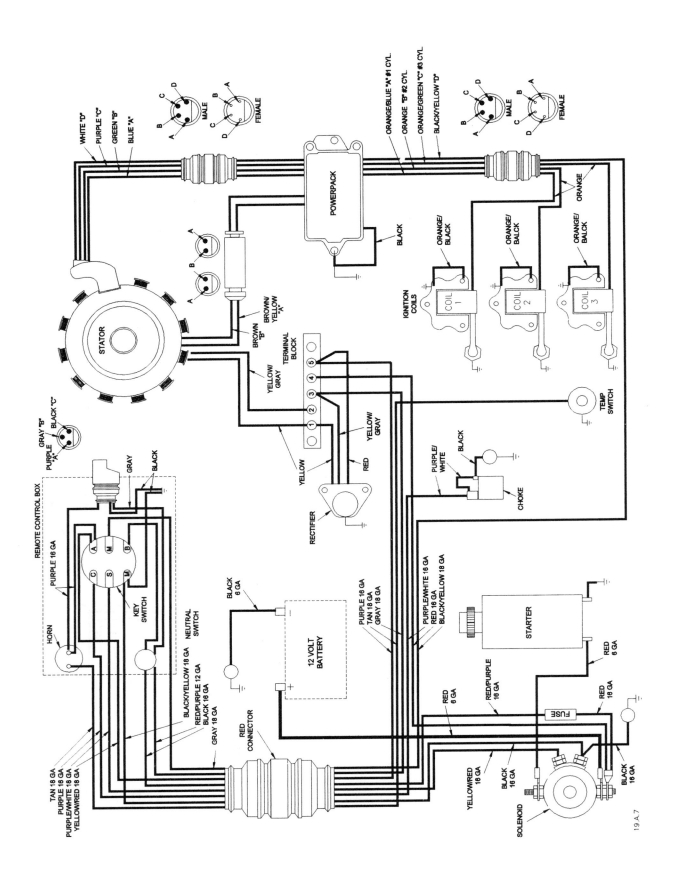

Wire Identification — 60 hp 1985 and on; 70 hp 1979 and on; 75 hp 1979 thru 1982

WIRE INDENTIFICATION   A-21

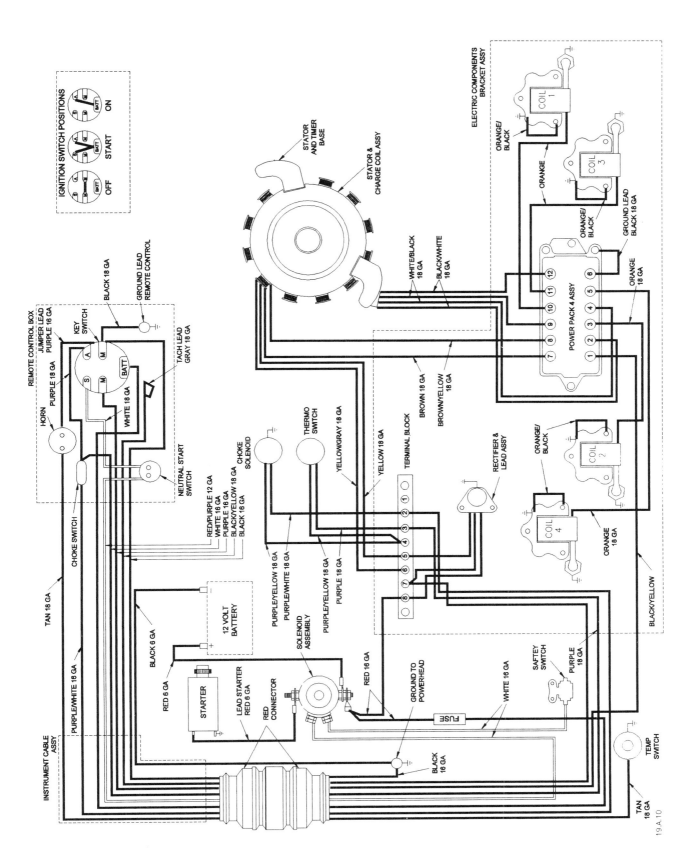

Wire Identification — 85 hp, 115 hp, and 135 hp V4 — 1972 thru 1976 (The 1974 model does not have the safety switch.)

APPENDIX A-22

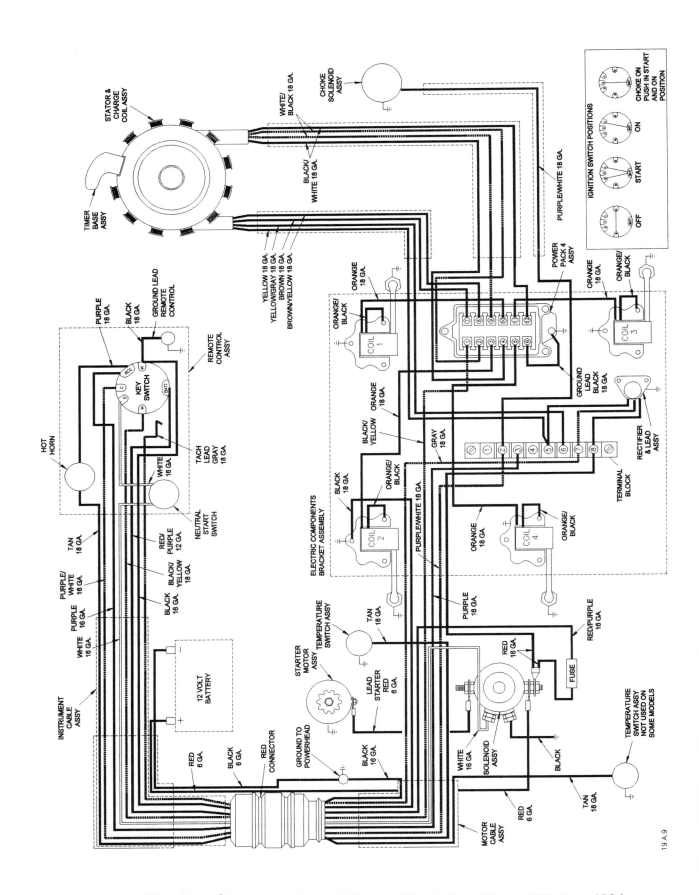

Wire Identification — 85 hp, 115 hp 1977; 140 hp V4 — 1977 thru 1984

WIRE IDENTIFICATION   A-23

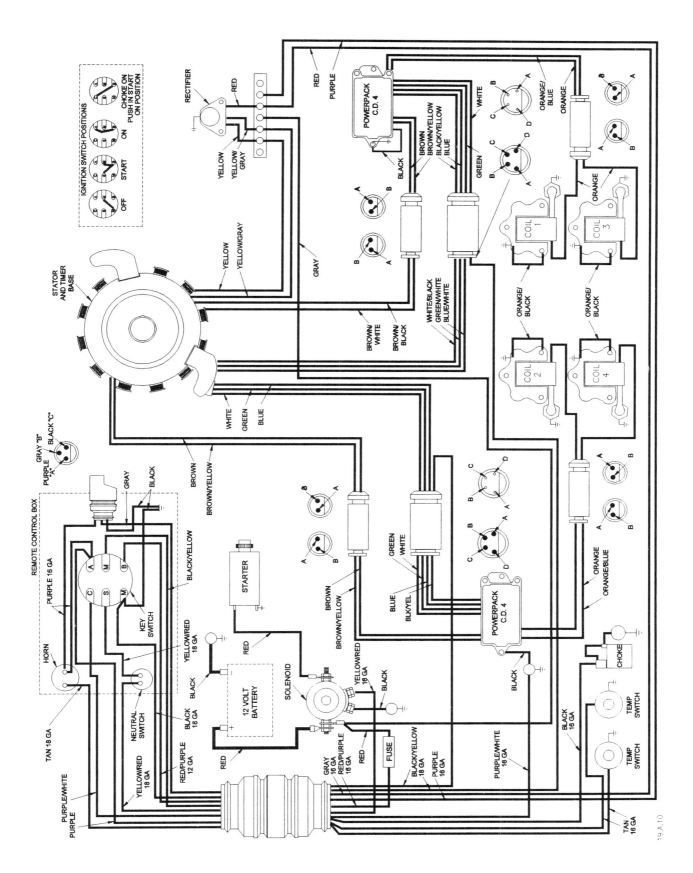

Wire Identification -- V4 Powerheads: 85 hp 1978 thru 1980; 90 hp 1981 and on; 100 hp 1979 and 1980; 110 hp 1985 thru 1989; 115hp 1978 thru 1984 and 1990 and on.

APPENDIX A-24

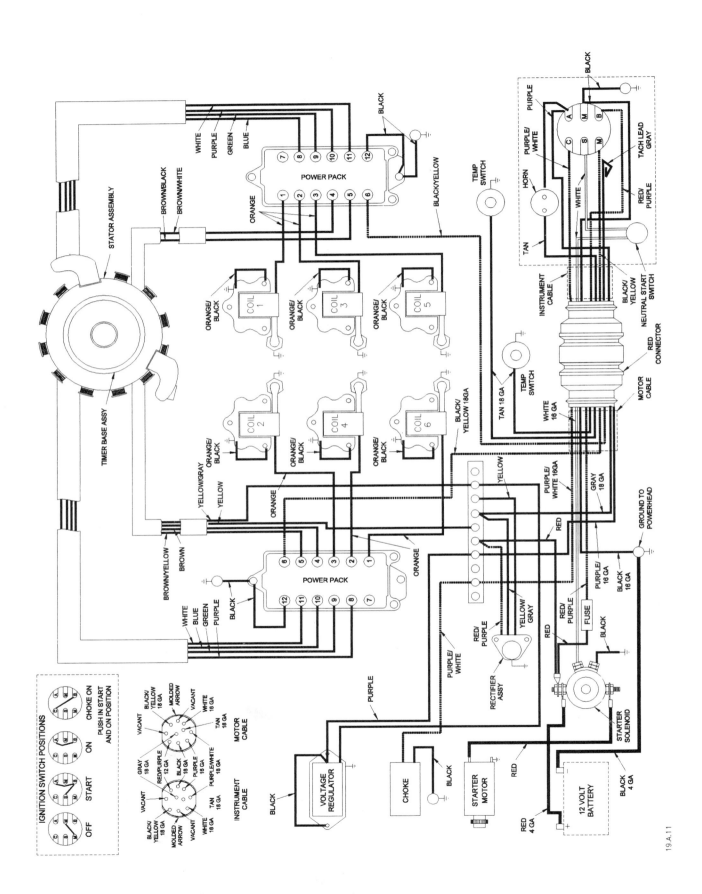

Wire Identification — 175 hp and 200 hp V6 — 1976 thru 1978; 235 hp V6 — 1978 only

WIRE IDENTIFICATION A-25

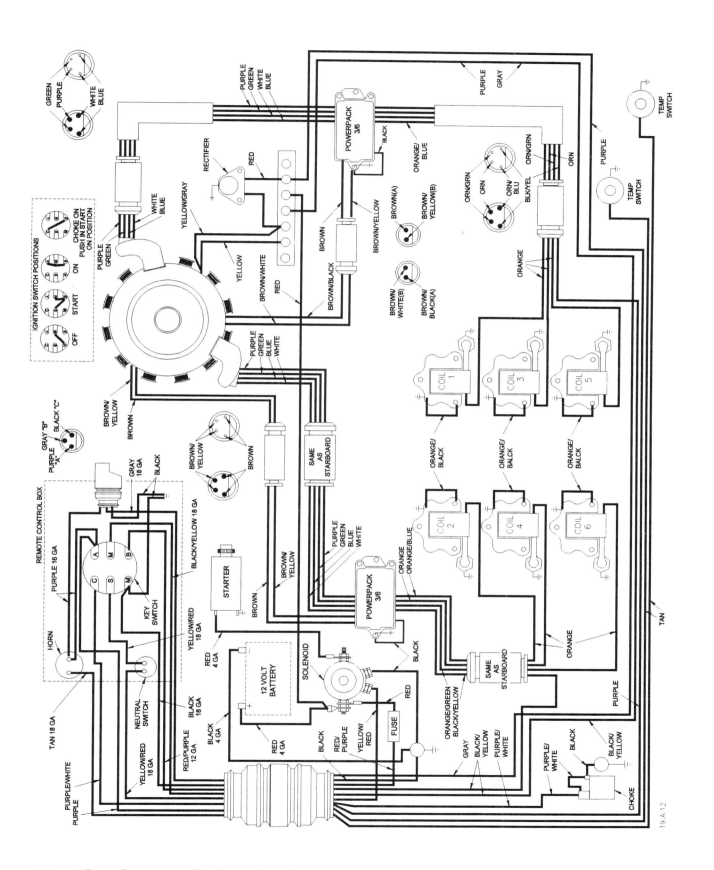

Wire Identification -- V6 Powerheads: 150 hp 1978 thru 1985; 175 hp 1979 thru 1985; 185 hp 1984 and 1985; 200 hp 1979 thru 1983; 235 hp 1979 thru 1985.

A-26 APPENDIX

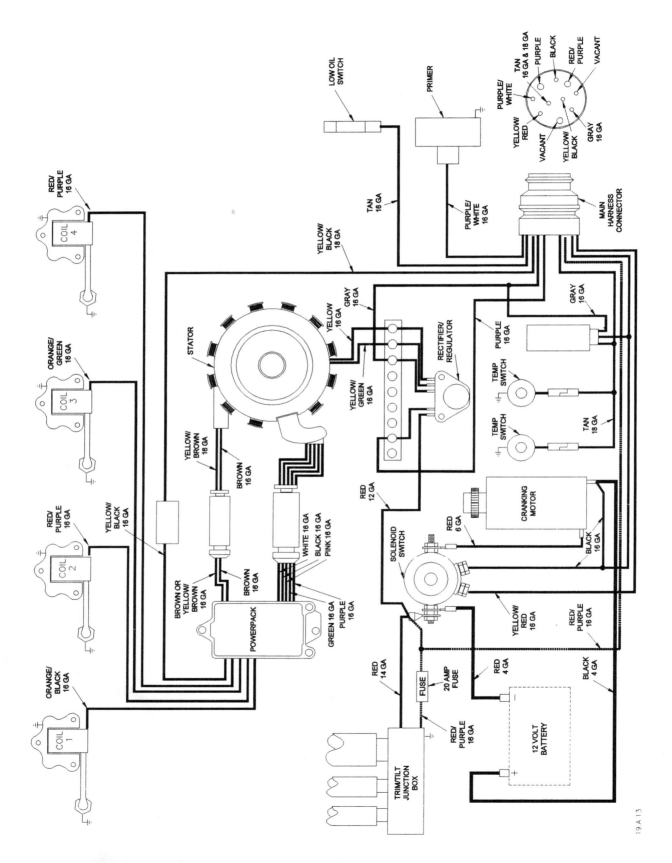

Wire Identification -- V4 Powerheads: 120 hp and 140 hp 1985 and on.

WIRE IDENTIFICATION A-27

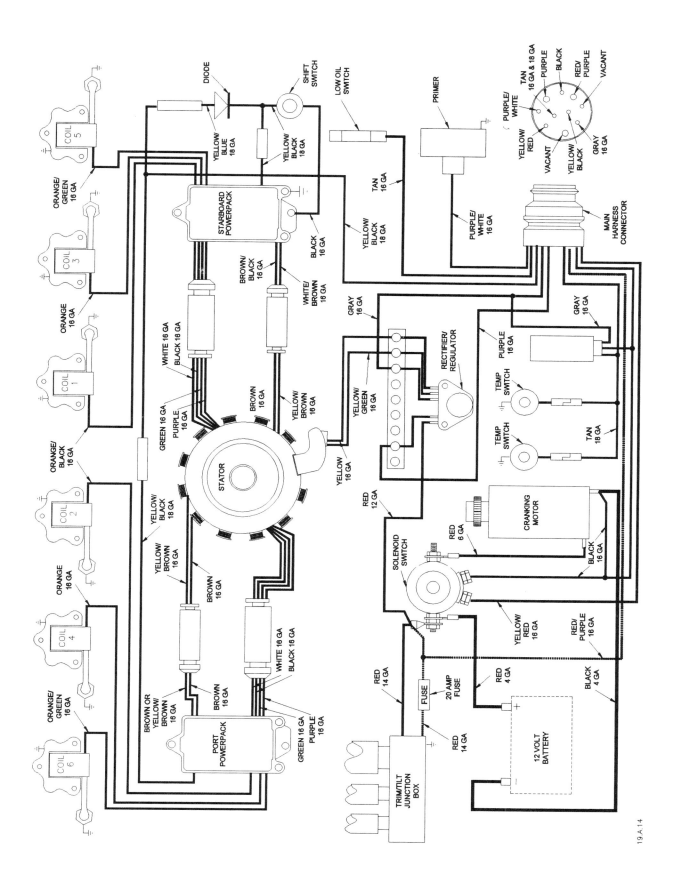

Wire Identification — V6 Powerheads 150 hp thru 225 hp 1986 and on

SelocOnLine

SelocOnLine is a maintenance and repair database accessed via the Internet.
Always up-to-date, skill level and special tool icons, quick access buttons to wiring diagrams, specification charts, maintenance charts, and a parts database.
Contact your local marine dealer or see our demo at www.seloconline.com

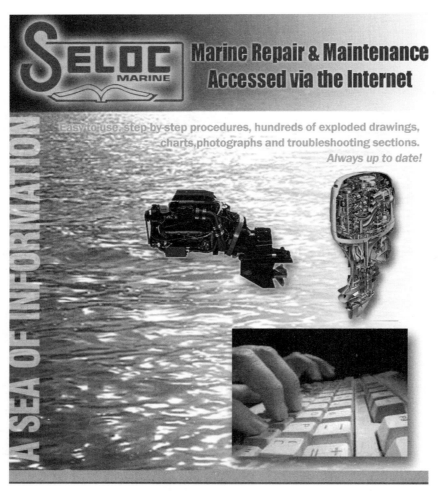

**Evinrude / Force / Honda / Johnson / Mariner / Mercruiser
Mercury / OMC / Suzuki / Volvo / Yamaha / Yanmar**

Purchase price provides access to one single year / model engine and drive system. The purchase of this product allows the user unlimited access for three years.